NMR 31

Basic Principles and Progress

Editors: P. Diehl E. Fluck H. Günther
R. Kosfeld J. Seelig

Solid-State NMR II
Inorganic Matter

Guest-Editor: B. Blümich

With contributions by
D. Brinkmann, G. Engelhardt, C. Jäger,
H. Koller, M. Mali, H. Pfeifer, A. Sebald

With 94 Figures and 24 Tables

Springer-Verlag Berlin Heidelberg GmbH

ISBN 978-3-642-50051-0 ISBN 978-3-642-50049-7 (eBook)
DOI 10.1007/978-3-642-50049-7

Library of Congress Catalog Card Number 93-9522

Originally published by Springer-Verlag Berlin Heidelberg New York in 1994.

Softcover reprint of the hardcover 1st edition 1994

Typesetting: Thomson Press (India) Ltd, New Delhi

51/3020 - 5 4 3 2 1 0 - Printed on acid-free paper

Preface

Solid-State NMR is a branch of Nuclear Magnetic Resonance which is presently experiencing a phase of strongly increasing popularity. The most striking evidence is the large number of contributions from Solid-State Resonance on NMR meetings, approaching that of liquid-state Resonance. Important progress can be observed in three areas: Methodical developments, applications to inorganic matter, and applications to organic matter. These developments are intended to be captured in three volumes of this series, each of them being devoted to more or less one of these areas.

The present volume on Solid-State NMR II is devoted to inorganic matter. The first chapter surveys recent developments in ^{29}Si-NMR by selected examples elucidating subtle structural details of crystalline and amorphous materials. Chapter two reviews NMR spectroscopy of structure and properties of zeolites with respect to adsorbate-adsorbent properties, a topic of fundamental importance in heterogeneous catalysis. Chapter three is devoted to MAS and CP/MAS NMR of less common spin-1/2 nuclei like ^{77}Te, ^{125}Te, ^{113}Cd, and ^{119}Sn. MAS of quadrupolar nuclei is particularly rewarding if spinning sidebands of the satellite transitions are analysed. Such experiments are treated in chapter four. The fifth chapter deals with NMR and NQR investigations of the electronic structure of high-temperature superconductors.

Of course, the topics chosen in all three Solid-State NMR volumes by no means cover the entire area of Solid-State NMR, but it is hoped that they treat an attractive cross-section of todays research. Particular thanks goes to the authors for their pleasant cooperation and, most importantly, for writing the contributions. Springer-Verlag has been very helpful in its assistance and editorial supervision.

Aachen, August 1993

B.Blümich
R.Kosfeld

Guest-Editor

Prof. Dr. Bernhard Blümich
Lehrstuhl für Makromolekulare Chemie,
RTWH Aachen, Worringer Weg 1,
D-52056 Aachen, FRG

Table of Contents

Table of Contents of Volume 30

Solid-State NMR I - Methods

^{29}Si NMR of Inorganic Solids

Günter Engelhardt and Hubert Koller

Institute of Chemical Technology I, University of Stuttgart, Pfaffenwaldring 55,
70550 Stuttgart, FRG

Table of Contents

NMR Basic Principles and Progress, Vol. 31

^{29}Si NMR spectroscopy, originally applied to study silicate solutions and silicon-organic compounds in the liquid phase, has recently developed into an important and potential tool for the investigation of crystalline and amorphous solids. New and powerful NMR techniques, such as magic-angle spinning, cross-polarization, two-dimensional methods, and measurements at high and low temperatures permit the routine-like registration of highly resolved solid-state ^{29}Si NMR spectra of a wide range of silicates and aluminosilicates, in particular zeolites and related materials, minerals, glasses, amorphous silicas, clays, and cements, but also of silicon-containing ceramics and other materials. Detailed information on the local structure around the silicon atom can be derived from the spectra which makes ^{29}Si NMR a valuable complement to X-ray diffraction techniques monitoring the periodic long-range order. The present article surveys recent developments in the experimental technique and structural interpretation of the spectra and demonstrates by selected examples the high potential of modern solid-state ^{29}Si NMR spectroscopy in elucidating subtle structural details of crystalline and amorphous materials.

1 Introduction

Since the first systematic study of silicates and aluminosilicates by high-resolution solid-state ^{29}Si NMR was published in 1980 [1], the application of this method has grown rapidly and is fast becoming a powerful tool in the structural characterization of a wide range of silicates and other silicon-containing materials. The development is closely related to the great progress made in the last decade in the instrumentation, techniques and methods for the registration and evaluation of the ^{29}Si NMR spectra and their structural interpretation. Nowadays, high magnetic field strengths and a wide selection of experimental techniques are available by modern solid-state NMR spectrometers. Since ^{29}Si is a spin 1/2 nucleus, magic angle spinning (MAS) yields simple spectra with complete averaging of the chemical-shift anisotropy and is, therefore, the most important technique for the measurement of highly resolved ^{29}Si NMR spectra of microcrystalline or amorphous solids. In proton containing samples, line broadening due to dipolar ^{29}Si-^{1}H interactions can be removed by high power proton decoupling, and cross-polarization (CP) may be used for signal enhancement and the detection of protons in close proximity to the silicon atom, e.g. in SiOH groupings. In addition, 2D ^{29}Si MAS NMR techniques such as COSY and INADEQUATE have recently been applied to study three-dimensional framework connectivities. All these experiments can, in principle, be performed at a wide range of variable temperatures which opens the possibility of investigating phase transitions and other temperature-induced structural transformations. In summary, the development of modern NMR techniques permit the routine-like registration of highly resolved solid-state ^{29}Si NMR spectra and has, therefore, greatly contributed to the efficient application and high capacity of ^{29}Si NMR in structural studies of a large scale of solid materials.

At the same time, considerable progress has been made in the detailed interpretation of the ^{29}Si NMR spectra. It was shown even in the initial work on ^{29}Si MAS NMR of silicates and aluminosilicates and has been further confirmed in numerous subsequent studies that the ^{29}Si chemical shift is most sensitive to the immediate chemical and structural surrounding of the silicon atoms. Chemically and/or crystallographically inequivalent Si sites can thus be detected by distinct resonances in the spectra. Moreover, empirical and theoretical relations have been established between the chemical shift and the kind, number and structural arrangement of the nearest and second nearest neighbour atoms of the silicon atom. ^{29}Si MAS NMR can thus provide detailed information on the local structure around the silicon atoms and is, therefore, a valuable complement to the X-ray or neutron diffraction techniques which monitor the periodic long-range order of crystalline materials. ^{29}Si MAS NMR can also be successfully applied to amorphous or highly disordered solids the study of which by diffraction techniques is notoriously difficult. It should be emphasized, however, that, at least for crystalline structures, the combination of ^{29}Si NMR and XRD results is often a prerequisite for the reliable interpretation of the

NMR spectra and, on the other hand, particularly appropriate for a more complete and detailed description of the structure.

The majority of the numerous papers published so far on the application of solid-state ^{29}Si NMR deal with crystalline and amorphous silicates and aluminosilicates, in particular with zeolites and related materials, glasses, minerals and amorphous silica, but an additional number of ^{29}Si NMR studies on cements, layer silicates (clays), silicon-containing ceramics and other materials have appeared in the literature. Several review articles have been published during the last few years summarizing the recent work on solid-state ^{29}Si NMR of zeolites [2–5], glasses [6–9], minerals [6, 10], and ceramics [11]. In addition, a comprehensive review of the fundamentals and applications of ^{29}Si NMR to silicates and zeolites considering the literature up to 1986 has been presented in a book [12]. Extensive compilations of ^{29}Si chemical shift data of silicates and aluminosilicates are given, e.g., in refs. [10, 12].

In the present article a concise review will be given of the current state and ongoing developments in the field, with emphasis on the most recent results. To this end Sect. 2 of this article will summarize the progress in the experimental ^{29}Si NMR techniques, Sect. 3 gives a brief review of the empirical and theoretical interpretation of the ^{29}Si chemical shift, and Sect. 4 surveys the specific information available from ^{29}Si NMR on structure and composition of solid silicates, aluminosilicates and other silicon containing materials. Finally, selected examples of recent applications of ^{29}Si NMR will be considered in Sect. 5.

2 Experimental Techniques

2.1 Introduction

Although single-crystal [13, 14] or static powder ^{29}Si NMR [15–18] can provide most valuable information on local bonding and symmetry in solids, these techniques have only been applied sparingly. Complete single-crystal studies are time-consuming and require, at least for work with ^{29}Si in its natural abundance (4.7%), relatively large crystals (usually about 5 mm in diameter) which are normally not available. Static powder studies yield direct information on the anisotropy of the ^{29}Si chemical shift but suffer from broad and often overlapping resonances and need also long measuring times. The overwhelming majority of solid-state ^{29}Si NMR studies have been performed using magic-angle spinning (MAS) of powders and, where appropriate, in combination with high-power proton decoupling and cross-polarization (CP). Since substantial gains in sensitivity and resolution are achieved with increasing strength of the static magnetic field, B_0, it is generally advantageous to measure the ^{29}Si NMR spectra at high field, e.g., at $B_0 \geqq 7\,T$, corresponding resonance frequencies of

$\nu_0(^{29}Si) \geqq 59.6$ MHz. In what follows, recent developments in the experimental techniques of solid state ^{29}Si NMR will be briefly considered with emphasis on MAS, CP, two-dimensional methods, and application of high temperatures.

2.2 Magic-Angle Spinning

The main reason for line broadening in ^{29}Si NMR of solid silicates and aluminosilicates, is, in general, shielding (chemical-shift) anisotropy which can be fully averaged by magic-angle spinning. Since for ^{29}Si the chemical-shift anisotropy does not, in general, exceed and is mostly smaller than 100 ppm, narrow isotropic line shapes without or with only weak spinning side bands can be observed at magnetic field strengths B_0 up to 9.4 T by MAS applying spinning frequencies of about 4–6 kHz. Such spinning rates are easily attainable by standard MAS probes with rotors of 7 mm diameter. If necessary, even spinning rates up to 15 kHz can be achieved if smaller rotors (4 mm diameter) are used. Although fast MAS completely averages chemical shift anisotropy, the three components of the shielding tensor can be retrieved from the detailed analysis of MAS sideband patterns [19] registered at moderate spinning rates [20]. MAS also averages weak dipolar interactions to distant protons or other NMR-active nuclei, e.g. ^{27}Al. Stronger dipolar couplings of ^{29}Si to nearby protons can be removed by high-power proton decoupling.

2.3 Cross-Polarization

In addition to MAS and dipolar decoupling, which by the line-narrowing effect result in a considerable improvement of the signal-to-noise ratio of the spectra, the intensities of the ^{29}Si NMR peaks may be substantially enhanced by use of the ^{1}H-^{29}Si cross-polarization (CP) technique, provided that ^{1}H nuclei capable of polarization transfer to ^{29}Si are present in the sample. In silicates and related materials the most likely candidates for CP are protons of SiOH groups, but under certain conditions also protons of water or organic molecules adsorbed in porous materials, or functional groups bonded at the surface may give effective CP enhancements. Another advantage of the CP technique is that significantly shorter repetition times for spectra accumulation can be applied since the magnetization recovery of the protons is governed by the longitudinal relaxation time T_{1H} rather than T_{1Si}, the former being generally much shorter. However, because the CP efficiency is determined by the cross-relaxation rate T_{HSi}^{-1} which depends strongly on the Si$\cdots$H distance ($T_{HSi}^{-1} \propto r_{HSi}^{-6}$), the resulting intensity enhancement in the CP MAS spectra may be rather different for distinct Si sites. Therefore, ^{29}Si CP MAS NMR spectra are not quantitatively reliable in the sense of single-pulse MAS spectra, and detailed studies of the CP characteristics by contact time variation and relaxation measurements (T_{HSi}, $T_{1\rho H}$) are required to derive quantitative information from the spectra [21–24]. As an

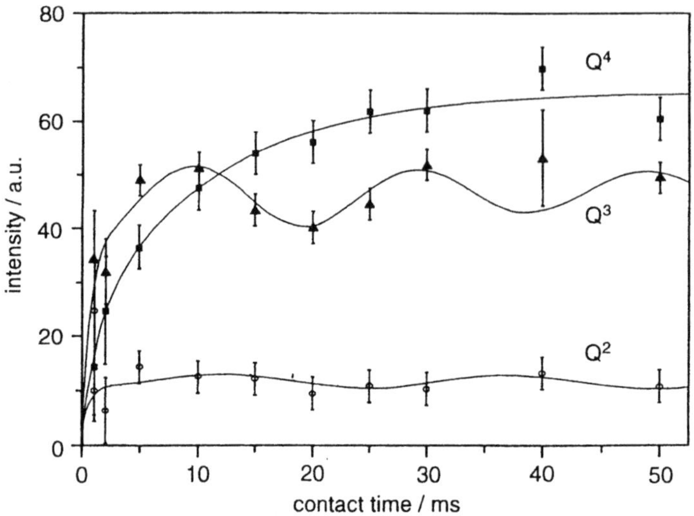

Fig. 1. Integrated ^{29}Si CP MAS intensities of the Q^2, Q^3 and Q^4 signals of Aerosil 200 (Degussa) as a function of the contact time. (Reprinted with permission from Ref. 22. Copyright 1990 Taylor & Francis)

example, Fig. 1 shows the integrated ^{29}Si CP MAS intensities as a function of the contact time for the central silicon atoms in the three different $Si(OSi)_n(OH)_{4-n}$ (denoted conventionally by Q^n with $n = 2$–4 [25]) structural groups of Aerosil 200 (Degussa) [22]. At short contact times the Q^3 signal may be more intense than that for Q^4 even though the concentration of the latter group is much higher. Besides the different intensity evolution for the distinct Si sites, an oscillation of the intensity is observed for the Q^2 and Q^3 signals which has been ascribed to a magnetization transfer from hydroxyl groups of neighbouring silica centres. In addition to CP, ^1H-^{29}Si dipolar dephasing experiments have been applied to examine type and distribution of structural groups in the neigbourhood of single (Q^3) or geminal (Q^2) OH groups at the surface of amorphous silica [26, 27].

2.4 *Two-Dimensional Techniques*

While the application of 2D NMR techniques in solution studies is well established and widely used, e.g., to determine connectivities between atoms within molecular structures [28, 29], corresponding applications to solids have found only limited use so far. There have been, however, several successful studies on the connectivities of tetrahedral Si sites in high-silica zeolites (see [4] and references therein), the octameric trimethylsilyl silicate Q_8M_8 [30], and glasses [31] using the well known COSY [32] or INADEQUATE [33] pulse sequences. In these studies scalar coupling interactions in the ^{29}Si-O-^{29}Si connections are used to establish the three-dimensional connectivities of the silicate framework. ^{29}Si isotopic enrichment of the samples reduces considerably the measuring times but 2D ^{29}Si MAS NMR spectra were also obtained from natural abundance materials, making the method quite generally applicable.

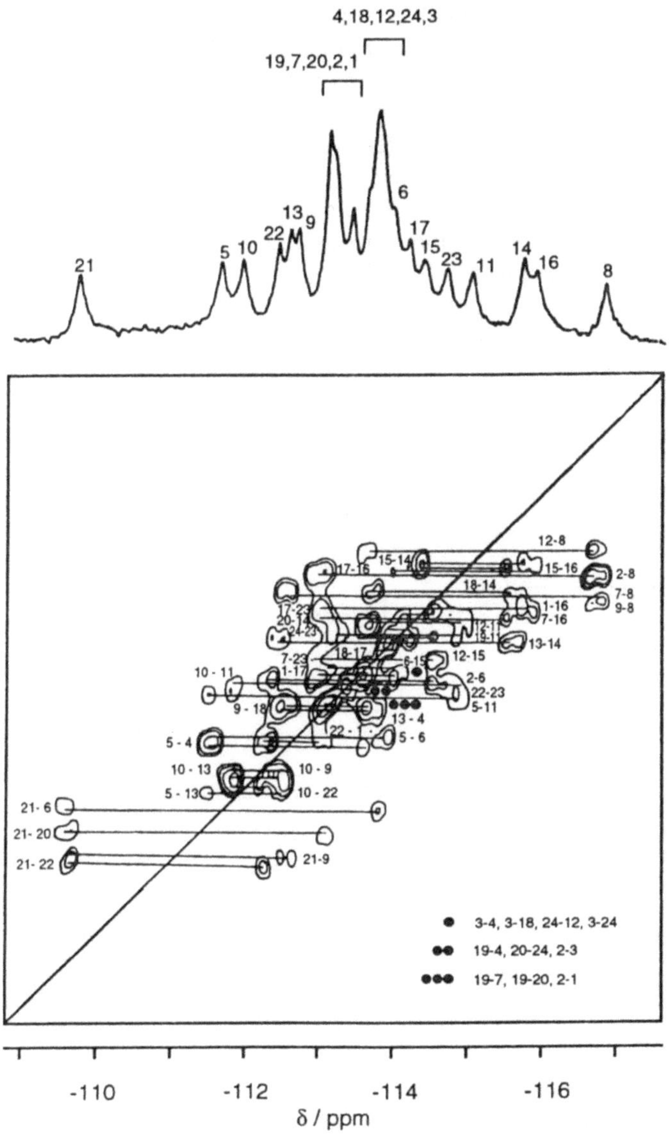

Fig. 2. Contour plot of an ^{29}Si MAS 2D INADEQUATE experiment on high-silica zeolite ZSM-5 at 300 K with a 1D MAS NMR spectrum above. The numbers indicate the SiOSi connectivities detected using the site numbering of Ref. 104. (Reprinted with permission from Ref. 34. Copyright 1990 American Chemical Society)

Compared to COSY, the INADEQUATE experiment has a number of advantages (e.g., better S/N, no diagonal peaks, no spinning sidebands) and will usually be the experiment of choice. As an example, Fig. 2 presents the 2D INADEQUATE ^{29}Si MAS NMR spectrum of the monoclinic form of highly siliceous ZSM-5 zeolite [34], a complex silica structure with 24 crystallographically distinct Si sites. Of the 48 expected SiOSi connectivities, 38 are clearly observed in the spectrum. 2D spin diffusion experiments [35] and several modifications of the conventional COSY sequence have also been applied, e.g., replacing the initial 90° pulse by a CP sequence for proton containing materials, COSY 45 (double quantum filtered COSY) [35], J-scaled COSY [36] and SUPERCOSY [37]. A two-dimensional ^{1}H-^{29}Si CP MAS NMR heteronuclear correlation experiment has been proposed where mixing is achieved by CP [38]. By this technique specific information on SiOH groups of surface species of silicas and zeolites are obtained. The pulse sequences of the various 2D techniques mentioned above are schematically depicted in Fig. 3.

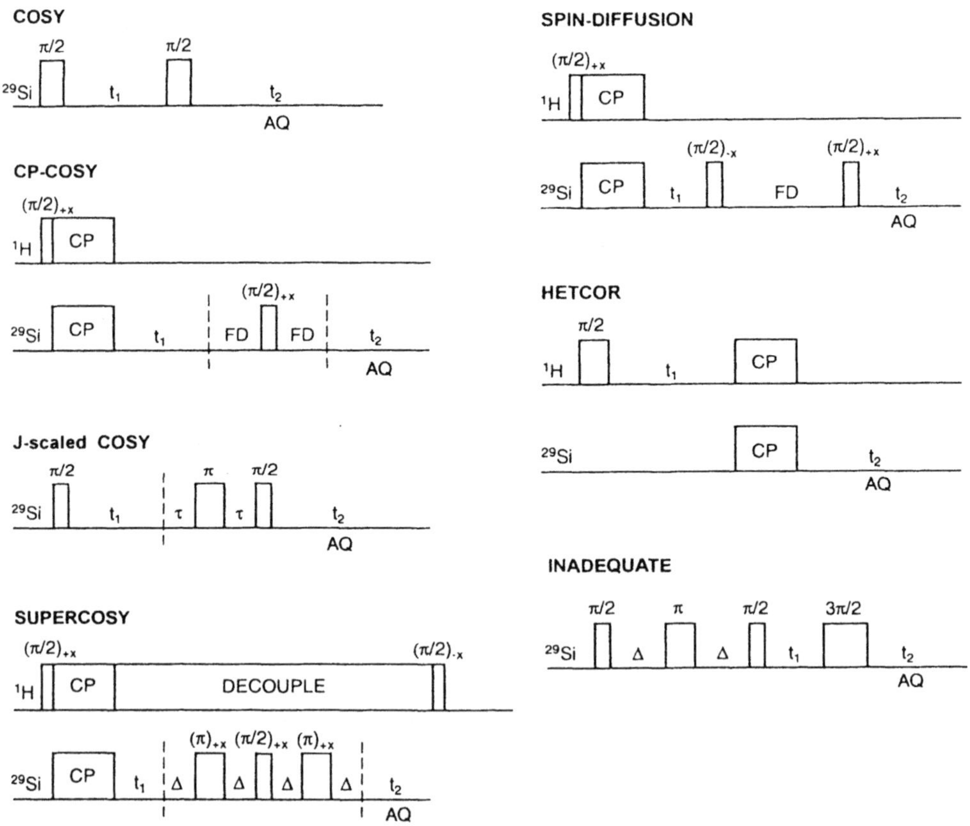

Fig. 3. Schematic representation of the pulse sequences used in the various two-dimensional ^{29}Si NMR experiments (adapted from Ref. 29 (COSY), Ref. 35 (CP-COSY, SPIN-DIFFUSION), Ref. 36 (J-scaled COSY), Ref. 37 (SUPERCOSY), Ref. 38 (HETCOR), and Ref. 102 (INADEQUATE))

2.5 *Measurements at High Temperatures*

The ability of ^{29}Si MAS NMR studies to give information at varying temperatures about dynamic processes and structural transformations in solid materials and their melts is well documented (see, e.g., [39–43]). However, the extension of the range of temperature applicable in NMR experiments is still a technical challenge [44]. In principle, MAS NMR experiments can be performed with most commercially available MAS probes in the temperature range of about −100 to +200 °C with cooling or heating the sample via the flowing gas stream used for sample spinning. However, owing to low gas heat capacities this technique rapidly becomes inefficient at higher temperatures and may further involve difficulties in cooling the rf coil and other parts of the probe. Nevertheless, gas heating has been used recently up to 600–700 °C in a special design of a MAS probe [45]. Laser heating can eliminate many of these difficulties, and a laser-heated high-temperature MAS probe for temperatures up to about 700 °C has been developed where the heating is confined to the sample container (Bruker Analytische Meßtechnik GmbH, Rheinstetten). Even higher temperatures have been reached in static NMR experiments and the upper limit has recently been extended to well above 2000 °C by a laser heated system [46]. Temperatures up to 1250 °C obtained by resistive heating (molybdenum furnace windings) of the sample have been applied to static ^{29}Si NMR investigations of solid and molten silicates [47].

3 Interpretation of ^{29}Si Chemical Shifts in Solids

3.1 *Introduction*

Most of the structural information that can be derived from ^{29}Si NMR spectra of solids arise from chemical shifts. Since the latter is dependent on the electronic environment of the ^{29}Si nucleus, it should be related to the type, position and bonding characteristics of the surrounding atoms. As mentioned above, the most detailed information can, in principle, be obtained if the three principal components of the ^{29}Si shielding tensor, i.e. the chemical shift anisotropies are known [48]. This has been demonstrated in a few papers applying single-crystal [13, 14] or static-powder ^{29}Si NMR [15–18] to silicates, and by theoretical considerations [16, 49, 50]. However, to avoid the difficulties in spectral analysis of broad and overlapping anisotropic line shapes, most of the solid-state ^{29}Si NMR work is done by use of the MAS technique which averages the distinct components of the shift anisotropy to the single mean value of the isotropic chemical shift, δ.

3.2 ^{29}Si Chemical Shifts of Silicates and Correlation with Structure Parameters

For silicate structures and related systems, many empirical relationships between the isotropic ^{29}Si chemical shift and structural features have been established, and, at least in part, substantiated by theoretical considerations. (The earlier work up to about 1986 was reviewed in Chapter IV of Ref. [12].) In particular, an approximately 10 ppm high-field shift is observed for the formation of a SiOSi connection at a given $SiO_{4-n}(OSi)_n$ environment (Q^n), while each replacement of Si with Al in the second coordination sphere yields a low-field shift of about 5 ppm [1]. Moreover, linear correlations have been established between δ and the mean SiOT bond angles, α, [51–55], non-bonded Si–Si distances, d_{SiSi} [56, 57], SiO bond lengths, d_{SiO}, [15, 58], and lattice parameters for cubic structures, a_0 [53, 54], derived from the X-ray structures of a large number of framework silicates and aluminosilicates. The correlations indicate decreasing δ values (i.e., shifts to high field) with decreasing d_{SiO} and increasing α, d_{SiSi} and a_0. These dependencies could be rationalized by a simple theoretical model (see [59] and references therein) combining semi-empirical [60] or ab initio [61, 62] quantum-chemical calculations with the general shielding theory. In short, the model shows that δ decreases almost linearly with the net charge at the silicon atom which depends on number and type of T-atoms ($T = Si, Al$) in the $SiO_{4-n}(OT)_n$ environments and on the s-hybridization of the oxygen atoms in the four O–Si bonds defining the SiO_4 tetrahedron. Since the oxygen s-hybridization, ρ, is related to α by $\rho = \cos\alpha/(\cos\alpha - 1)$, and reciprocal to d_{SiO}, the observed correlation with larger shielding at increasing α and decreasing d_{SiO} is confirmed by the theoretical model. Linear correlations between δ and d_{SiO} have also been observed for a number of compounds containing six-fold coordinated Si in SiO_6 environments [63], and for monomer SiO_4 tetrahedra in a series of calcium monosilicates [64]. However, an opposite slope (i.e., decreasing δ with increasing d_{SiO}) has been obtained for the latter case which shows that the relationships derived for framework silicates can not be simply applied to nesosilicates with monomer SiO_4 anions.

The bond angle dependence of δ for silica polymorphs has also been discussed by a point charge model describing the influence of the second coordination sphere on the chemical shift [65]. More sophisticated quantum chemical methods have been applied by several authors to calculate ^{29}Si chemical shifts of silicates and related systems [50, 66–68] but the results are mostly confined to small molecules or clusters and are not directly applicable to the general interpretation of the large body of experimental data.

A simple relationship has been derived between the structures of a large number of silicate minerals and ^{29}Si chemical shifts which is based on the magnetic susceptibility anisotropy of the bonds to second nearest neighbours modified by the bond valence and the angle at the bridging oxygen [69]. Only the valence, position and orientation of bonds between cations and oxygens of the silicate tetrahedron lying within 4 Å of the Si atom are used in this correlation.

The fit of 124 Si sites yields a correlation coefficient of 0.99 and a root-mean-square deviation of 0.66 ppm. Calculations on model structures show an almost linear relationship between δ and the cation-oxygen distance while the actual effect on δ increases with increasing angle at the oxygen atom. Another approach applicable to crystalline, glassy and gel silicates relates δ to a parameter P which takes into account the electronegativity and structural description (a graph quantity) of the silicate units as well as the ionic potential of the cation [70]. An exponential dependence of δ on P is found which, however, is difficult to explain.

4 Survey of Structural and Chemical Information Available from ^{29}Si NMR of Solids

4.1 Introduction

In this chapter the information on local structure and chemical composition that can be derived from the ^{29}Si NMR spectra of solids will be summarized from a general point of view. Unless otherwise stated the following considerations refer to ^{29}Si MAS NMR spectra of microcrystalline samples recorded at high magnetic field strengths and under optimum conditions for maximum spectral resolution and reliable line intensities.

4.2 Number and Population of Si Sites in Non-Equivalent Environments

The most direct and rather fundamental information which follows immediately from the number of distinct resonances observed in the ^{29}Si MAS NMR spectra is the number of chemically and/or crystallographically inequivalent Si sites present in the sample. Moreover, from the normalized peak intensities the relative populations of the various sites can be determined. The number of inequivalent sites that can be distinguished in the spectra depends on the relationship between spectral resolution (line width) and chemical shift difference between the distinct sites. For highly crystalline samples with well ordered structure usually very narrow lines (line widths of a few Hz) are observed, and subtle differences in the structural surrounding of the corresponding Si sites may be detected. Figure 4 shows the ^{29}Si MAS NMR spectrum of highly crystalline silicalite, a microporous silica polymorph containing 24 crystallographically distinct Si sites per unit cell. 20 well resolved peaks appear in the spectrum, and a total of 24 Si sites follows from the normalized integrated peak intensities [39]. In contrast, amorphous, glassy or ill-crystallized samples with highly disordered structures show, in general, broad and overlapping resonances rendering the interpretation

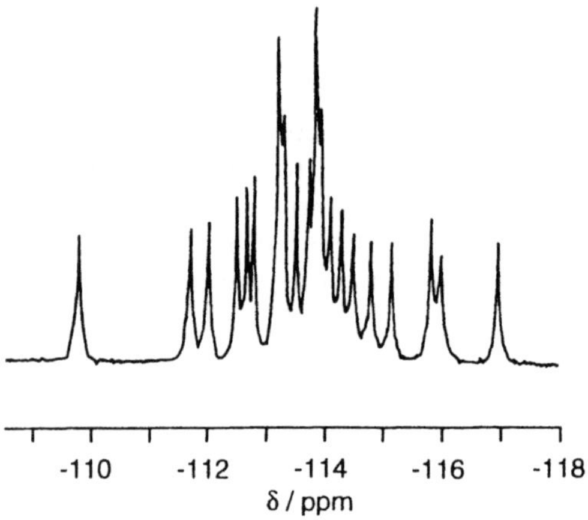

Fig. 4. ^{29}Si MAS NMR spectrum of high-silica zeolite ZSM-5 (Silicalite) (Reprinted with permission from Ref. 39. Copyright 1988 American Chemical Society)

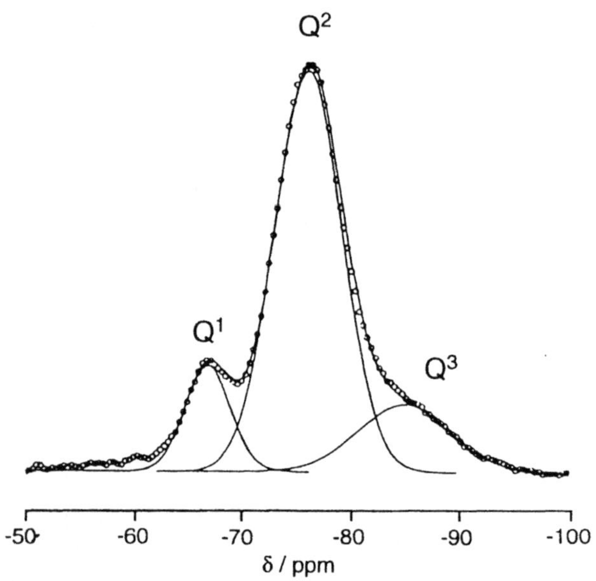

Fig. 5. ^{29}Si MAS NMR spectrum of sodium metasilicate glass showing peak separation by Gaussian functions into three component lines for the distinct Q^n environments (Reprinted with premission from Ref. 71. Copyright 1991 Elsevier Science Publishers)

of the spectra difficult. However, if sensible assumptions can be made on line positions and line shapes, those spectra may be analysed by deconvolution and line fitting procedures. As an example, Fig. 5 displays the ^{29}Si MAS NMR spectrum of sodium metasilicate glass [71]. The broad spectral pattern can be clearly separated into three resonances corresponding three different Si environments present in the glass.

4.3 Coordination Number of Si

The vast majority of silicates and other inorganic silicon compounds contain four-fold coordinated silicon but a number of structures with five- and six-

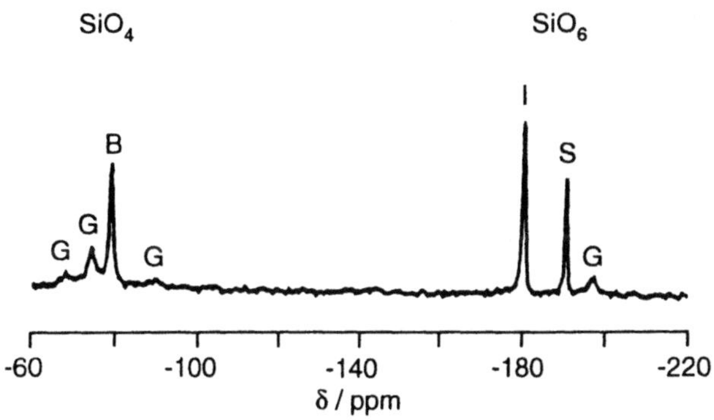

Fig. 6. ^{29}Si MAS NMR spectrum of a high-pressure $MgSiO_3$ phase showing lines of four- (*left*) and six-fold coordinated (*right*) silicon in garnet (*G*), β-Mg_2SiO_4 (*B*), ilmenite (*I*) and stishovite (*S*). (Reprinted with permission from Ref. 74a. Copyright 1991 American Association for the Advancement of Science)

coordinated Si are known. ^{29}Si MAS NMR can clearly distinguish between the different Si coordination numbers owing to large differences in the ^{29}Si chemical shifts. While δ-values between -60 and -120 ppm are typical of four-coordinated SiO_4 in silicates [12], the resonances of octahedral SiO_6 coordinations, as observed, e.g., in thaumasite, stishovite [63], $MgSiO_3$-rich garnets, perovskites and other high-pressure silicate minerals [72–74], and phosphosilicate glasses [31, 75] and gels [76], appear in the range between -170 and -220 ppm. Moreover, weak lines at about -150 ppm has recently been attributed to five-coordinated SiO_5 in quenched high-pressure alkali silicate glasses and crystalline $CaSiO_3$ samples [74b, 77]. The clear distinction between four- and six-coordinated Si is demonstrated in Fig. 6 by the ^{29}Si MAS NMR spectrum of a high-pressure $MgSiO_3$ sample which exhibits several resonances that could be attributed to SiO_4 and SiO_6 environments of distinct silicate phases present in this material [74].

4.4 Local Chemical Environment of Si

Effect of nearest neighbour atoms. The ^{29}Si chemical shift depends sensitively on type and structural arrangement of the first and second nearest neighbour atoms at the silicon atom. Detailed information on the local Si environment can thus be derived from the ^{29}Si NMR spectra. The effect on δ of the type of the atoms directly bound to the Si atom is illustrated in Fig. 7 which summarizes the δ-ranges observed for various $SiX_{4-n}Y_n$ coordinations with $X, Y = O, N, Si$ and C in solid materials, such as silicates [12], silicon nitrides [78, 79], YSiON phases [80], SiAlON ceramics [81], silicon carbides [82, 83], SiCO ceramics [84], and crystalline silicon [79]. Although the different coordinations have shift ranges with considerable overlap, the observed tendencies in the shift

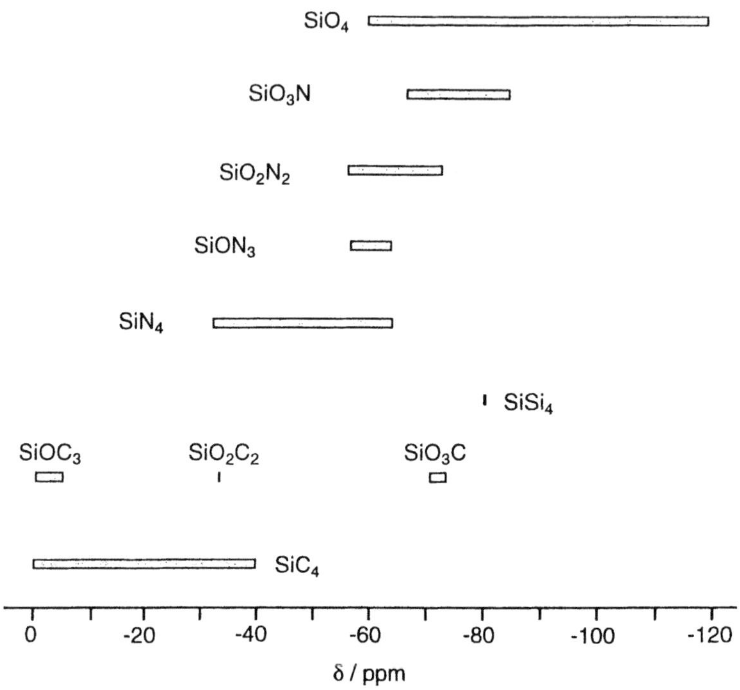

Fig. 7. Ranges of ^{29}Si chemical shifts of $SiX_{4-n}Y_n$ coordinations with X, Y = O, N, Si and C

variation together with some knowledge on the kind and chemical composition of the material may be helpful in the characterization of the immediate chemical surrounding of the silicon atom.

Effect of second nearest neighbour atoms. Second nearest neighbour effects on the ^{29}Si chemical shift has been particularly well studied for silicates, aluminosilicates and related materials (for a comprehensive review see [12]). As mentioned in Sect. 3.2, there are especially two characteristic features which affect the ^{29}Si chemical shift of a SiO_4 group: characteristic high-field shifts are observed with increasing number of SiOT bridges ($T = Si, Al$) formed by the SiO_4 tetrahedron (degree of SiO_4 polymerization), and typical low-field shifts follow from the replacement of Si with Al in the second coordination sphere of the central silicon with a given number of SiOT bridges (degree of tetrahedral Al substitution). Similar shift effects have been observed for tetrahedral Ga substitution in gallosilicates. Typical ranges of δ for the central Si atoms of the five $Si(O^-)_{4-n}(OSi)_n$ environments in silicates (conventionally denoted by Q^n), the five $Si(OSi)_{4-m}(OAl)_m$ environments in framework aluminosilicates (Q^4(mAl) or Si(nAl) [85]) and the four $Si(OH)(OSi)_{3-m}(OAl)_m$ environments (Q^3(mAl)) in layer aluminosilicates [12, 86] derived from a large body of experimental data are displayed in Figs. 8, 9 and 10, respectively. As in the case of the nearest neighbours (Fig. 7), the shift ranges of the different environments overlap but

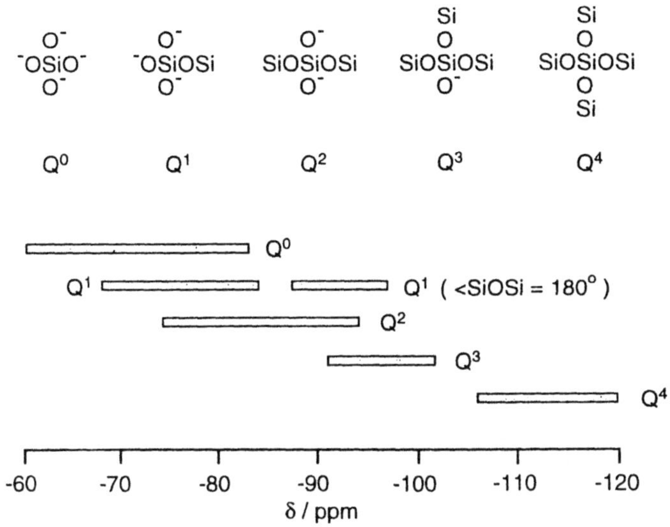

Fig. 8. Ranges of ^{29}Si chemical shifts of $Si(O^-)_{4-n}(OSi)_n$ coordinations (Q^n) in silicates

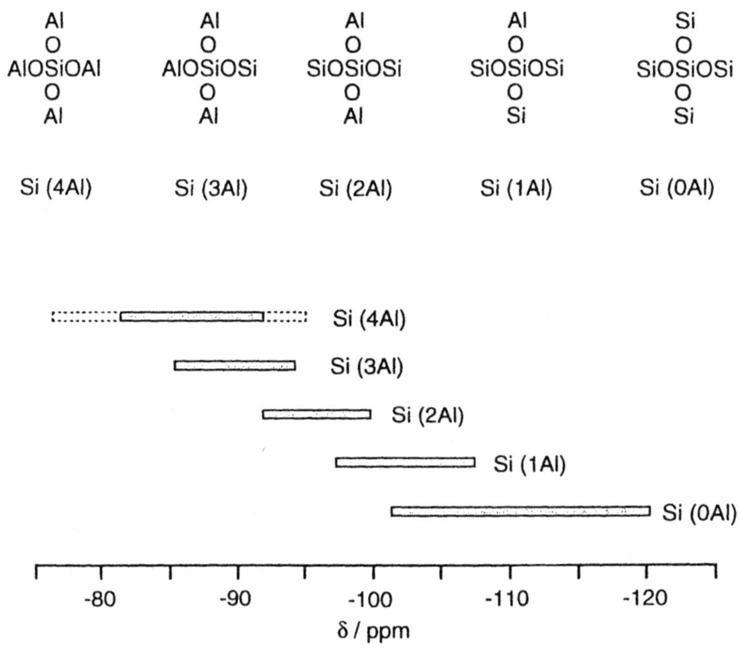

Fig. 9. Ranges of ^{29}Si chemical shifts of $Si(OSi)_{4-n}(OAl)_n$ coordinations (Si(*n*Al)) in framework aluminosilicates. The *dotted lines* for Si(4Al) designate the δ range observed for 1:1 aluminosilicate sodalites with different cage fillings [53, 54]

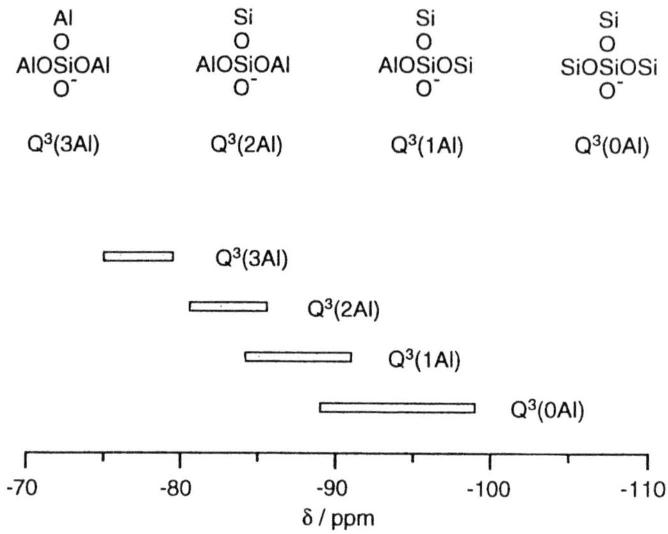

Fig. 10. Ranges of ^{29}Si chemical shifts of $Si(OH)(OSi)_{3-m}(OAl)_m$ coordinations (Q^3(mAl)) in layer aluminosilicates

the assignment of well separated peaks in the ^{29}Si MAS NMR spectrum of a certain silicate or aluminosilicate to the corresponding Si environments is, in general, feasible. Thus, the presence of the distinct types of structural units and, from peak intensities, their quantitative distribution in the structure can be determined from the ^{29}Si MAS NMR spectrum.

4.5 Si/Al Atomic Ratio of the Aluminosilicate Framework

Provided that the ^{29}Si NMR spectrum of an aluminosilicate is correctly interpreted in terms of the different Q^n(mAl) units, the quantitative ratio of tetrahedral Si and Al in the structure can be directly calculated from the peak intensities according to the equation $Si/Al = \Sigma I_{n,m}/\Sigma(m/n)I_{n,m}$, where $I_{n,m}$ are the intensities of the Q^n(mAl) peaks and summation is from $m = 0$ to $m = n$ (i.e., from 0–3 in layer aluminosilicates and from 0–4 in framework aluminosilicates) [12]. The above equation is independent of the specific structure of the aluminosilicate and includes the aluminium which is substitutionally incorporated into the tetrahedral framework but excludes any non-framework aluminium frequently present, e.g., in layer aluminosilicates and in chemically or thermally treated zeolites. If the total Si/Al ratio of such samples is known from the overall chemical composition, the proportion of non-framework Al can readily be determined.

4.6 Si, Al Ordering in the Aluminosilicate Framework

Information about the distribution of Si and Al atoms on the various tetrahedral sites of the aluminosilicate framework can be derived by comparing the relative Q^n(mAl) populations obtained from ^{29}Si NMR peak intensities with model-generated populations. The latter populations may be derived, e.g., from ordering schemes which are constructed by distributing the respective number of Si and Al atoms on the *T*-sites of a certain repeating unit from which the whole structure can be formed. It should be noted, however, that the NMR results reflect the local environments of the Si atoms averaged throughout the framework and the whole sample volume and do not, therefore, necessarily imply any simple long-range ordering. Nevertheless, specific Si, Al distributions or, less specifically, the degree of dispersion of Al atoms over the framework has been deduced from the ^{29}Si MAS NMR spectra of a number of zeolites and layer aluminosilicates with different tetrahedral Si/Al ratios. A review of the strategies and methods applied is given in Chap. V of Ref. [12], for recent applications see, e.g., [87–93].

4.7 Crystallographically Inequivalent Si Sites

The broad shift ranges observed for Si atoms residing in identical first and second neighbour chemical environments (see Figure 7–10) indicate that in addition to the chemical surrounding other structural features affect the ^{29}Si chemical shift. These are, in particular, different bonding geometries of the local structure around the Si atom. Distinct lines can thus be observed in the ^{29}Si MAS NMR spectra for Si atoms in chemically equivalent environments but characterized by different bond angles and/or bond lengths, i.e., occupying crystallographically non-equivalent sites in the structure. Pertinent examples are the two lines for the SiN_4 environment in α-Si_3N_4 [78], the three lines for SiC_4 units in the SiC polytypes 6H and 15R [82], and, in particular, the line splittings in the ^{29}Si MAS NMR spectra of a variety of silica polymorphs and high-silica zeolites [4, 12, 57] (e.g., the 20 lines observed for the 24 crystallographically inequivalent Q^4 sites of silicalite [39], see Fig. 4). The number and relative occupancies of crystallographically distinct Si sites can thus be obtained from the spectra and may be directly related to the results of diffraction experiments. Moreover, for framework silicates the mean SiOT bond angles of the different Si sites can be estimated from δ applying the quantitative correlations between δ and α considered in Sect. 3.2. These correlations reveal that δ changes by about 0.6 ppm for a 1° change in α. Since, for highly crystalline materials, a peak-to-peak resolution of 0.1 ppm can be achieved in the ^{29}Si MAS NMR spectra [39], differences in the mean bond angles of about 0.2° may be detected. On the other hand, if the mean bond angles are known from the X-ray structure, δ can be calculated which may be helpful in the correct assignment of multi-line spectra [55, 59, 94, 95].

4.8 Characterization of SiOH Groups in Silicate Materials

It was shown in Sect. 2.3 that application of the CP technique enhances selectively the signal intensity of Si nuclei near protons and that the enhancement factor depends on the Si···H distance and the CP contact time applied. At short contact times, intense lines of SiOH (e.g., in Q^3 or Q^2 structural groups) are observed, while the signals of Si nuclei with larger Si···H distances (e.g., Q^4 adjacent to Q^3) are strongly reduced or do not appear (e.g., Q^4 surrounded only by other Q^4 groups) in the CP MAS spectra. An even more detailed discrimination of the different Si environments can be achieved by application of CP with variable contact times [22–24, 38, 95, 96]. Figure 11 shows the ^{29}Si MAS NMR spectra of a silica gel registered with and without CP [23]. The considerable decrease of the intensity of the Q^4 line at about -110 ppm and the relative enhancement of the Q^3 (-110 ppm) and Q^2 line (-90 ppm) in the CP MAS spectrum is obvious. In addition, the resonance of the Q^4 groups in the CP spectrum is shifted by about 1.5 ppm to low field from that in the single pulse MAS spectrum, indicating different chemical shifts for "surface" Q^4 (adjacent to Q^3 and Q^4) and "bulk" Q^4 (surrounded by other Q^4 only) environments [23]. However, as mentioned in Sect. 2.3, the CP MAS spectra are not quantitatively reliable. But once the line width and position of less abundant species have

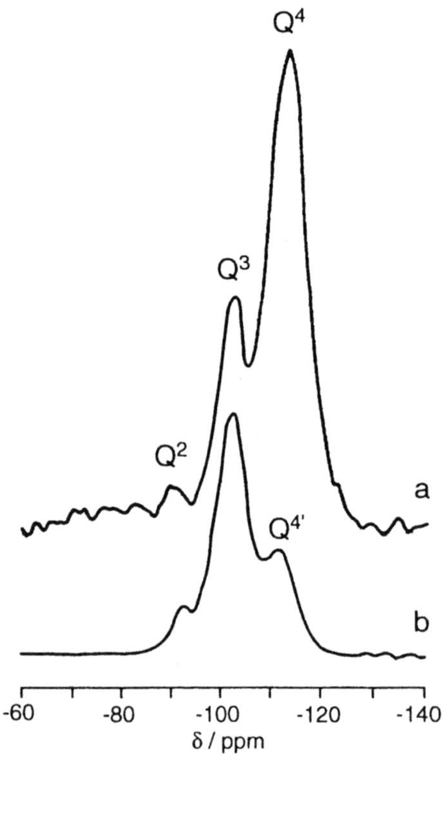

Fig. 11a, b. ^{29}Si MAS (**a**) and ^{29}Si CP MAS (**b**) NMR spectra of a silica gel. (Reprinted with permission from Ref. 23. Copyright 1990 Elsevier Science Publishers)

been determined from the CP MAS spectrum, this information may be used for the quantitation of weak and overlapping lines in the single pulse MAS spectrum by line fitting procedures [96, 97].

4.9 Si-O-Si Framework Connectivities

The direct determination of bonding connectivities between the various crystallographically inequivalent SiO_4 sites of a three-dimensional silicate framework has recently become feasible by application of two-dimensional ^{29}Si MAS NMR experiments such as COSY and INADEQUATE [4] (see Sect. 2.4). The latter experiment has been proved to be particularly useful since only double quantum coherence signals, i.e., only signals due to ^{29}Si nuclei coupled via scalar spin interactions in the ^{29}Si-O-^{29}Si connection are observed in the 2D plot (see Figs. 2 and 12b). The F2 dimension is the normal ^{29}Si chemical shift scale while the F1 domain represents the double-quantum frequencies of the resonances. Connected signals occur equally spaced on either side of the diagonal of the plot at the same frequencies in F2 as the corresponding lines from the simple 1D experiment. Other than INADEQUATE, 2D plots from COSY experiments have the normal ^{29}Si chemical shift scale in both F1 and F2 and the connectivities appear by cross-peaks. However, large intensities along the diagonal occur

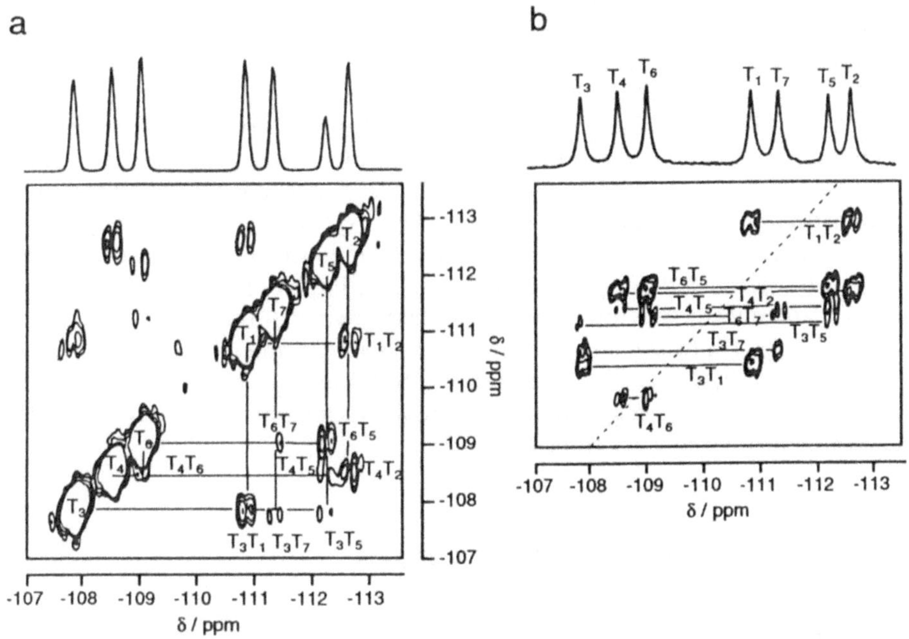

Fig. 12a, b. Contour plots of ^{29}Si MAS 2D COSY (**a**) and INADEQUATE (**b**) experiments on high-silica zeolite ZSM-12 with the F2 projection (**a**) and 1D MAS NMR spectrum (**b**) *above*. The *T*-site assignments and TOT connectivities are indicated in the spectra. (Reprinted with permission from Ref. 98. Copyright 1990 American Chemical Society)

which may obscur cross-peaks close to the diagonal. Figure 12a shows the 2D contour plot of a COSY experiment and Fig. 12b that of an INADEQUATE experiment on siliceous zeolite ZSM-12 [4, 98]. The ZSM-12 structure has seven crystallographically distinct Si sites with equal occupancies in the unit cell which are reflected by the seven lines of equal intensity in the 1D MAS NMR spectrum shown above the 2D plot of Fig. 12b. (The lower intensity of the T_5 line in the F2 projection on top of Fig. 12a arises from the much longer spin-lattice relaxation at this site). All of the nine connectivities expected for the known ZSM-12 structure are observed in the 2D spectra. Some ambiguity appear for that between T_4 and T_6 in the COSY experiment but is clearly resolved in the INADEQUATE spectrum. The lines may be assigned unambiguously from the expected connectivities by trial and error, but a partial assignment from other data may also be used as a starting point. Such information may be gained, e.g., from the expected intensity distribution of the lines, the above mentioned correlations between δ and SiOSi bond angles, or from different relaxation behaviour of particular sites (see, e.g., the above mentioned T_5 site of zeolite ZSM-12). So far two-dimensional ^{29}Si NMR experiments have been applied preferably to high-silica zeolite materials with known structures (e.g., zeolites ZSM-39 [35, 99], ZSM-12 [4, 98], ZSM-5 [34], ZSM-11 [100], ZSM-22 [98], DD3R [101], and mordenite [36]), and additional details of some of the structures have been revealed. In the case of unknown structures, 2D experiments will make it possible to select between different structural models, or to prove or disprove proposed structures or crystallographic space groups as, e.g., shown recently for the low-temperature form of zeolite ZSM-11 [100] and for zeolite ZSM-23 [102].

^{29}Si COSY MAS NMR spectroscopy has also been applied to ^{29}Si enriched sodium silicate glasses and sodium phosphosilicate glasses [31]. The cross-peaks observed in the COSY spectra of sodium silicate glasses suggest that a substantial proportion of the Q^2 and Q^3 as well as Q^3 and Q^4 silicate sites in the glass are interconnected on a nearest neighbour level. In contrast, no cross-peaks were observed for the sodium phosphosilicate glass, implying that the Q^4 and six-coordinate silicate sites present in this glass are not linked directly together. Interestingly, a slow exchange between the silicon Q^3 and Q^4 environments has been detected by 2D chemical-exchange ^{29}Si NMR spectroscopy of potassium tetrasilicate above its glass transition (highly viscous melt) not present in the glass just below the glass transition [103].

5 Applications: Two Selected Examples

5.1 Introduction

The huge number of papers published in recent years on applications of ^{29}Si NMR spectroscopy to inorganic materials renders impossible a complete survey of the literature in the field within the framework of this article. However, to illustrate further the high potential and strategies of modern solid-state ^{29}Si NMR spectroscopy in investigating subtle structural details of crystalline as well as amorphous systems, two more or less arbitrarily selected examples will be briefly considered in the following two sections. The first example has been selected to demonstrate the unique possibilities of the combined application of ^{29}Si MAS NMR and diffraction methods in obtaining a most detailed and complete description of complex crystal structures, while the second example is chosen to show that ^{29}Si NMR may be used as a bridge between crystalline, amorphous (glassy) and liquid structures, i.e. between the structure found by diffraction techniques and that which exists in other phases where diffraction cannot be applied.

5.2 *Combined Application of ^{29}Si MAS NMR and XRD: High-Silica Zeolite ZSM-5 as a Case in Point*

Zeolite ZSM-5, which is of particular interest because of its high catalytic activity and shape selective sorptive properties, has a very complex structure [104] and its study by ^{29}Si NMR spectroscopy is a demanding test for the reliability of this technique. It is advantageous to perform those studies with highly siliceous

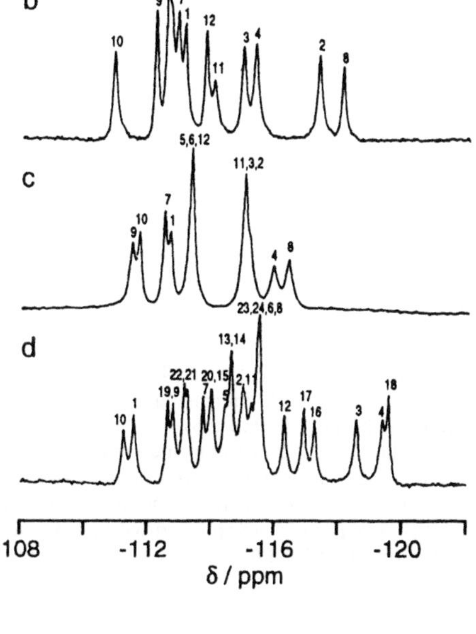

Fig. 13a–d. ^{29}Si MAS NMR spectra of high-silica zeolite ZSM-5 at 300 K (**a**), loaded with 2 molecules *p*-xylene per u.c. (**b**), at 403 K (**c**), loaded with 8 molecules *p*-xylene per u.c. (**d**). The assignments of the individual resonances are indicated. (Reprinted with permission from Ref. 34. Copyright 1990 American Chemical Society)

ZSM-5 samples (silicalite) of high crystallinity, since the ^{29}Si MAS NMR spectra of these materials are not affected by the Al content and lattice defects and thus show extremely narrow lines (see Figs. 4, 13, 14). XRD reveals that at ambient temperature ZSM-5 exists in a monoclinic form (space group $P2_1/n$) with 24 independent Si sites in the 96 Si atom unit cell, while at high temperature or loading the zeolite with two molecules of *p*-xylene/u.c. a reversible phase transition occurs to an orthorhombic form (space group Pnma) characterized by 12 crystallographically distinct Si sites per u.c. (see [104] and literature cited therein). Loading the high-silica ZSM-5 with 8 molecules *p*-xylene/u.c. induces a change of the orthorhombic Pnma symmetry to the likewise orthorhombic space group $P2_12_12_1$ with 24 distinct Si sites in the asymmetric unit [105]. The ^{29}Si MAS NMR spectra shown in Fig. 13 are in full agreement with these conclusions. While the spectra of the monoclinic form (Fig. 13a) and of the orthorhombic form containing 8 molecules *p*-xylene/u.c. (Fig. 13d) reveal the presence of 24 distinct Si sites, the spectra of the high-temperature (Fig. 13c) and the low-loaded (2 molecules *p*-xylene/u.c.) orthorhombic forms (Fig. 13b) show only 12 crystallographically inequivalent sites [34]. Detailed variable temperature studies of the unloaded sample demonstrate that the thermally induced change from the monoclinic 24 to the orthorhombic 12 Si-site form occurs over a very small temperature range of only about 2° [39]. However, Fig. 14 shows that the spectral changes with increasing *p*-xylene loadings are gradual, i.e., the different forms are highly ordered and coexist at intermediate

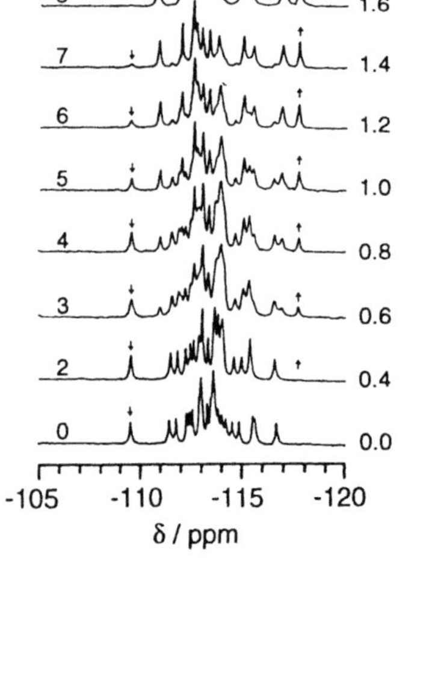

Fig. 14. ^{29}Si MAS NMR spectra at 295 K of high-silica zeolite ZSM-5 loaded with increasing amounts of *p*-xylene as indicated. The ↓ symbol shows the disappearance of a resolved characteristic of the monoclinic symmetry (24 T sites per u.c.) while the ↑ symbol indicates the appearance of one characteristic of the orthorhombic symmetry (12 T site per u.c.). (Reprinted with permission from Ref. 39. Copyright 1988 American Chemical Society)

conversions [34, 106]. It has further been shown that the effect of temperature on the spectra depends critically on the level of *p*-xylene loading of the samples, i.e., on the sorbate-induced initial symmetry at ambient temperature. The general conclusions derived from the combined effects of temperature and *p*-xylene concentration on the spectra have been summarized in the "phase diagram" presented in Fig. 15 [39]. It appears that in the low-temperature, low-concentration range, the unit cell has 24 crystallographically inequivalent Si sites of equal occupancy, while in the high temperature, high-concentration domain the unit cell contains 12 crystallographically distinct Si sites, again equally occupied, and mixtures of both phases exist in the intermediate range. Similar but even more complex effects on the ZSM-5 structure have been observed for other sorbate molecules, e.g., acetylacetone, dimethyl sulfoxide, toluene, pyridine, or benzene [39]. These conclusions agree, in general, with those from XRD studies but the substantial changes observed in the NMR peak positions demonstrate that high-resolution solid-state NMR is more sensitive to local site geometries than diffraction measurements. It should be noted, however, that the high sensitivity of the ^{29}Si MAS NMR spectra against temperature variation and/or sorbate loading is not universal for all zeolite structures, but has also been

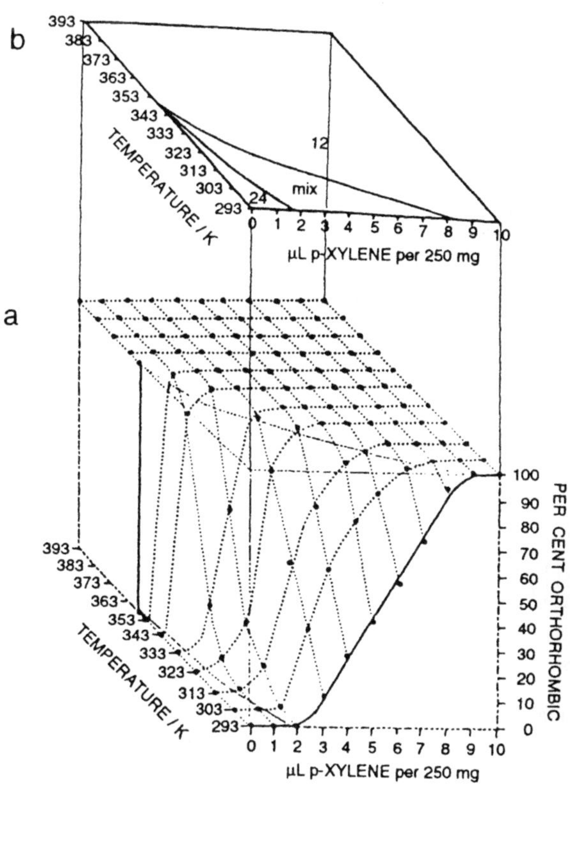

Fig. 15a. Three-dimensional representation of the effects of *p*-xylene loading and temperature on the crystal symmetry of high-silica zeolite ZSM-5 as derived from ^{29}Si MAS NMR data (*solid circles*). **b** is a two-dimensional projection of **a** showing the relationship of the orthorhombic (12 T sites per u.c.) and monoclinic (24 T sites per u.c.) phases with varying temperature and concentration of *p*-xylene. (Reprinted with permission from Ref. 39. Copyright 1988 American Chemical Society)

observed for some other microporous high-silica materials, e.g., zeolite ZSM-11 [107], titanosilicalite TS-1 [108], and dodecasil-3C (ZSM-39) [43, 109].

To achieve a deeper understanding of the local structural changes occurring upon heating or sorbate loading of the ZSM-5 samples, the detailed assignment of the numerous lines to the distinct Si sites of the ZSM-5 framework structure is required. To this end, two different approaches have been used. In the first approach, the δ-values of the different lines were calculated from the mean SiOSi bond angles of the X-ray structure using the correlation considered in Sect. 3.2, and compared with the experimental data [55]. Figure 16a shows a schematic presentation of the calculated spectrum of monoclinic ZSM-5 which agrees well with the experimental one presented in Fig. 16b. The assignment of the lines to the corresponding Si sites is indicated in Fig. 16a using the numbering of the X-ray structure [104]. Good agreement between the calculated and experimental data has also been obtained for the orthorhombic form. The second approach is based on the site connectivities derived from 2D INADEQUATE

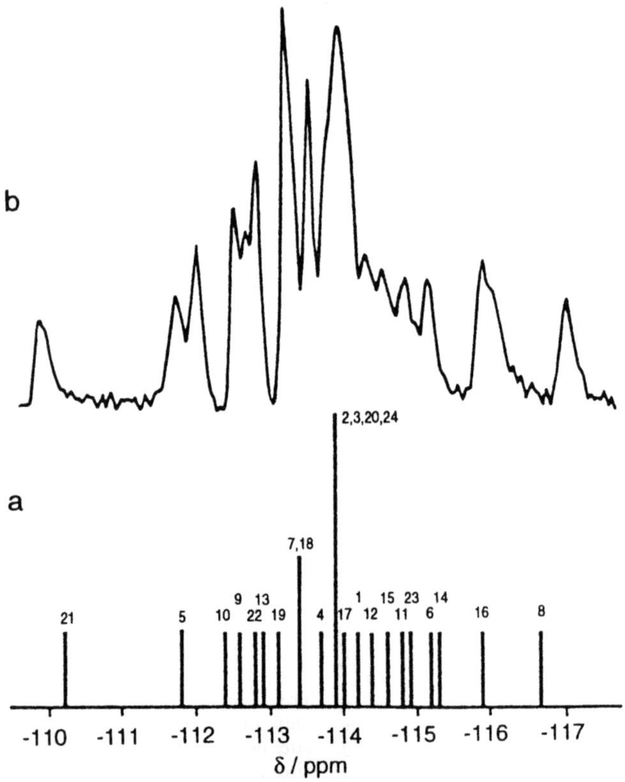

Fig. 16a. Calculated and **b** experimental ^{29}Si MAS NMR spectra of monoclinic high-slica zeolite ZSM-5. The *T*-site assignments are indicated in **a**. (Reprinted with permission from Ref. 55. Copyright 1990 Butterworth-Heinemann)

spectra (see Sect. 4.8 and Fig. 2) which yield immediately the site attribution of the lines indicated in the spectra of Fig. 13 [34]. The latter method is more universally applicable since it does not depend on the refined XRD structures of the various materials which are often not available. Comparison of the line assignments for the monoclinic ZSM-5 phase as given in Fig. 16a and Fig. 13a reveals some differences, particularly in the more crowded central region of the spectrum. This is probably due to the ambiguities introduced in the assignment via the bond angle/chemical shift correlation by the errors of the XRD data [55].

5.3 Hydrous Sodium Silicates: Crystals, Melts and Glasses

Type and distribution of the silicate species and/or Q^n structural units present in crystalline and amorphous silicate materials and their transformations upon de- and rehydration, melting, and glass formation can advantageously be studied by ^{29}Si MAS NMR. In agreement with the known crystal structures, single and narrow lines are observed e.g. in the spectra of a series of crystalline sodium silicates with chemical shifts typical of monomer silicate anions (Q^0) in $Na_2H_2SiO_4 \cdot nH_2O$ ($n = 4, 5, 7, 8$), $Na_3HSiO_4 \cdot nH_2O$ ($n = 5, 2$) [40, 110] and NaH_3SiO_4 [1], of dimer anions (Q^1) in $Na_6Si_2O_7$ [111], of middle groups (Q^2) in the single-chain silicate Na_2SiO_3 [111], and of branching groups (Q^3) in the layer silicate $Na_2Si_2O_5$ (α- and δ-form, the β-form shows two resonances due to crystallographical inequivalence [20]). In contrast, broad resonances appear in the ^{29}Si MAS NMR spectra of hydrous sodium silicate glasses, e.g., of composition $Na_2Si_2O_5 \cdot nH_2O$ [112] or $Na_2Si_4O_9 \cdot nH_2O$ [97] owing to overlapping lines of several Q^n environments present in the glass structure. However, type and relative proportions of the distinct Q^n coordinations could be determined by careful interpretation of the overlapping resonances and subsequent line deconvolution. A detailed Q^n speciation of the glasses could thus be given which shows clearly the increasing depolymerisation of the glass network with increasing water content.

The structural reorganization occurring in hydrous sodium silicates upon crystallization, melting, and glass formation has been monitored by in situ ^{29}Si MAS NMR measurements. This is demonstrated in Fig. 17 by a selection of spectra obtained from variable-temperature ^{29}Si MAS NMR experiments of the system $Na_2O \cdot SiO_2 \cdot 9H_2O$ [40, 110]. While crystalline $Na_2H_2SiO_4 \cdot 8H_2O$ at 298 K shows the expected single line of monomer silicate anions (Q^0), a number of new lines appears in the melt at 330 K, indicating that a major part of the monomer species is transformed into a variety of distinct oligomer anions. The presence of dimer (Q_2^1), cyclotrimer (Q_3^2), cyclotetramer (Q_4^2) and prismatic hexamer (Q_6^3) silicate anions (see Fig. 17) follows from the chemical shifts of the lines [40]. Interestingly, the spectrum of the melt is practically identical with that of an aqueous solution of the same composition prepared from water glass solution and NaOH. Recrystallization of the melt can be

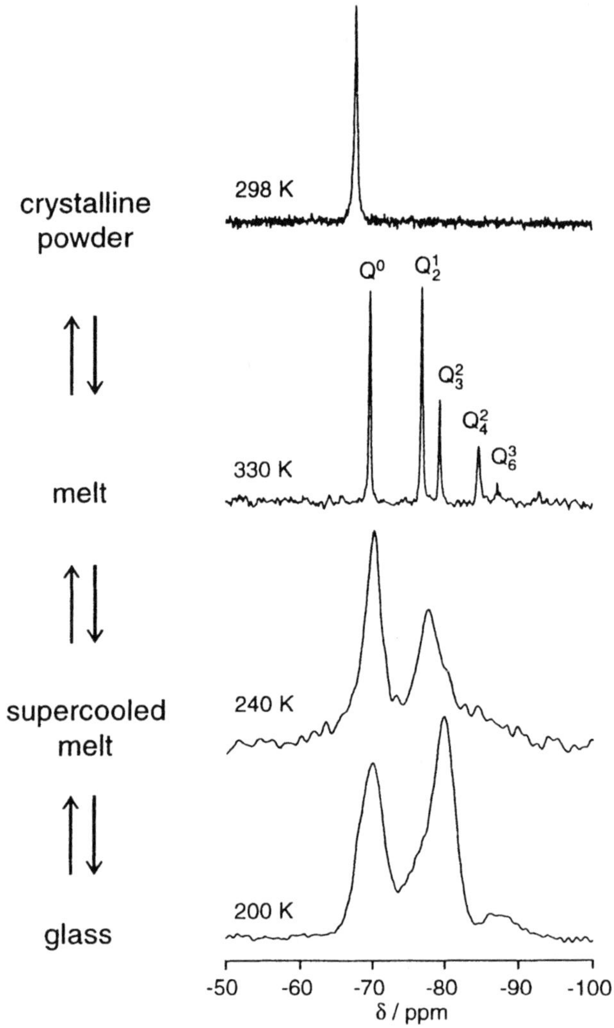

Fig. 17. In situ variable temperature ^{29}Si MAS NMR spectra of the crystalline powder (298 K), melt (330 K), supercooled melt (240 K), and glass (220 K) of composition $Na_2O \cdot SiO_2 \cdot 9H_2O$. The line assignment to monomer (Q^0), dimer (Q^1_2), cylotrimer (Q^2_3), cyclotetramer (Q^2_4), and prismatic hexamer silicate anions (Q^3_6) is indicated in the spectrum of the melt

achieved by cooling the sample slowly to about 270 K, and the single-line spectrum of the initial crystalline product appears again. However, if the melt is cooled quickly to 240 K, no recrystallization occurs and the highly viscous supercooled melt shows two broad lines and an even broader shoulder at the high-field side of the spectrum, the intensity distribution of which correspond approximately to that of the lines of the monomer and oligomer silicate species, respectively, in the melt at 330 K (Fig. 17). The line broadening may be due to freezing of the mobility and dynamic exchange of the silicate anions [113]. Further cooling below the glass transition temperature of 227 K [114] yields remarkable changes of the spectrum (see Fig. 17). The relative intensity of the line in the range of the dimer and cyclotrimer anions is now strongly enhanced and its maximum is shifted distinctly to higher field. This leads to the conclusion that a redistribution of the silicate anions occurs at the glass transition with a higher proportion of oligomer anions and, possibly, some preference of the cyclotrimer silicate species. It should be noted that all spectral changes shown in Fig. 17 are fully reversible, and that the composition of the sample is not changed during the NMR experiments. The spectrum of the crystalline sample and the glass were measured by the CP MAS technique, while single pulse MAS has been applied for the melt at 330 K and 240 K. (In fact, selection of the technique applied permits us to discriminate against the spectra of the crystalline solid and the liquid melt in the course of melting: While CP MAS yields the spectrum of the solid, single pulse MAS with short pulse intervals yields the spectrum of the liquid phase [40]). CP MAS experiments with variable contact times and single pulse MAS measurements of the deuterated sample reveal that the CP MAS spectrum of the glass is quantitatively reliable. Moreover, the results of corresponding investigations at compositions $Na_2O \cdot SiO_2 \cdot nH_2O$ with $n = 5$ and 6 are very similar to those described above for the system with $n = 9$.

6 References

1. Lippmaa E, Mägi M, Samoson A, Engelhardt G, Grimmer A-R (1980) J. Am. Chem. Soc. 102: 4889
2. Engelhardt G (1989) Trends in Anal. Chem. (TrAC) 8: 343
3. Klinowski J (1991) Chem. Rev. 91: 1459
4. Fyfe CA, Feng Y, Grondey H, Kokotailo GT, Gies H (1991) Chem. Rev. 91: 1525
5. Engelhardt G (1991) Stud. Surf. Sci. Catal. 58: 285
6. Kirkpatrick RJ (1988) In: Hawthorne FC (ed) Spectroscopic methods in mineralogy and geology. Mineralogical Society of America, Washington DC, p 341 (Reviews in Mineralogy, vol 18)
7. Grimmer A-R (1988) Exp. Techn. Physik 36: 81
8. Dupree R (1988) Exp. Techn. Physik 36: 315
9. Eckert H (1992) Progr. NMR Spectr. 24: 159
10. Wilson MA (1987) NMR Techniques and applications in geochemistry and soil chemistry. Pergamon, Oxford, p95; Stebbins JF (1993) In: Ahrens TH, (ed) Handbook of physical constants. American Geophysical Union, Washington DC
11. Hatfield GR, Carduner KR (1989) J. Mater. Sci. 24: 4209
12. Engelhardt G, Michel D (1987) High-resolution solid-state NMR of silicates and zeolites, John Wiley, Chichester
13. Weiden N, Rager H (1985) Z. Naturforsch. A40: 126
14. Spearing DR, Stebbins JF (1989) Am. Mineral. 74: 956
15. Smith KA, Kirkpatrick RJ, Oldfield E, Henderson DM (1983) Am. Mineral. 68: 1206
16. Grimmer A-R (1985) Chem. Phys. Lett. 119: 416
17. Grimmer A-R, v.Lampe F (1986) J. Chem. Soc. Chem. Commun. 219
18. Bray PJ, Emerson JF, Lee D, Feller SA, Bain DL, Feil DA (1991) J. Non-Cryst. Solids 129: 240
19. Herzfeld J, Berger J (1980) J. Chem. Phys. 73: 6021
20. Heidemann D, Hübert C, Schwieger W, Grabner P, Bergk K-H, Sarv P (1992) Z. Anorg. Allg. Chem. 617: 169
21. Pfleiderer B, Albert K, Bayer E, van de Ven L, de Haan J, Cramers C (1990) J. Phys. Chem. 94: 4189

22. Walther KL, Wokaun A, Baiker A (1990) Mol. Phys. 71: 769
23. Tuel A, Hommel H, Legrand AP, Chevallier Y, Morawski JC (1990) Colloids and Surfaces 45: 413
24. Leonardelli S, Facchini L, Fretigny C, Tougne P, Legrand AP (1992) J. Am. Chem. Soc. 114: 6412
25. Engelhardt G, Jancke H, Hoebbel D, Wieker W (1974) Z. Chem. 14: 109
26. Chuang I-S, Kinney DR, Bronnimann CE, Zeigler RC, Maciel GE (1992) J. Phys. Chem. 96: 4927
27. Legrand AP, Taibi H, Hommel H, Tougne P, Leonardelli S (1993) J. Non-Cryst. Solids, in the press
28. Ernst RR (1987) Chimia 41: 323
29. Kessler H, Gehrke M, Griesinger C (1988) Angew. Chem. 100: 507
30. Benn R, Grondey H, Brevard C, Pagelot A (1988) J. Chem. Soc. Chem. Commun. 102
31. Knight CTG, Kirkpatrick RJ, Oldfield E (1990) J. Non-Cryst. Solids 116: 140
32. Muller L, Kumar A, Ernst RR (1975) J. Chem. Phys. 63: 5490
33. Bax A, Freeman R, Frenkiel TA, Levitt MHJ (1981) J. Magn. Reson. 43: 478
34. Fyfe CA, Grondey H, Feng Y, Kokotailo GT (1990) J. Am. Chem. Soc. 112: 8812
35. Fyfe CA, Gies H, Feng Y (1989) J. Am. Chem. Soc. 111: 7702
36. Kolodziejski W, Barrie PJ, He H, Klinowski J (1991) J. Chem. Soc. Chem. Commun. 961
37. Kolodziejski W, Klinowski J (1992) Solid State NMR 1: 41
38. Vega AJ (1988) J. Am. Chem. Soc. 110: 1049
39. Fyfe CA, Strobl H, Kokotailo GT, Kennedy GJ, Barlow GE (1988) J. Am. Chem. Soc. 110: 3373
40. Koller H, Engelhardt G, Felsche J (1990) J. Chem. Soc. Chem. Commun. 371
41. Spearing DR, Farnan I, Stebbins JF (1992) Phys. Chem. Minerals 19: 307
42. Engelhardt G, Sieger P, Felsche J (1993) Anal. Chim. Acta, in the press
43. Tse JS, Desando M, Ripmeester JA, Handa YP (1993) J. Am. Chem. Soc. 115: 281
44. Stebbins JF (1991) Chem. Rev. 91: 1353
45. Doty D (1990) Abstracts, Rocky Mountain Conference in Analytical Chemistry, American Chemical Society, Washington DC, p 32
46. Taulelle F, Coutures JP, Massiot D, Rifflet JP (1989) Bull. Magn. Reson. 11: 318
47. Farnan I, Stebbins JF (1990) J. Am. Chem. Soc. 112: 32
48. Grimmer A-R (1993) In: Tossell JA (ed) Nuclear magnetic shieldings and molecular structure. Kluwer, Dordrecht, p 191
49. Vogel Ch, Wolff R, Radeglia R (1989) Z. Phys. Chem. (Leipzig) 270: 1073
50. Tossell JA (1991) Phys. Chem. Minerals 17: 654
51. Radeglia R, Engelhardt G (1985) Chem. Phys. Lett. 114: 28
52. Newsam JM (1987) J. Phys. Chem. 91: 1259
53. Weller MT, Wong G (1988) J. Chem. Soc. Chem. Commun. 1103
54. Engelhardt G, Luger S, Buhl, J Ch, Felsche J (1989) Zeolites 9: 182
55. Engelhardt G, van Koningsveld H (1990) Zeolites 10: 650
56. Ramdas S, Klinowski J (1984) Nature 308: 521
57. Alma-Zeestraten NCM, Dorrepaal, J. Keijsper J, Gies H (1989) Zeolites 9: 81
58. Grimmer AR, Radeglia R (1984) Chem. Phys. Lett. 106: 263
59. Engelhardt G (1989) Stud. Surf. Sci. Catal. 52: 151
60. Wolff R, Radeglia R, Vogel Ch (1990) J. Phys. Chem. Solids 51: 123
61. Wolff R, Radeglia R, Sauer J (1986) J. Mol. Struct. (Theochem) 139: 113
62. Wolff, R, Radeglia R, Vogel Ch, Sauer J (1989) J. Mol. Struct. (Theochem) 183: 223
63. Grimmer A-R, v.Lampe F, Mägi M (1986) Chem. Phys. Lett. 132: 549
64. Skibstedt J, Hjorth J, Jakobsen HJ (1990) Chem. Phys. Lett. 172: 279
65. Sternberg U (1988) Mol. Phys. 63: 249
66. Tossell JA, Lazzeretti P (1987) J. Chem. Phys. 112: 205
67. Malkin VG, Zhidimorov, GM (1990) Zeolites 10: 207
68. van Wazer JR, Ewig CS, Ditchfield R (1989) J. Phys. Chem. 93: 2222
69. Sheriff BL, Grundy HD, Hartman JS (1991) Eur. J. Mineral. 3: 751
70. Prabakar S. Rao KL, Rao CNR (1991) Chem. Phys. Lett. 183: 176
71. Maekawa H, Maekawa T, Kawamura K, Yokokawa T (1991) J. Non-Cryst. Solids 127: 53
72. Phillips BL, Howell DA, Kirkpatrick RJ (1992) Am. Mineral. 77: 704
73. Kirkpatrick RJ, Howell D, Phillips BL, Cong X-D (1991) Am. Mineral. 76: 673

74. (a) Stebbins JF, Kanzaki M (1991) Science 251: 294; (b) Kanzaki M, Stebbins JF, Xue X (1992) In: Syono Y, Manghnani MH (eds) High-pressure research: Applications to earth and planetary sciences, Terra Scientific, Tokyo/American Geophysical Union, Washington DC, p 89
75. Dupree R, Holland D, Mortuza MG, Collins JA, Lockyer MWG (1988) J. Non-Cryst. Solids 106: 403
76. Szu S-P, Klein LC, Greenblatt M (1992) J. Non-Cryst. Solids 143: 21
77. Stebbins JF (1991) Nature 351: 638
78. Carduner KR, Blackwell CS, Hammond WB, Reidinger F, Hatfield GR (1990) J. Am. Chem. Soc. 112: 4676
79. Carduner KR, Carter RO, Milberg ME, Crosbie GM (1987) Anal. Chem. 59: 2794
80. Dupree R, Lewis MH, Smith ME (1988) J. Am. Chem. Soc. 110: 1083
81. Dupree R, Lewis MH, Smith ME (1989) J. Am. Chem. Soc. 111: 5125
82. Hartman JS, Richardson MF, Sheriff BL, Winsborrow BG (1987) J. Am. Chem. Soc. 109: 6059
83. Wagner GW, Na B-K, Vannice MA (1989) J. Phys. Chem. 93: 5061
84. Lipowitz J, Turner GL (1988) Polym. Prep. 29: 74
85. Lippmaa E, Mägi M, Samoson A, Tarmak M, Engelhardt G (1981) J. Am. Chem. Soc. 103: 4992
86. Weiss CA, Altaner SP, Kirkpatrick RJ (1987) Am. Mineral. 72: 935
87. Massiani P, Fajula F, Figueras F, Sanz J (1988) Zeolites 8: 332
88. Takaishi T (1988) J. Chem. Soc. Faraday Trans. 1 84: 2967
89. Herrero CP, Sanz J, Serratosa JM (1989) J. Phys. Chem. 93: 4311
90. Herrero CP (1991) J. Chem. Soc. Faraday Trans. 87: 2837
91. Herrero CP (1991) J. Phys. Chem. 95: 3282
92. Herrero CP (1990) Chem. Phys. Lett. 171: 369
93. Phillips BL, Kirkpatrick RJ, Carpenter MA (1992) Am. Mineral. 77: 484
94. Newsam JM, Melchior MT, Beyerlein RA (1988) Mat. Res. Soc. Symp. Proc. 111: 125
95. Rocha J, Welch MD, Klinowski J (1991) J. Am. Chem. Soc. 113: 7100
96. Farnan I, Kohn SC, Dupree R (1987) Geochim. Cosmochim. Acta 51: 2869
97. Kümmerlen J, Merwin LH, Sebald A, Keppler H (1992) J. Phys. Chem. 96: 6405
98. Fyfe CA, Feng Y, Gies H, Grondey H, Kokotailo GT (1990) J. Am. Chem. Soc. 112: 3264
99. Fyfe CA, Gies H, Feng Y (1989) J. Chem. Soc. Chem. Commun. 1240
100. Fyfe CA, Feng Y, Grondey H, Kokotailo GT, Mar A (1991) J. Phys. Chem. 55: 3747
101. Fyfe CA, Gies H, Feng Y, Grondey H (1990) Zeolites 10: 278
102. Fyfe CA, Grondey H, Feng Y, Kokotailo GT, Ernst S, Weitkamp J (1992) Zeolites 12: 50
103. Farnan I, Stebbins JF (1990) J. Non-Cryst. Solids 124: 207
104. van Koningsveld H, Jansen JC, van Bekkum H (1990) Zeolites 10: 235
105. van Koningsveld H, Tunstra F, van Bekkum H, Jansen JC (1989) Acta Cryst. B45: 423
106. Fyfe CA, Feng Y, Grondey H, Kokotailo GT (1990) J. Chem. Soc. Chem. Commun. 1224
107. Fyfe CA, Gies H, Kokotailo GT, Pasztor C, Strobl H, Cox DE (1989) J. Am. Chem. Soc. 111: 2474
108. Tuel A, Ben Taarit Y (1992) J. Chem. Soc. Chem. Commun. 1578
109. Strobl H, Fyfe CA, Kokotailo GT, Pasztor CT (1987) J. Am. Chem. Soc. 109: 4733
110. Koller H, Engelhardt G (1993) (unpublished results)
111. Mägi M, Lippmaa E, Samoson A, Engelhardt G, Grimmer A-R (1984) J. Phys. Chem. 88: 1518
112. Uchino T, Sakka T, Ogata Y, Iwasaki M (1992) J. Phys. Chem. 96: 7308
113. Engelhardt G, Hoebbel D (1984) J. Chem. Soc. Chem. Commun. 1416
114. Felsche J, Ketterer B, Schmid RL (1984) Thermochim. Acta 77: 109

NMR of Solid Surfaces

Harry Pfeifer

Fachbereich Physik der Universität Leipzig, Linnéstraße 5, 04103 Leipzig, FRG

Table of Contents

After a short review of the structure and properties of zeolites and a discussion of two basic problems connected with an application of NMR spectroscopy, namely the signal-to-noise ratio and the measurement of chemical shifts in adsorbate-adsorbent systems, three main topics will be treated: The study of surface OH groups which may act as proton donators (Brönsted acid sites), of extra-framework aluminium species as typical electron-pair acceptors (Lewis acid sites) in zeolites, and of molecular transport. These topics have been chosen because they are of special importance for an understanding of the elementary processes controlling heterogeneous catalytic reactions.

NMR Basic Principles and Progress, Vol. 31

1 Introduction

Experimental and theoretical studies of the interaction between molecules and solid surfaces are of importance not only from the standpoint of basic research but also with respect to their application to heterogeneous catalysis and sorption. In general, the experimental studies may be grouped into three main directions of research:

(1) The use of technical adsorbents and catalysts. In these studies however, it cannot be excluded that the structure and chemical composition of the solid surfaces will change under the influence of the external atmosphere so that in contrast to statements about the macroscopic process it is very difficult or even impossible to derive unambiguous information about the elementary processes.

(2) Experiments performed with molecules adsorbed on single crystal surfaces in vacuum. The area of these surfaces amounts to a few mm^2 and the "vacuum" to residual gas pressures of not more than 10^{-8} Pa. This technique allows a direct study of the elementary processes taking place on solid surfaces of definite structure and composition [1–3].

(3) The use of highly porous crystals with a definite structure and definite chemical composition (zeolites) so that the elementary processes taking place in the single pores, which are of molecular dimension, can be studied spectroscopically, viz. by the signal of a large number of identical systems.

After a short review of the structure and properties of zeolites and a discussion of two basic problems connected with an application of NMR spectroscopy, namely the signal-to-noise ratio and the measurement of chemical shifts in adsorbate-adsorbent systems (Sect. 2), three further topics will be treated: Surface OH groups which may act as proton donators (Brönsted acid sites) (Sect. 3), extra-framework aluminium species (Sect. 4) as typical electron-pair acceptors (Lewis acid sites) in zeolites, and molecular transport (Sect. 5). These topics have been selected because they are of special importance for an understanding of the elementary processes controlling heterogeneous catalytic reactions.

For more general reviews or reviews on topics which are not treated in the present paper, the reader is referred to the following references: Surface Phenomena Investigated by Nuclear Magnetic Resonance [4], NMR of Organic Compounds Adsorbed on Porous Solids [5], NMR Studies of Simple Molecules on Metal Surfaces [6], High-Resolution Solid-State NMR of Silicates and Zeolites [7], Comparison of Analytical Techniques for the Determination of Silicon and Aluminium Content in Zeolites [8], NMR Spectroscopy and Zeolite Chemistry [9], ^{129}Xe NMR Study of Adsorbed Xenon: A New Method for Studying Zeolites and Metal-Zeolites [10, 18], Multinuclear Solid-State NMR in Silicate and Zeolite Chemistry [11], Applications of NMR Spectroscopy to the Study of Zeolite Synthesis [12], Methodologies to Establish the Structure and Composition of New Zeolitic Molecular Sieves [13], NMR of Zeolites, Silicates and Solid Catalysts [14], Nuclear Magnetic Resonance Spectroscopy in Studies of Catalysts [15], Solid-State NMR Studies of Molecular Sieve Catalysts [16], One- and Two-Dimensional High-Resolution Solid-State NMR Studies of Zeolite Lattice Structures [17], ^{1}H NMR Magic-Angle Spinning (MAS) Studies of Heterogeneous Catalysis [19], NMR in Studies of Zeolites [20].

2 Zeolites and NMR Spectroscopy

2.1 *Structure and Properties of Zeolites*

Zeolites are porous inorganic crystallites built from TO_4 tetrahedra [21]–[26]. In the case of the original zeolites, T represents silicon or aluminium while for the so-called SAPOs which are silicon substituted aluminium phosphates ($ALPO_4$s), T stands for Si, Al, and P. The original zeolites can be represented by the formula

$$(SiO_2)_n(AlO_2)^-M^+$$

where $n \geqq 1$ denotes the silicon-to-aluminium ratio of the zeolite framework which is built from corner-sharing SiO_4 and AlO_4 tetrahedra. Through the spatial arrangement of these tetrahedra, regular systems of intracrystalline cavities and channels of molecular dimensions are formed. Since the radius of the oxygen is distinctly larger than that of the cations (O^{2-}: 0.132 nm; Al^{3+}: 0.051 nm; Si^{4+}: 0.042 nm) the intracrystalline surface exposed to adsorbed molecules consists entirely of oxygen. The negative charge of the zeolite framework $(SiO_2)_n(AlO_2)^-$ and hence the concentration of the exchangeable cations M^+ which are located on extra-framework sites [27] can be made equal to zero by enhancing the silicon-to-aluminium ratio n so that in the limiting case $n = \infty$, zeolites exhibit a perfect homogeneous and neutral solid surface formed by oxygen atoms. The cations M^+ neutralizing the electric charge of the framework can be exchanged for other cations of equivalent net charge (e.g., $1/2\,M^{2+}$). An exchange with ammonium ions followed by a heat treatment leads to the formation of bridging OH groups

$$\overset{NH_4^+}{\equiv Al\text{-}O^-\text{-}Si\equiv} \rightarrow \overset{H}{\equiv Al\text{-}O\text{-}Si\equiv} + NH_3\uparrow$$

which are known to be able to protonate adsorbed molecules (Brönsted acid sites). On the other hand, steaming of zeolites by a fast heating of hydrated zeolites (deep-bed treatment, cf. Sect. 3.3) or by a treatment with wet air at elevated temperatures leads to the formation of extra-framework aluminium species which may act as Lewis acid sites (electron-pair acceptors).

In the last 40 years more than 30 natural zeolites and more than 100 synthetic zeolites have been identified. A typical example of large-pore zeolites is the so-called zeolite Y or faujasite which is denoted as zeolite X if n assumes values less than 1.4. In this type of zeolites large cavities (supercages) with an inner diameter of 1.16 nm are tetrahedrally interconnected via windows (12-oxygen rings) with an inner diameter of 0.74 nm. Through this arrangement of the large cavities, an equal number of small cavities (cubooctahedra) are formed with an inner diameter of 0.66 nm which are accessible from the large cavities through windows (6-oxygen rings) of only 0.25 nm inner diameter. The unit cell of this zeolite contains 8 large and 8 small cavities and is built up by 384 oxygen atoms and 192 T atoms. The four windows to the large cavities of the small-pore zeolite A are formed by 8-oxygen rings with an inner diameter of 0.41 nm and 0.5 nm for Na-A and Ca-A, respectively. The medium-pore zeolite ZSM-5 (for $n = \infty$ it is denoted as silicalite) consists of intersecting channels formed by 10-oxygen rings with a medium inner diameter of 0.54 nm. The concentration of the large cavities or intersections for the zeolites mentioned here is $4.18 \times 10^{20}\,g^{-1}$ for a silicon-to-aluminium ratio $n = \infty$. For finite values of n, the concentration per gramme decreases slightly in dependence on n and on the sort of exchangeable cations (M, see above) but in most cases $4 \times 10^{20}\,g^{-1}$ may be taken as an approximate value. Zeolites are largely used in industry for a wide range of applications including drying through adsorption, separation of gases and of hydrocarbons ("molecular sieves"), ion exchange (additives to washing agents) and catalysis. Catalysis based on zeolite Y containing bridging OH groups have a cracking activity which is orders of magnitude greater than that of the formerly used silica-alumina catalysts. The synthetic zeolite ZSM-5 is very useful for many applications since its high silicon content ($n \geqq 15$) gives it high thermal stability, while the channel diameters lead to quite striking effects of shape selectivity including the ability to synthesize gasoline from methanol (MTG = methanol-to-gasoline process).

Summarized, zeolites share the following five properties: (1) Well-defined crystalline structure, (2) large internal surface areas since 4×10^{20} supercages/g with ca. $2.5 \times 10^{-18}\,m^2$ inner surface per supercage correspond to a specific surface area of ca. $1000\,m^2/g$, (3) uniform pores, (4) good thermal stability, and (5) adjustable concentration of Brönsted acid sites, that make them attractive not only for practical application but also for basic research including NMR spectroscopy inspite of its relatively small signal-to-noise ratio compared with other spectroscopic methods.

2.2 *NMR Signal-to-Noise Ratio*

The amplitude U of the NMR signal induced in the coil of the tuned receiver circuit by the precessing magnetization M_0 immediately after a short intense $\pi/2$ pulse (free induction decay) is given by the Faraday law:

$$U = \mu_0 \omega_0 M_0 Q n A \eta \qquad (1)$$

where μ_0 denotes the permeability of vacuum,

$$\omega_0 = \gamma B_0 \tag{2}$$

the resonance frequency, γ the gyromagnetic ratio of the resonating nuclei and B_0 the intensity of the constant magnetic field. Q is the quality factor of the circuit, nA the total coil area crossed by the magnetic flux, and η the filling factor which can be approximated by the ratio of the cross sections of the sample and of the coil. The magnetization M_0 of the sample containing N spins can be shown (Curie's law) to be proportional to the product $\gamma^2 I(I+1)B_0 N/T$, where I is the spin quantum number of the resonating nuclei and T the absolute temperature of the sample. During a measuring time T_m the experiment can be repeated z times with

$$z \leqq T_m/T_1 \tag{3}$$

since the time interval between two succeeding experiments must be larger than the longitudinal relaxation time T_1 in order to restore the thermal equilibrium of the magnetization before the application of the next $\pi/2$ pulse. Hence, it follows for the total NMR signal after z accumulations

$$U_s = zU \propto (T_m/T_1) Q n A \eta N \gamma^3 I(I+1) B_0^2/T. \tag{4}$$

Neglecting other sources of noise than the thermal noise of the circuit one may write for the total noise after z accumulations

$$U_n \propto (T_m/T_1)^{1/2} (R_p T \Delta\nu)^{1/2} \tag{5}$$

with the electrical resistance R_p of the circuit at resonance and the band width $\Delta\nu$ of the receiver. Furthermore, we assume that the electrical losses are caused by the electrical resistance R_s of the coil for which we can write

$$R_s \propto n \cdot \sqrt{A} \cdot \sqrt{\omega_0} \tag{6}$$

regarding the influence of the skin effect upon the effective cross section of the wire. Denoting the self inductance of the coil by L with

$$L \propto n^2 A \tag{7}$$

and taking into consideration the relation

$$Q = R_p/(\omega_0 L) = \omega_0 L/R_s \tag{8}$$

it follows for the total signal-to-noise ratio

$$U_s/U_n \propto K N I(I+1) \gamma^{5/2} B_0^{3/2} T^{-3/2} (\Delta\nu)^{-1/2} (T_m/T_1)^{1/2} \tag{9}$$

with the constant

$$K = \eta (QA)^{1/2} \tag{10}$$

which depends on the construction of the probehead. To increase the signal-to-noise ratio, the band width $\Delta\nu$ should be as small as possible (cf. Eq. (9)). Its lower limit is given by the condition that all Fourier components of the

NMR signal which decays with the time constant T_2 (transverse relaxation time) must be transmitted:

$$\Delta\nu_{\min} \approx T_2^{-1} \tag{11}$$

so that Eq. (9) may be rewritten as

$$(U_s/U_n)_{\max} \propto NI(I+1)\gamma^{5/2}(B_0/T)^{3/2}(T_m T_2/T_1)^{1/2}. \tag{12}$$

Hence, it follows for the minimum number of nuclei $N_{\min}$ which are necessary to yield a reasonable value (e.g. 10) of the signal-to-noise ratio

$$N_{\min} \propto (\gamma/\gamma_p)^{1/2} r^{-1}(T/B_0)^{3/2}(T_1/T_2 T_m)^{1/2} \tag{13}$$

with the relative signal intensity

$$r = 4(\gamma/\gamma_p)^3 I(I+1)/3. \tag{14}$$

γ_p denotes the gyromagnetic ratio of the proton. Values for the spin quantum number I, for γ/γ_p which is the ratio of the resonance frequencies for a given magnetic field, for the relative signal intensity r, and for the natural abundance n.a. of various nuclei which are of interest in NMR studies of solid surfaces have been collected in Table 1. Data for other nuclei can be found, e.g., in Ref. [28]. In order to make use of Eq. (13), some empirical results for the values of $N_{\min}$ shall be given, assuming a signal-to-noise ratio of 10, a measuring time $T_m = 600$ s, a magnetic field intensity $B_0 = 7.04598$ T corresponding to a proton resonance frequency $\omega_0/2\pi = 300$ MHz, and a temperature $T = 300$ K:

(1) ^{1}H MAS NMR of surface OH groups. In this case typical values of T_1 and T_2 (MAS enhanced value) are ca. 1 s and ca. 1 ms, respectively, and one finds experimentally

$$N_{\min} \approx 10^{18}. \tag{15}$$

Table 1. Values for the spin quantum number I, the resonance frequency in MHz at a field of 2.34866 T, the relative signal intensity (cf. Eq. (14)) and the natural abundance (n.a.) of various nuclei

nucleus	I	$100\gamma/\gamma_p$	r	n.a.
^{1}H	1/2	100	1	99.98%
^{7}Li	3/2	38.863	0.29	92.58%
^{11}B	3/2	32.084	0.17	80.42%
^{13}C	1/2	25.144	0.0159	1.108%
^{15}N	1/2	10.133	0.00104	0.37%
^{17}O	5/2	13.557	0.0291	0.037%
^{23}Na	3/2	26.451	0.0925	100%
^{27}Al	5/2	26.057	0.21	100%
^{29}Si	1/2	19.865	0.00784	4.67%
^{31}P	1/2	40.481	0.0663	100%
^{129}Xe	1/2	27.660	0.0212	26.44%
^{195}Pt	1/2	21.499	0.00994	33.8%

For 0.2 g of a zeolite with typically 4×10^{20} cavities per gram, this value corresponds to ca. 0.01 OH groups per cavity, and for 0.2 g of a catalyst with a specific surface area of S m^2/g to ca. $5/S$ OH groups per nm^2.

(2) ^{13}C NMR of adsorbed carbon monoxide enriched to 100% in ^{13}C. In this case typical values of T_1 and T_2 are ca. 0.1 s and 20 ms, respectively, and one finds experimentally

$$N_{min} \approx 4 \times 10^{18} \tag{16}$$

corresponding to ca. 0.05 CO per cavity or ca. 20/S carbon monoxide molecules per nm^2 of the catalyst.

(3) ^{27}Al MAS NMR of extra-framework aluminium species. These are characterized by large differences in their line widths and therefore of the respective values for the transverse relaxation time. Octahedrally coordinated extra-framework aluminium which can be observed in hydrated zeolites reveals the smallest line widths with not more than 300 Hz or $T_2 \geqq 10^{-3}$ s. The minimum concentration of these species which has been determined experimentally at 70 MHz for 0.2 g of hydrated ZSM-5 is of the order of 0.05 sites per channel intersection [29] corresponding to $N_{min} \approx 4 \times 10^{18}$. Values for the longitudinal relaxation time of extra-framework aluminium species are not known. It can be assumed that they are not greater than the values for framework aluminium which are between 0.3 and 70 ms depending on the temperature and degree of hydration [30]. Since for other extra-framework aluminium species line widths of the order of 30 kHz have been observed (see, e.g., [31]), it can be estimated by the use of Eq. (13) that N_{min} for these species is not much smaller than 4×10^{20} corresponding to as much as 1 extra-framework aluminium per channel intersection which is NMR invisible. The only way to reduce this value is the application of ultra high ($\Omega/2\pi \geqq 18$ kHz) MAS speeds [31] since even through the application of the more advanced ^{27}Al DOR NMR method the line widths of extra-framework condensed aluminium species of low symmetry cannot be reduced remarkably (see Sect. 4).

2.3 Measurement of Chemical Shifts

As can be found in textbooks on solid state physics, the local magnetic field acting upon a nucleus i is given by

$$B_i = \{1 + (4\pi/3 - \alpha_i + k_{Li})\chi_{vi} - \sigma_i\}B_0 \tag{17}$$

where B_0 denotes the intensity of the applied constant magnetic field, χ_{vi} and α_i the volume susceptibility and the demagnetization factor of the sample, respectively, $k_{Li}\chi_{vi}B_0$ the contribution to the local magnetic field resulting from the magnetic dipoles inside the hypothetical Lorentzian sphere and σ_i the chemical shielding of the nucleus. Experimentally, instead of the absolute value of the resonance frequency

$$\omega_i = \gamma B_i \tag{18}$$

another quantity, viz. the so-called apparent chemical shift δ^{app} is determined by a comparison of ω_i with the resonance frequency

$$\omega_r = \gamma B_r \tag{19}$$

of a reference:

$$\delta_i^{app} = (\omega_r - \omega_i)/\gamma B_0. \tag{20}$$

With respect to the question which reference should be used in the case of adsorbate-adsorbent systems like zeolites there exist two possibilities: (1) The reference sample contains the same zeolite specimen but loaded with a suitable sort of molecules (external reference), or the sample under study is loaded additionally with a small amount of these molecules itself (internal reference). In the latter case it must be proved by additional experiments that the loading of the sample does not affect the resonances to be studied. (2) The reference sample contains only the reference material either in the gaseous or liquid or solid state.

For an application of the first method, the zeolite must be unloaded or loaded only to such a degree that it does not change the total volume susceptibility χ_v of the samples. χ_v is given by Wiedemann's relation

$$\chi_v/(p_a\rho_a + p_z\rho_z) = p_a\chi_{va}/\rho_a + p_z\chi_{vz}/\rho_z \tag{21}$$

where ρ_a and p_a denote the density and the relative mass of the adsorbate, respectively. ρ_z and $p_z = 1 - p_a$ are the corresponding quantities for the zeolite. Apparently, the condition $\chi_v \approx \chi_{vz}$ can be realized in any case by a suitably small value of p_a. For the reference sample, a relation quite analogous to Eq. (17) holds:

$$B_r = \{1 + (4\pi/3 - \alpha_r + k_{Lr})\chi_{vr} - \sigma_r'\}B_0 \tag{22}$$

where $\alpha_r = \alpha_i = \alpha$ (the same shape of both samples) and $\chi_{vr} = \chi_{vi} = \chi_{vz}$ (the same zeolite specimen with $p_a \ll 1$). σ_r' denotes the chemical shielding of the reference molecules r in the adsorbed state. The essential point is now, that molecules such as methane, cyclohexane or tetramethylsilane have been found for which the chemical shielding of the ^1H nuclei is the same in the gaseous and adsorbed states [32]. For practical application methane has to be preferred due to its smallness. Using

$$\sigma_{rH}' = \sigma_{rH} \tag{23}$$

it follows from Eqs. (17)–(20) and (22):

$$\delta_{iH} \equiv \sigma_{rH} - \sigma_{iH} = \delta_{iH}^{app} + (k_{Lr} - k_{Li})\chi_v. \tag{24}$$

About the Lorentzian factors k_{Lr} and k_{Li}, no reliable predictions can be made in the case of heterogeneous systems. Assuming however an isotropic distribution of the magnetic dipoles inside the hypothetical Lorentzian sphere, k_L can be put equal to zero and Eq. (24) simplifies to

$$\delta_{iH} \equiv \sigma_{rH} - \sigma_{iH} = \delta_{iH}^{app} \tag{25}$$

where σ_{rH} is the chemical shielding of the reference molecule (e.g., methane) in the gaseous state. For other nuclei than ^{1}H like ^{13}C or ^{15}N no molecules have been found for which the chemical shielding is the same in the gaseous and the adsorbed states. Even for "inert" molecules like those mentioned above, there will be in general a contribution of van der Waals' interaction to the chemical shielding. The order of magnitude of this contribution can be estimated from the resonance shift between the gaseous and liquid states [33]. Its value is relatively large if the nuclei under study are found on the periphery of the molecule. Some typical examples are given in Table 2. With respect to these shifts, the influence of the susceptibility upon the local field (cf. Eq. (17)) is relatively small: Using the Faraday method, the diamagnetic, paramagnetic, and ferromagnetic contributions to the susceptibility of various zeolite specimens have been measured [37]. Some results are presented in Table 3. The concentration of the iron impurities were determined by emission spectroscopy, and the specimens studied were either zeolites synthesized in laboratory by S. P. Shdanov (S) [38] under conditions of controlled purity, or commercial zeolites (C). By definition, the volume susceptibility χ_v, the mass susceptibility χ_m, and the molar susceptibility χ_M are related by

$$\chi_v = \rho\chi_m = (\rho/M)\chi_M \tag{26}$$

Table 2. Values of the ^{13}C and ^{15}N NMR shift between the liquid and gaseous states for various molecules. Values are in ppm and positive if the resonance of the liquid is shifted to higher frequencies

Molecule	NMR shift	Reference
$\underline{C}H_3-\underline{C}H_3$	2.23	[33]
$\underline{C}H_2=\underline{C}H_2$	5.30	[33]
$(\underline{C}H_3)_3CH$	2.4	[34]
$(CH_3)_3\underline{C}H$	0.6	[34]
$C_5H_5\underline{N}$	−8.9	[35]
$\underline{N}H_3$	18	[36]

Table 3. Experimental values for the diamagnetic (χ_{md}), the paramagnetic (χ_{mp}), and the ferromagnetic (χ_{mf}) contributions to the magnetic mass susceptibility (χ_m) of various zeolite specimens with different concentration of paramagnetic impurities. The measurements were performed at 295 K and with a magnetic field of intensity $B_0 = 1$ T [37]

zeolite	Si/Al	Fe^{3+}/ppm	$\chi_{md} \times 10^9$/ $m^3\,kg^{-1}$	$\chi_{pm} \times 10^9$/ $m^3\,kg^{-1}$	$\chi_{mf} \times 10^9$/ $m^3\,kg^{-1}$
Na-X (S)	1.37	<5	−0.395	0	0
Na-Y (S)	1.76	40	−0.390	0.008	0
Na-Y (S)	2.26	110	−0.385	0.02	0
Na-Y (S)	2.69	280	−0.420	0.05	0.025
Na-X (C)	1.2	550	−0.350	0.1	0.090
Na-Y (C)	2.6	500	−0.340	0.08	0.080

where ρ denotes the density (the value of the apparent density for zeolites is ca. 500 kg/m^3) and M the molar mass in kg/mol. Considering these results, the mass susceptibility χ_m of pure zeolites is ca. -0.4×10^{-9} m^3/kg and does not depend on the temperature T or on the intensity B_0 of the applied magnetic field. This value is in reasonable agreement with the molar susceptibility of -29.6×10^{-12} m^3/mol for SiO_2 [28] which corresponds (cf. Eq. (26)) to a mass susceptibility of -0.49×10^{-9} m^3/kg. In contrast, for the commercial zeolite containing ca. 550 ppm Fe^{3+} ions as impurities, the susceptibility depends on both the temperature and the intensity of the magnetic field: For $B_0 = 1$ T, the mass susceptibility increases from -0.3×10^{-9} m^3/kg to -0.2×10^{-9} m^3/kg if the temperature decreases from about 550 K to 170 K. At room temperature the mass susceptibility of this specimen increases by a factor of 1.3 if B_0 decreases from 1 T to 0.5 T. For the usual form of the NMR samples, viz. cylindrical tubes oriented perpendicular to the magnetic field B_0 with a value of about 3 for the ratio between the filling height and the diameter of the the sample, the demagnetization factor α is ca. 1.8π [37] so that it follows for the relative contribution of the susceptibility to the local field (cf. Eq. (17))

$$\{B_i - B_0(1-\sigma_i)\}/B_0 = (-1.47 - k_L)\chi_v. \tag{27}$$

Assuming for simplicity $k_L = 0$, this contribution amounts to 0.3 ppm for pure zeolites and to 0.1 ppm for the commercial zeolite Na-Y (C) at room temperature and $B_0 = 1$ T. Hence, for other nuclei than ^1H like ^{13}C or ^{15}N where the difference of the chemical shielding between the gaseous and adsorbed states amounts to several ppm [5], [7] the second method, namely the use of a sample containing only the reference material either in the gaseous or liquid or solid state is recommended.

3 Surface OH Groups

3.1 *Strength of Acidity and Chemical Shift*

The elementary step of a Brönsted acid catalyzed reaction is the proton transfer from a surface OH group ZOH to the adsorbed molecule M:

$$ZOH + M \rightleftarrows ZO^- + MH^+. \tag{28}$$

Therefore, apart from a possible rate determining influence of diffusion (see Sect. 5), the efficiency of a catalyst will be determined by three independent parameters for each sort of OH groups: (1) The strength of acidity as defined qualitatively by the ability to protonate an adsorbed molecule, (2) the concentration, and (3) the accessibility of the respective hydroxyls. Apparently, the ability to protonate an adsorbed molecule will depend both on the properties of the acidic site ZOH and on the molecule M. In order to define a quantity which characterizes the protonation ability of the ZOH group independent of the particular molecule, the reaction described by Eq. (28) is decomposed into the deprotonation of the ZOH group:

$$ZOH \rightleftarrows ZO^- + H^+ \tag{29}$$

and the protonation of the molecule M:

$$M + H^+ \rightleftarrows MH^+. \tag{30}$$

Denoting the standard Gibbs free energy change for the deprotonation (Eq. (29)) by ΔG^0_{DP}, the so-called strength S_a of gas phase acidity of the ZOH group is defined by

$$S_a = (\Delta G^0_{DP})^{-1} \tag{31}$$

so that increasing values of S_a correspond to decreasing values of ΔG^0_{DP} as it should be. In order to compare the strength of the gas phase acidity with the deprotonation energy ΔE_{DP} of a ZOH group, a quantity which follows from quantum chemical calculations, one must take into consideration that ΔG^0_{DP} is the sum of three terms:

$$\Delta G^0_{DP} = \Delta E_{DP} + \Delta E^0_{DP} + \Delta G^t_{DP}. \tag{32}$$

ΔE^0_{DP} denotes the change of the zero point energy and ΔG^t_{DP} the change of Gibbs' free energy resulting from the conversion of the three vibrational degrees of freedom of the proton in the ZOH group into its three translational degrees of freedom after it has left this group. In Ref. [39] it has been shown that the zero-point energy change ΔE^0_{DP} is fairly constant and that the contribution of ΔG^t_{DP} can be neglected. Therefore, according to Eq. (31) a change of ΔE_{DP} should be directly related to a change of the reciprocal value of the strength of gas phase aciditiy.

With respect to nuclear magnetic resonance spectroscopy the question arises whether the chemical shift of the proton resonance δ_H of a ZOH group can be related to its strength of acidity S_a. For such a comparison however, the OH groups must be isolated which requires for gaseous samples simply a measurement at low pressure while in the case of surface OH groups three conditions must be fulfilled: (1) The measurements have to be performed with evacuated samples since adsorbed molecules may strongly shift the proton resonance of a ZOH group to higher values of δ_H. This shift depends both on the strength of acidity and on the particular molecule and may be of the order of 10 ppm (see Sect. 3.4). (2) The concentration of the OH groups must be not too high in order to exclude a direct interaction of neighbouring hydroxyls. This condition is fulfilled in general for surface OH groups of zeolites. (3) Additional electrostatic interactions of the OH groups with the framework as e.g. caused by the formation of a hydrogen bond to other oxygen atoms of the surface must be excluded: If one plots the chemical shift $\delta_H(r_{oo})$ of OH groups measured in various solid samples as a function of the O-H$\cdots$O distance r_{oo}, one finds a nearly linear dependence which can be approximated by

$$\delta_H(r_{oo}) = \delta_H + s \cdot (300 - r_{oo}/\text{pm}) \tag{33}$$

for the interval

$$240\,\text{pm} < r_{oo} < 300\,\text{pm}. \tag{34}$$

δ_H denotes the chemical shift of the isolated OH group ($r_{oo} \geqq 300\,\text{pm}$), and for the slope s a value of 0.26 ppm/pm and 0.38 ppm/pm follows from data published in [40] and [41], respectively. It is this dependence which leads to different values of the chemical shift for the bridging OH groups pointing into large and small cavities of zeolites (lines *b* and *c*, respectively, see Sect. 3.3).

With regard to a relation between δ_H and the strength of acidity S_a of isolated OH groups, the results of two different experimental studies shall be mentioned:

(1) Experimental results for molecules in the gaseous state. Values of δ_H which have been measured [42] for seven aliphatic alcohols (methanol, ethanol, *n*-propanol, *i*-propanol, *tert*-butanol, *i*-butanol, neopentanol), 2,2,2-trifluoroethanol, and benzyl alcohol in the gaseous state under conditions (ca. 10^4 Pa, 148.6 °C) where hydrogen bonding has little effect on chemical shifts, correlate well with experimental results [43], [44] for the standard Gibbs free energy change of deprotonation ΔG^0_{DP}. The slope of the line derived from a linear regression [42] is given by

$$d(\Delta G^0_{DP})/d(\delta_H) = -43 \pm 2\,\text{kJ ppm}^{-1}\,\text{mol}^{-1} \tag{35}$$

with a correlation coefficient of 0.99.

(2) Experimental results for isolated bridging OH groups of zeolites with different silicon-to-aluminium ratio (n). It is apparent that an enhancement of the electronegativity of the framework of a zeolite will decrease the electron density of the OH bond of a surface OH group which should lead to an increased strength of acidity. In Fig. 1 values for Sanderson's intermediate electronegativity S_m are plotted together with experimental values for δ_H in dependence on the silicon-to-aluminium ratio n. The intermediate electronegativity is defined as the geometric mean of the atomic electronegativities [45], so that it follows for zeolites of composition $HAlO_2(SiO_2)_n$

$$S_m = (S_H \cdot S_{Al} \cdot S_O^{2n+2} \cdot S_{Si}^n)^{1/(3n+4)} \tag{36}$$

with $S_H = 3.55$, $S_{Al} = 2.22$, $S_O = 5.21$, $S_{Si} = 2.84$ as the atomic electronegativities [45]. The nearly identical functional dependence of S_m and δ_H is another experimental proof for the usefulness and sensitivity of δ_H as a measure for the strength of acidity of surface OH groups. With respect to the absolute values of δ_H plotted in Fig. 1 it should be mentioned that these are smaller by about 0.3 ppm with respect to a former paper [46]. This difference is due to the influence of the susceptibility of the zeolite powder which has not been regarded in [46] and which could be eliminated by the use of methane as an inner standard [47] (see Sect. 2.3).

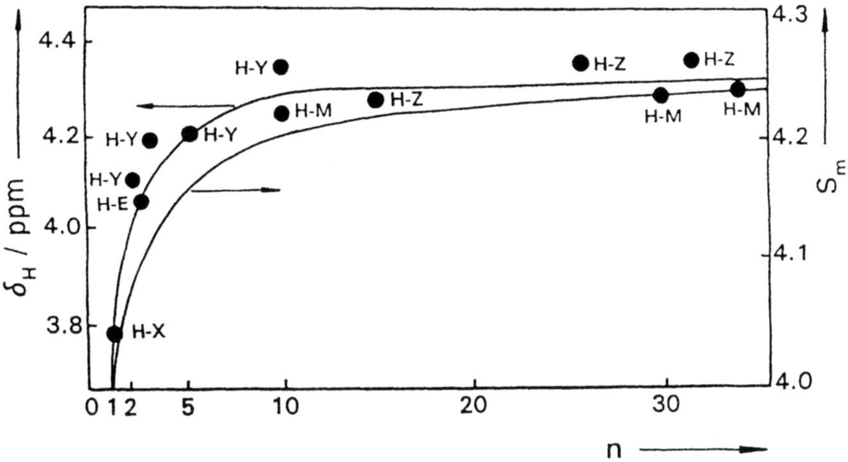

Fig. 1. Values for the intermediate electronegativity S_m and the chemical shift δ_H of isolated bridging OH groups (line *b*, corresponding to the HF band in IR spectroscopy) in dependence on the silicon-to-aluminium ratio *n* for various zeolites [51]

Table 4. Results of ab initio SCF calculations of the chemical shielding σ_H and the deprotonation energy ΔE_{DP} for various molecules and model clusters of surface OH groups [48]. For comparison in the last column experimental values for the chemical shift $\delta_H = \sigma_{TMS} - \sigma_H$ are given. (*) The chemical shielding calculated for methanol to 31.9 ppm and the shift of methanol relative to gaseous TMS has been found experimentally to be 0.02 ppm [42]

	ΔE_{DP}/kJ mol^{-1}	δ_H/ppm	$\delta_H^{(*)} = 31.92 - \sigma_H$/ppm	δ_H^{exp}/ppm
H_2O	1700	30.9	1.02	0.3(*g*); 4.8(1)
H_3O^+	734	23.5	8.42	19.2 (1) [49]
CH_3OH	1677	31.9	0.02	0.02 [42]
$CH_3OH_2^+$	804	25.7	6.22	5.3 – 9.1 [50]
terminal SiOH				1.8 – 2.3 [51]
H_3SiOH	1561	31.0	0.92	
$Si(OH)_4$	1563	29.7	2.22	
terminal AlOH				(– 0.5) – 1 [51]
H_2AlOH	1615	29.7	2.22	
$Al(OH)_3$	1636	30.7	1.22	
H_3POAlH_2OH	1494	31.2	0.72	
terminal POH				1.5 – 4 [52]
H_3AlOPH_2OH	1285	28.7	3.22	
$OP(OH)_3$	1438	28.5	3.42	
terminal BOH				2.5 ± 0.2 [52]
$B(OH)_3$	1571	29.0	2.92	
Si/Al-bridging OH groups				3.8 – 4.4 [51]
$H_3SiOHAlH_3$	1329	28.9	3.02	
$H_3SiOHAl(OH)_3$	1353	27.8	4.12	

Qualitatively, the results of both experimental studies which lead to the conclusion that increasing values of the chemical shift δ_H correspond to increasing values of the strength of acidity S_a of the OH group can be explained as follows: An increasing strength of acidity which is tantamount to a decreasing value of the deprotonation energy should correspond to an increase of the net atomic charge (i.e. to a decrease of the electron density) of the hydrogen atom. On the other hand, the chemical shielding σ_H of the ^1H nucleus will decrease with an increase of the net atomic charge so that the chemical shift δ_H of the OH group which is defined by

$$\delta_H = \sigma_{TMS} - \sigma_H \tag{37}$$

with σ_{TMS} denoting the chemical shielding of the ^1H nucleus in tetramethylsilane, should increase with increasing strength of acidity. In a recent paper, J. Sauer et al. [48] have performed ab initio SCF calculations of both the chemical shielding σ_H and the deprotonation energy ΔE_{DP} for various molecules and model clusters of surface OH groups. The latter include models for terminal surface OH groups ($H_3Si\underline{OH}$ and $Si(\underline{OH})_4$ for SiOH; $H_2Al\underline{OH}$, $Al(\underline{OH})_3$ and $H_3POAlH_2\underline{OH}$ for AlOH; $OP(\underline{OH})_3$ and $H_3AlOPH_2\underline{OH}$ for POH) and for

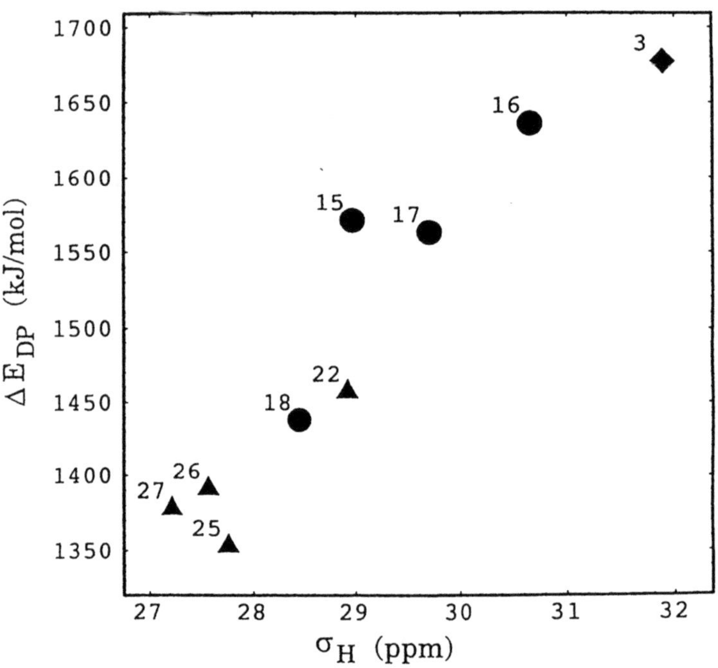

Fig. 2. Dependence of the deprotonation energy ΔE_{DP} on the shielding constant σ_H for molecules that are relevant as models of surface OH groups [48]: $CH_3O\underline{H}$ (3), $B(OH)_3$ (14), $Al(OH)_3$ (15), $Si(OH)_4$ (16), $OP(OH)_3$ (17), $H_3POAlPH_2O\underline{H}$ (19), $H_3SiO\underline{H}\cdot BH_3$ (22), $H_3SiO\underline{H}\cdot AlH_3$ (23), $H_3SiO\underline{H}\cdot Al(OH)_3$ (24)

Si/Al-bridging OH groups ($H_3Si\underline{OH}AlH_3$ and $H_3Si\underline{OH}Al(OH)_3$). Some of the results are collected in Table 4 and compared with experimental values for the chemical shift δ_H. After a comparison of all results (not shown in Table 4) including most different molecules with XOH groups, the authors of Ref. [48] come to the conclusion that there is no general linear dependence between δ_H and ΔE_{DP}. The reason is that the contributions of the orbitals other than the OH bond orbital which are localized on the O atom of the OH group, namely the lone pairs and the OX bond orbital, are significant and not constant within the set of molecules studied. It is the variation of these non-OH-bond contributions which spoils the linear dependence between δ_H and ΔE_{DP} for some XOH groups. However, for sets of hydroxyl groups in similar bonding environments and especially for surface hydroxyl groups which are responsible for the Brönsted acidity of catalysts, a linear relation has been found. This is due to the fact that these OH groups are all bonded to cations (B, Al, Si or P) whose first coordination sphere consists of oxygen atoms only. Figure 2 shows the data for those models studied in [48] that have the property of a complete oxygen coordination around the cations. As a reference also the methanol molecule with $\Delta E_{DP} = 1677\,\mathrm{kJ\,mol^{-1}}$ and $\sigma_H = 31.9$ ppm has been included. It appears that these points (except methanol) are reasonably close to a straight line with a slope of $84 \pm 12\,\mathrm{kJ\,mol^{-1}\,ppm^{-1}}$ so that the chemical shift δ_H can be used in fact as a reliable and sensitive measure for the strength of acidity of surface OH groups.

3.2 Resolution of ^1H MAS NMR Spectra

It has been shown in Sect. 3.1 that the isotropic chemical shift δ_H of isolated surface OH groups may be taken as a quantitative measure for the strength of gas phase acidity S_a. Therefore, the accuracy of such a measurement of S_a depends on the residual line width $\Delta\nu^{MAS}$ of the resonance line (central line in the case of a spinning sideband pattern) of the corresponding ^{1}H MAS NMR signal. For a quantitative discussion we shall use the line narrowing factor v defined by

$$v = (M_2/M_2^{MAS})^{1/2} \tag{38}$$

where M_2 denotes the second moment of the static line (non-spinning sample) and M_2^{MAS} the second moment of the corresponding central line of the MAS spectrum. For spin-1/2 nuclei v may be determined by the following six factors:

(1) An inhomogeneity of the constant magnetic field B_0. For a probehead designed with non-magnetic materials the residual line widths can be made less than 0.001 ppm through a careful shimming of the magnet so that this influence is negligible in general.

(2) Misadjustment of the magic angle. If Θ denotes the angle between the axis of rotation and the direction of B_0, the line narrowing factor is given by

$$v = 2|1 - 3\cos^2\Theta|^{-1}. \tag{39}$$

In order to ensure $\upsilon > 100$, the deviation of Θ from the magic angle $\theta_m = \arccos(1/\sqrt{3})$ must be less than 0.4°.

(3) Anisotropy of the magnetic susceptibility. Assuming that the particles are spheriods and that their magnetic susceptibility can be described by an axial symmetric tensor (χ_p, χ_p, χ_s) with

$$|\chi_p + 2\chi_s| \gg 3|\chi_p - \chi_s| \tag{40}$$

it follows [53] for the line narrowing factor

$$\upsilon = |(\chi_p + 2\chi_s)(\chi_p - \chi_s)^{-1}|. \tag{41}$$

However, it could be shown experimentally by a comparison of the residual line widths measured for zeolites of cubic $(\chi_p = \chi_s)$ and orthorhombic $(\chi_p \neq \chi_s)$ symmetry that the anisotropy of the magnetic susceptibility is of no measurable influence upon the resolution of the ^1H MAS NMR signals of the surface OH groups [54].

(4) Thermal motion. It is well-known (see, e.g., [55]) that the protons, especially those of bridging hydroxyl groups are not fixed in space. The mean residence time τ of a briding hydroxyl proton at the oxygen atom where it experiences a magnetic dipole–dipole interaction with the neighbouring aluminium nucleus depends on residual amounts of ammonium and on the temperature. Typical values of this quantity τ may be as small as 0.5 ms at room temperature. Assuming that the magnetic dipole–dipole interaction between ^1H and ^{27}Al is dominant and that the jump lengths are large compared to the $^1H-^{27}Al$ distance of the bridging OH groups so that the mean residence time τ is identical with the correlation time of the magnetic dipole–dipole interaction, the envelope $\Phi^{MAS}(t)$ of the ^1H MAS NMR free induction decay can be shown [56] to be given by

$$\Phi^{MAS}(t) = \exp\{-(M_2/3)[2F(\Omega, t) + F(2\Omega, t)]\} \tag{42}$$

where

$$\begin{aligned} F(\Omega, t) = {} & t\tau(1 + (\Omega\tau)^2)^{-2} \\ & + \tau^2((\Omega\tau)^2 - 1)(1 + (\Omega\tau)^2)^{-2}[1 - \exp(-t/\tau)\cos\Omega t] \\ & - 2\Omega\tau^3(1 + (\Omega\tau)^2)^{-2}\exp(-t/\tau)\sin\Omega t. \end{aligned} \tag{43}$$

Ω denotes the spinning rate of the sample and M_2 the second moment (in s^{-2}) of the proton resonance line for the rigid lattice $(\tau = \infty)$ and $\Omega = 0$. For the special case of a static sample $(\Omega = 0)$, Eqs. (42) and (43) yield

$$\Phi(t) = \exp\{-M_2\tau[t + \tau\exp(-t/\tau) - \tau]\} \tag{44}$$

which is known as the Anderson-Weiss formula [57]. If the condition $\tau \gg t$ is fulfilled, where t is of the order of the observation time, i.e., for the rigid lattice, Eqs. (42) and (43) simplify to

$$\Phi^{MAS}(t) = \exp\{-(M_2/3\Omega^2)[2(1 - \cos\Omega t) + (1 - \cos 2\Omega t)/4]\}. \tag{45}$$

This formula has been derived by Schneider et al. [58]. Under the condition $t \gg \tau$ which can be written as

$$t \approx (\pi \Delta \nu^{\mathrm{MAS}})^{-1} \gg \tau \tag{46}$$

Eqs. (42) and (43) yield an exponential decay corresponding to a Lorentzian line with a line width (FWHM)

$$\Delta \nu^{\mathrm{MAS}} = (M_2/3\pi)\{2\tau[1+(\Omega\tau)^2]^{-1} + \tau[1+(2\Omega\tau)^2]^{-1}\} \tag{47}$$

which agrees with the result of Andrew et al. [59]. For $\Omega \neq 0$, the line width given by Eq. (47) exhibits a maximum as a function of the correlation time at $\Omega\tau \approx 1$ and a monotonous decrease of the value of $\Delta\nu^{\mathrm{MAS}}$ both at smaller (line narrowing caused by thermal motion) and at larger (line narrowing caused by MAS) values of the correlation time. In this latter case it follows

$$\Delta \nu^{\mathrm{MAS}} = (3/4\pi) M_2/(\tau \Omega^2) \tag{48}$$

so that the residual line width must strongly depend on the spinning rate Ω if it is controlled by thermal motion. In general, the MAS NMR spectrum does not consist of a single Lorentzian line. Instead it comprises a central line and spinning sidebands. The whole pattern is described quantitatively by the Fourier transform of Eq. (42), which, however, can be performed only numerically. In Fig. 3 typical spectra are shown which have been calculated for $\Omega/2\pi = 3$ kHz, a static line width $(M_2 2 \ln 2)^{1/2}/\pi = 8$ kHz and values of the correlation time τ varying between 5 µs and 10 ms [56]. The essential point is now that an inspection of such spectra shows that the spinning sidebands must exhibit a much larger line width (FWHM) than the central line if the residual line width is controlled by thermal motion. In the case of bridging OH groups the FWHM of the first

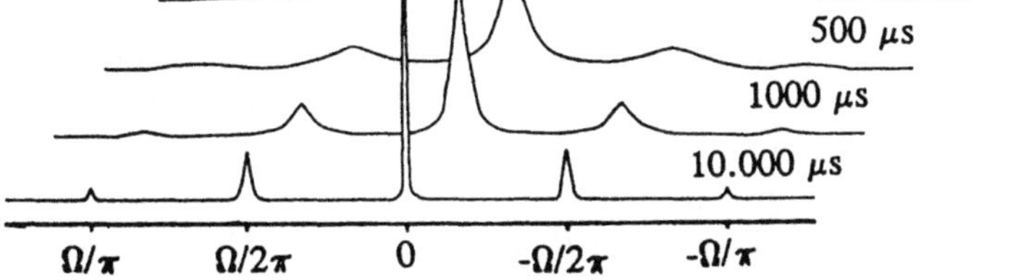

Fig. 3. Calculated ^1H MAS NMR spectra for dominating inhomogeneous magnetic dipole–dipole interaction (for bridging OH groups: ^1H-^{27}Al) and large translational jumps of the proton. Speed of rotation $\Omega/2\pi = 3$ kHz, static line width $(M_2 2\ln 2)^{1/2}/\pi = 8$ kHz, and values of τ varying between 5 µs and 10 ms [56]

spinning sideband should be approximately twice as large as the FWHM of the central line for $\tau = 0.5\,\mathrm{ms}$ and $\Omega/2\pi = 3\,\mathrm{kHz}$ [60]. This however has not been observed [61] so that also thermal motion must be excluded as a dominant source of the residual line width at room temperature.

(5) Heteronuclear magnetic dipole–dipole interaction with quadrupole nuclei. The MAS NMR signals of $I = 1/2$ nuclei in solids show a characteristic deformation of the line shape, mostly observed as a line broadening, if the nuclei are magnetically coupled to a nucleus with spin $S > 1/2$ (see [62, 63, 64]). Since for bridging OH groups the magnetic dipole–dipole interaction with the neighbouring ^{27}Al nucleus ($S = 5/2$) is the dominating line-broadening interaction for the static ^{1}H NMR signal [65, 66], this effect may be significant. The line broadening due to the magnetic dipole–dipole interaction of a spin $I = 1/2$ with a neighbouring quadrupole bearing nucleus (spin $S > 1/2$) has been calculated numerically by Böhm et al. [63] under two assumptions: (a) the electric field gradient tensor (EFGT) at the site of the spin S is of axial symmetry and (b) the symmetry axis of the EFGT is parallel to the I–S distance vector $\mathbf{r}$. The result can be approximated by a simple relation if the NMR frequency ν_L of the ^{27}Al nuclei is greater than their quadrupole frequency ν_Q. This condition is fulfilled in most experiments with superconducting magnets and one may write for the line narrowing factor

$$v = 1.1(\nu_L/\nu_Q). \tag{49}$$

In the general case of an asymmetric EFGT (asymmetry parameter η) and an arbitrary orientation of $\mathbf{r}$ with regard to the EFGT (angle ϑ and ϕ) it follows [61] for the line narrowing factor v instead of Eq. (49)

$$v \approx 0.9[S\sqrt{F(\vartheta,\phi,\eta)}]^{-1}(\nu_L/\nu_Q) \tag{50}$$

with

$$F(\vartheta,\phi,\eta) = (2/1215)\{2 + \eta^2 + 1.5(1 - \eta^2/3)(\cos\vartheta)^2 - \eta(\sin\vartheta)^2\cos 2\phi + (73/4)[1 - 3(\cos\vartheta)^2 - \eta(\sin\vartheta)^2\cos 2\phi]^2\}. \tag{51}$$

The value of $F(\vartheta,\phi,\eta)$ depends strongly on the orientation of $\mathbf{r}$ relative to the principal axes of the EFGT. Its maximum value of 154/1215 is achieved for $\vartheta = \phi = 0$ and $\eta = 1$. By a measurement at different values of B_0, however, it is possible to decide by the use of Eq. (50) whether the residual line width is determined by the magnetic dipole–dipole interaction with the aluminium nuclei of the bridging OH groups. For a zeolite H-ZSM-5 with a silicon-to-aluminium ratio $n = 15$ which may be taken as a typical example for catalysts with a low concentration of bridging OH groups (catalysts of type *l*), the line narrowing factor has been measured to be 34 and 23 at $B_0 \approx 7.05\,\mathrm{T}$ and $11.7\,\mathrm{T}$, respectively [54]. However, since ν_L is proportional to B_0, the line narrowing factor for 11.7 T should be 1.66 times greater than that for 7.05 T, and therefore we must conclude that the residual line width is not determined by the magnetic dipole–dipole interaction with the ^{27}Al nuclei. For the other limiting case of a high concentration of bridging OH groups (catalysts of type *h*) represented by a zeolite H-Y with a silicon-to-aluminium ratio $n = 2.6$, a similar result has been found [61]: Also for these zeolites the residual line width is not determined by the magnetic dipole–dipole interaction with the ^{27}Al nuclei if B_0 is not smaller than ca. 7 T.

(6) Combined influence of magnetic dipole–dipole interactions and of the anisotropy of the chemical shift. The second moment M_2 of the NMR signal of spin $I = 1/2$ nuclei without thermal motion contained in a static powder sample is given by (see, e.g., [67])

$$M_2 = M_2^{\mathrm{II}} + M_2^{\mathrm{IS}} + M_2^{\mathrm{CSA}} \tag{52}$$

with

$$M_2^{\mathrm{II}} = (3/5)(\mu_0/4\pi)^2\gamma_{\mathrm{I}}^4\hbar^2 I(I+1)N_{\mathrm{I}}^{-1}\Sigma_{\mathrm{i}\neq\mathrm{j}}r_{\mathrm{ij}}^{-6}, \tag{53}$$

$$M_2^{\mathrm{IS}} = (4/15)(\mu_0/4\pi)^2\gamma_{\mathrm{I}}^2\gamma_{\mathrm{S}}^2\hbar^2 S(S+1)N_{\mathrm{I}}^{-1}\Sigma_{\mathrm{j}}\Sigma_{\mathrm{k}}r_{\mathrm{jk}}^{-6}, \tag{54}$$

$$M_2^{\mathrm{CSA}} = (1/5)\gamma_{\mathrm{I}}^2 B_0^2\delta_{\mathrm{CSA}}^2(1+\eta_{\mathrm{CSA}}^2/3). \tag{55}$$

Here γ_{I}, N_{I}, δ_{CSA} and η_{CSA} denote the gyromagnetic ratio and the number of the resonanting nuclei, the anisotropy and the asymmetry parameter of their shielding tensor, respectively. γ_{S} is the gyromagnetic ratio, S the spin and N_{S} the number of the non-resonating nuclei. r_{ij} denotes the distance between two resonating spins and r_{jk} the distance between a resonating and a non-resonating spin. Using the concept of the average Hamiltonian [68], the influence of these interactions upon the residual line width of MAS NMR spectra has been calculated [67]. Under the assumption that the central line and the spinning sidebands can be separated, the following equation for the second moment M_2^{MAS} of the central line of the MAS spectrum was obtained

$$M_2^{\mathrm{MAS}} = \Omega^{-2}\{K_1 M_2^{\mathrm{II}}M_2^{\mathrm{II}} + K_2 M_2^{\mathrm{II}}M_2^{\mathrm{IS}} + K_3 M_2^{\mathrm{SS}}M_2^{\mathrm{IS}} + K_4 M_2^{\mathrm{II}}M_2^{\mathrm{CSA}}\}. \tag{56}$$

M_2^{SS} is defined analogously to Eq. (53) by interchanging I and S, and characterizes the strength of the homonuclear magnetic dipole–dipole interaction between the non-resonating nuclei. The coefficients $K_1 - K_4$ depend on geometrical parameters [67]. For isolated pairs of resonating nuclei without inter-pair interaction and also for linear chains without inter-chain interaction it can be shown that $K_1 = 0$, i.e., the residual line width vanishes. The same holds if there is no homonuclear magnetic dipole–dipole interaction ($M_2^{\mathrm{II}} = M_2^{\mathrm{SS}} = 0$). For isolated systems of three resonating spins at the corners of an equilateral triangle it follows $K_1 = 0.023$ and for a simple cubic lattice formed by the resonating spins $K_1 = 0.041$ [69]. Assuming that the shape of the central line of the MAS spectrum can be approximated by a Gaussian function, the residual line width is given by

$$\Delta\nu^{\mathrm{MAS}} = (M_2^{\mathrm{MAS}} 2\ln 2)^{1/2}/\pi \tag{57}$$

so that it will be inversely proportional to the spinning rate. This behaviour has been found experimentally by Wind [70] for the ^1H MAS NMR signal of a Torlon rotor. The presence of an additional heteronuclear magnetic dipole–dipole interaction and/or anisotropy of the chemical shielding may lead to a further enlargement of the residual line width. For a system of two resonating and one one-resonating nuclei, or one resonating and two non-resonating nuclei at the corners of an equilateral triangle, K_2 assumes the same value 0.087 [67]. If the shielding tensors for the spins of an isolated pair of resonating nuclei have different orientations, K_4 does not vanish, its maximum value, however, is only of the order of 0.1. For an application of this theory to bridging OH groups it is necessary to discuss separately zeolites with a low (catalysts of type *l*) and high (catalysts of type *h*) concentration of OH groups:

Catalysts of type l. For a zeolite H-ZSM-5 with a silicon-to-aluminium ratio $n = 15$, the experimentally determined residual line width $\Delta\nu^{MAS}$ is plotted in Fig. 4 as a function of the spinning rate $\Omega/2\pi$ [54]. Apparently, there is no dependence of the line width on the spinning rate in contrast to Eqs. (56) and (57). From this result one must conclude that the experimentally observed residual line width is not caused by the combined influence of magnetic dipole–dipole interactions including the anisotropy of the chemical shift. Considering also the results which were already discussed above (cf. (1)–(5)), the residual line width of the ^{1}H MAS NMR signal of bridging OH groups in catalysts of type *l* must be determined therefore by the distribution width of the isotropic value of the chemical shift: The observed value of $\Delta\nu^{MAS}/300\,\text{MHz} \approx 0.8\,\text{ppm}$ (cf. Fig. 4) must be interpreted as a natural limit for this catalyst which cannot be further reduced by an enlargement of the magnetic field B_0 or of the spinning rate Ω or by an application of combined rotation and multiple pulse sequences (CRAMPS-technique) [54].

Catalysts of type h. As a typical example a zeolite H-Y with a silicon-to-aluminium ratio $n = 2.6$ shall be considered. At first, it is necessary to estimate the values of M_2^{II}, M_2^{SS}, M_2^{IS} and M_2^{CSA}. From an analysis of the sideband pattern of the partially deuterated (in order to reduce the homonuclear magnetic dipole–dipole interaction) zeolite H-Y [71] it could be derived $M_2^{IS} \approx 4.5 \times 10^8\,\text{s}^{-2}$ and $M_2^{CSA} \approx 1.2 \times 10^8\,\text{s}^{-2}$. The value of M_2^{II} has been estimated to ca. $0.4 \times 10^8\,\text{s}^{-2}$

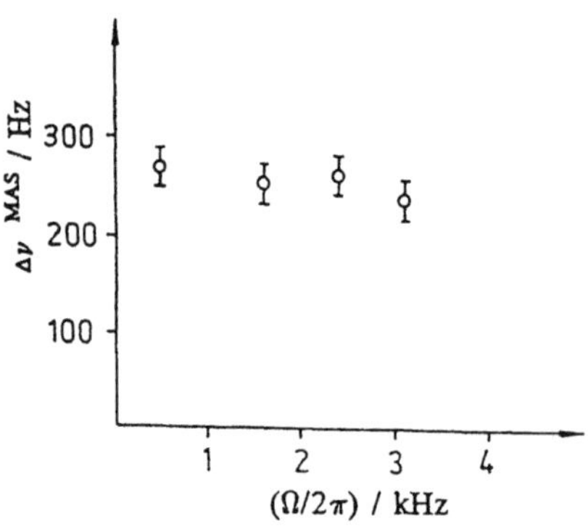

Fig. 4. Dependence of the residual line width $\Delta\nu^{MAS}$ on the spinning rate $\Omega/2\pi$ for the ^{1}H MAS NMR signal of bridging OH groups of a catalyst of type *l* (zeolite H-ZSM-5 with a silicon-to-aluminium ratio $n = 15$). Resonance frequency 300 MHz [54]

[66] from the static ^{1}H NMR spectrum of the same zeolite but non-deuterated. Since the density of aluminium atoms is approximately equal to the density of the bridging OH groups in shallow-bed treated (i.e., a non-dealuminated framework, see Sect. 3.3) zeolites, it follows from Eq. (53)

$$M_2^{II}/M_2^{SS} \approx (\gamma_I/\gamma_S)^4 I(I+1)/S(S+1) = 18.6 \tag{58}$$

or $M_2^{II} \gg M_2^{SS}$. Since M_2^{IS} is the largest contribution to M_2 and since K_1, K_3 and K_4 are of the same order or smaller than K_2, Eqs. (56) and (57) simplify to

$$\Delta\nu^{MAS} \approx (K_2 M_2^{II} M_2^{IS} 2 \ln 2)^{1/2}/(\Omega\pi). \tag{59}$$

In order to separate two lines which are ca. 1 ppm apart (e.g., the ^{1}H MAS NMR signals of bridging OH groups pointing into the large and into the small cavities

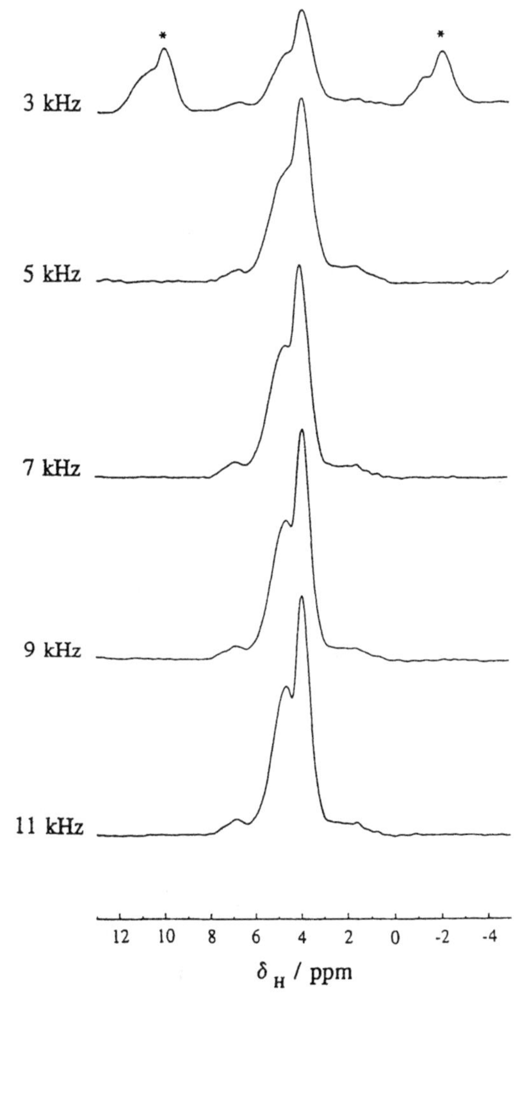

Fig. 5. ^{1}H MAS NMR spectra of a catalyst of type *h* (zeolite H-Y with a silicon-to-aluminium ratio $n = 2.6$) measured at a resonance frequency of 500 MHz with increasing values of the spinning rate $\Omega/2\pi$. * denotes spinning sidebands [61]

of H-Y zeolites), the line width must be smaller than 500 Hz at an ^1H NMR frequency of 500 MHz. Inserting the estimated values of M_2^{II} and M_2^{IS} together with $K_2 \approx 0.1$ into Eq. (59), the spinning rate $\Omega/2\pi$ must be approximately equal to or larger than 5 kHz. Experimental results are shown in Fig. 5 [61] and one may conclude that spinning rates of ca. 10 kHz are sufficient to eliminate the influence of the homonuclear magnetic dipole–dipole interaction on the residual line width of the ^1H MAS NMR spectra of catalysts of type *h*. The line width achieved under these conditions is then given by the distribution width of the isotropic value of the chemical shift (natural line width) which cannot be further reduced and which is achieved for catalysts of type *l* at much smaller values of the spinning rate.

Summarized, the following statements can be made:

(a) The natural line width for the signals of Brönsted acid sites (bridging OH groups) in the ^1H MAS NMR spectra of evacuated zeolite catalysts is given by the distribution width of the isotropic values of the chemical shift (distribution width of the strength of gas phase acidity).

(b) In the case of zeolites containing a low concentration of OH groups (catalysts of type *l*) as, e.g., H-ZSM-5, it is only necessary to apply constant magnetic fields of a sufficiently large intensity (B_0 not smaller than ca. 7 T) in order to achieve the natural line width.

(c) For zeolites with a high concentration of OH groups (catalysts of type *h*), as, e.g., non-dealuminated H-Y zeolites, in addition to a sufficiently strong magnetic field ($B_0 \geqq 7$ T) the homonuclear magnetic dipole–dipole interaction must be reduced in order to achieve the natural line width. This can be performed either by the use of high spinning rates (cf. Fig. 5) or by a partial deuteration of the OH groups [67] or by the application of a multiple pulse sequence during the period of the free induction decay (CRAMPS technique, see, e.g., Ref. [76] and Fig. 6).

3.3 *Evacuated Zeolites*

As was shown in Sect. 3.1, ^1H MAS NMR measurements of Brönsted acidity must be performed with evacuated samples. Since most of the zeolite catalysts are hygroscopic and fully hydrated in the as-synthesized form, it is necessary to remove the sorbed water which requires the application of relatively high temperatures usually in conjunction with evacuation of the sample. Through such a procedure however, the zeolite framework may be affected which mostly leads to a partial displacement of aluminium into a non-framework environment. This latter effect must be not disadvantageous since extra-framework aluminium species are a source of Lewis acidity (see Sect. 4) and hence of increased catalytic activity of a zeolite. Therefore it is reasonable to distinguish between two limiting procedures of dehydration: The shallow-bed treatment which should be used in order to evacuate the zeolite without affecting its framework, and the deep-bed treatment where the dehydration is combined with a generation of Lewis acid sites. In a typical shallow-bed treatment, which has been used – if not mentioned otherwise – in all experiments described here, the sample depth should be as small as possible (a few mm only). After a slow evacuation to a pressure below 1 Pa, the sample is heated slowly (10 K per hour) to the maximum temperature (typically 400 °C) where it is kept for 24 hours. Finally, the sample is cooled naturally over a period of several hours still under vacuum and sealed. In contrast, for a deep-bed treatment a deep layer (depth in excess of 20 mm) is heated relatively fast (100 K per hour) to the maximum temperature (self-steaming of the zeolite) where it is evacuated over a period of 24 hours. Finally, the sample is cooled under vacuum to room temperature and sealed.

The first highly resolved ^{1}H MAS NMR spectrum of a catalyst is shown in Fig. 6. Since it was a zeolite containing a high concentration of bridging OH groups (catalyst of type *h*) it was necessary to reduce the influence of the homonuclear magnetic dipole–dipole interaction which has been achieved in this experiment by the simultaneous application of a multiple pulse sequence (CRAMPS technique) which however was not mentioned in [72]. Another problem was the fast rotation of evacuated powder samples which requires a high degree of axial symmetry of the sealed glass ampoules. In the ^{1}H MAS NMR spectra of evacuated zeolites containing only oxygen, silicon and aluminium in the framework, five lines can be separated which have been denoted [72] as lines *a*, *b*, *c*, *d*, and *e*:

Line a which appears in the interval between ca. 1.8 and 2.3 ppm is caused by non-acidic (silanol) groups. In contrast to an older paper [73] it seems

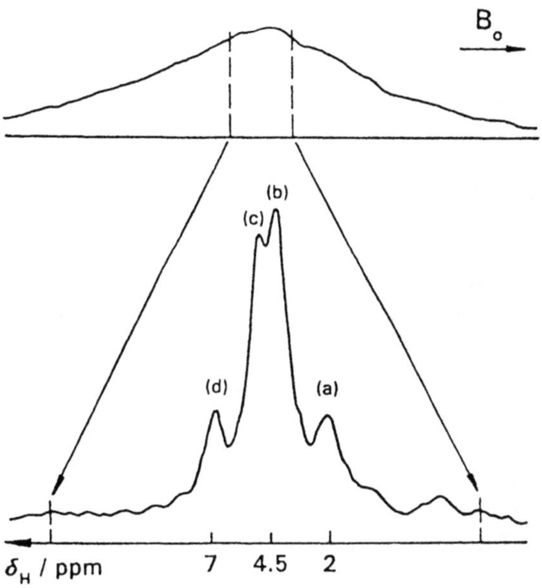

Fig. 6. Enhancement of spectral resolution for a typical catalyst of type *h* (shallow-bed treated (24 h at 300 °C) zeolite H-Y with a silicon-to-aluminium ratio $n = 2.6$) by spinning of the evacuated powder sample which is contained in a sealed glass ampoule about the magic angle and simultaneous application of a multiple pulse sequence (CRAMPS technique). *Top*: ^{1}H NMR spectrum of the static sample. *Bottom*: ^{1}H CRAMPS NMR spectrum with a spinning rate of 2.5 kHz. 270 MHz NMR frequency, room temperature [72]

necessary to mention that a distinction between single (SiOH) and geminal ($Si(OH)_2$) hydroxyl groups is not possible by this technique since the difference between the corresponding values of δ_H is less than 0.1 ppm [51]. This is in agreement with an ^1H CRAMPS study of silica gel prepared under various stages of dehyration where only three peaks at 1.7 ppm, 3.0 ppm and 3.5 ppm could be found which where identified as isolated (non-hydrogen bonded) silanols, hydrogen-bonded silanols and physisorbed water, respectively [97]. In ^{29}Si MAS NMR spectra however, well separated signals due to single and geminal silanols appear at about −100 ppm and about −90 ppm, respectively [98]. Unfortunately, the possibility of ascertaining these two species is limited to adsorbents built up only by silicon and oxygen since the insertion of other metal atoms as, e.g., aluminium leads to overlapping signals in the ^{29}Si MAS NMR spectra [7].

Line b at 3.8 to 4.4 ppm is ascribed to acidic OH groups (Brönsted acid sites) which are known to be of bridging type (SiOHAl). The value of δ_H increases with increasing silicon-to-aluminium ratio of the zeolite (cf. Fig. 1).

Line c at 4.8 to 5.6 ppm is also ascribed to acidic OH groups of the bridging type but under the influence of an additional electrostatic interaction as, e.g., in the case of formation of hydrogen bonds to neighbouring oxygen atoms (non-isolated OH groups). The same effect has been found for the stretching vibration of OH groups giving rise to the so-called LF band (3540 cm^{-1}) in addition to the HF band (3650 cm^{-1}). In a former paper [72] line *c* was ascribed tentatively to OH groups located in the large cavities which are known to be responsible for the HF band in the IR spectrum. In a following paper [46], however, it could be shown unambiguously by the use of deuterated pyridine as a probe molecule that the correlation

line $b \leftrightarrow$ HF band
line $c \leftrightarrow$ LF band

is the correct one. The fact that for the bridging OH groups pointing into the large and into the small cavities separate lines appear in the ^1H MAS NMR spectra (lines *b* and *c*, respectively) excludes the possibility of a fast proton exchange among the four oxygens around an aluminium atom of the zeolite framework.

Line d at 6.5 to 7.0 ppm is due to residual ammonium ions.

Line e at 2.5 to 3.6 ppm represents hydroxyl groups associated with extra-framework aluminium species. Due to the limited space available for these OH groups their δ_H-value will be affected by additional electrostatic interactions. Accordingly, for isolated AlOH groups the chemical shift is much smaller, with values in the interval between −0.5 and +1 ppm. In a recent paper, Harris et al. [87] have studied systematically H-Y zeolites containing large amounts of extra-framework aluminium. For dehydrated specimens they found a dramatic redistribution of the hydrogen nuclei within the sample over the course of the aging period (24 days) with non-framework aluminium species gaining OH groups at the expense of the other hydroxyl groups.

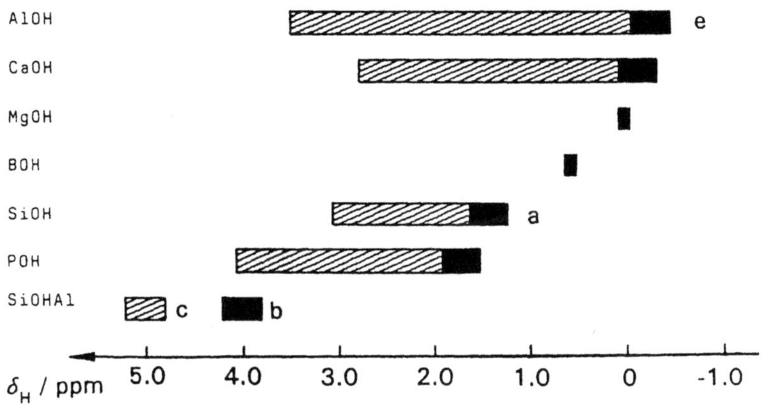

Fig. 7. Measured values for the ^{1}H NMR chemical shift of isolated (*black*) and interacting (*hatched area*) OH groups observed in zeolites [60]

In Fig. 7, measured values for the ^{1}H NMR chemical shift of isolated (black) and interacting (hatched area) OH groups appearing in zeolite catalysts of various type and composition are collected [60].

With respect to measurement of the concentration of hydroxyl groups, nuclear magnetic resonance spectroscopy has an extremely important advantage compared with infrared spectroscopy since the area of an ^{1}H MAS NMR signal (including the spinning sidebands which appear if the spinning rate is smaller than the width of the static line) is directly proportional to the concentration of the hydrogen nuclei contributing to this signal irrespective of their bonding state, so that any compound with a known concentration of hydrogen atoms can be used as a reference (mostly water). In Fig. 8, ^{1}H MAS NMR and infrared stretching vibration spectra are shown [51] for two differently synthesized specimens of SAPO-5. While the positions of the various signals in the NMR and IR spectra correspond to each other quite well and are in agreement with IR results published by other authors [74], there are dramatic differences in the relative intensities: From the IR spectrum of the first specimen one would erroneously conclude that the concentrations of bridging OH groups corresponding to line *b* and line *c* are approximately equal, and from the IR spectrum of the second specimen that the concentration of POH groups is about three times larger than that of the bridging OH groups corresponding to line *b*. Therefore, even the relative intensity of an OH stretching vibration band cannot be taken as a measure for the concentration of the respective hydroxyl groups in contrast to the intensity of an NMR signal. On the other hand there is a surprisingly good correlation between the positions of the various signals in the IR and ^{1}H MAS NMR spectra. Experimental results are collected in Table 5 and plotted in Fig. 9. With the exception of the cationic OH groups there is a nearly linear interdependence between the wave number ν_{OH} of the IR band and the chemical shift δ_H of the ^{1}H MAS NMR signal. The straight line approxi-

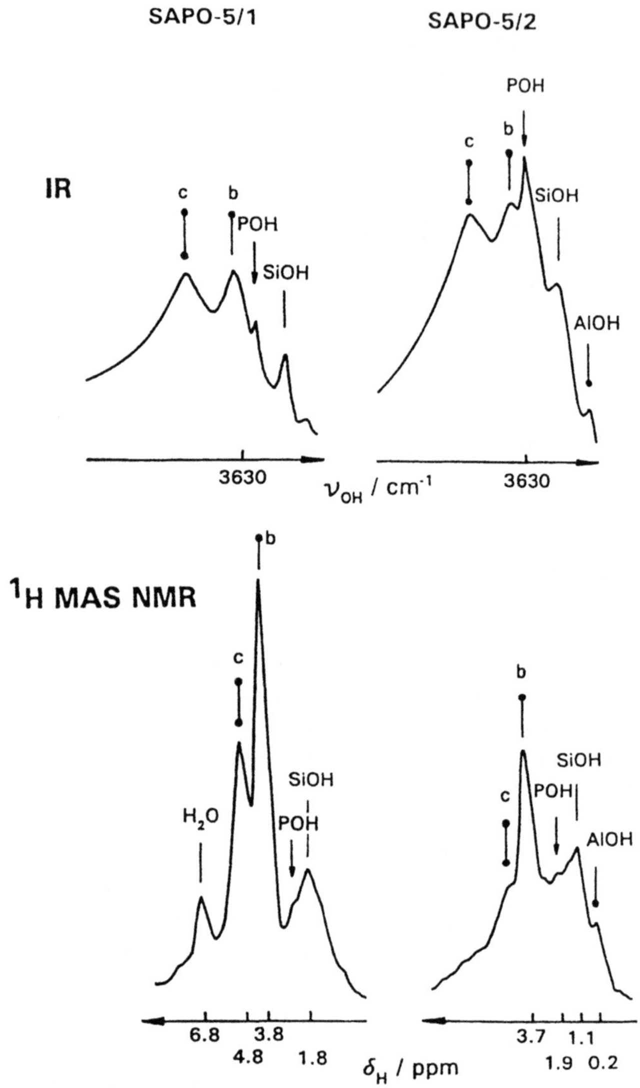

Fig. 8. IR stretching vibration (Digilab FTS-20) and ^{1}H MAS NMR spectra (Bruker MSL 300) of two differently synthesized specimens of SAPO-5 [51]

mating the experimental results is given by [75]

$$\nu_{OH}/\mathrm{cm}^{-1} = 3870 - 67.8 \cdot \delta_H/\mathrm{ppm}. \tag{60}$$

In Sect. 3.2 it was shown that the experimentally determined minimum line width of ca. 0.8 ppm for the ^{1}H MAS NMR signal of bridging OH groups is given by the distribution width of the isotropic value of the chemical shift. Therefore, the minimum line width set by the method itself must be less than

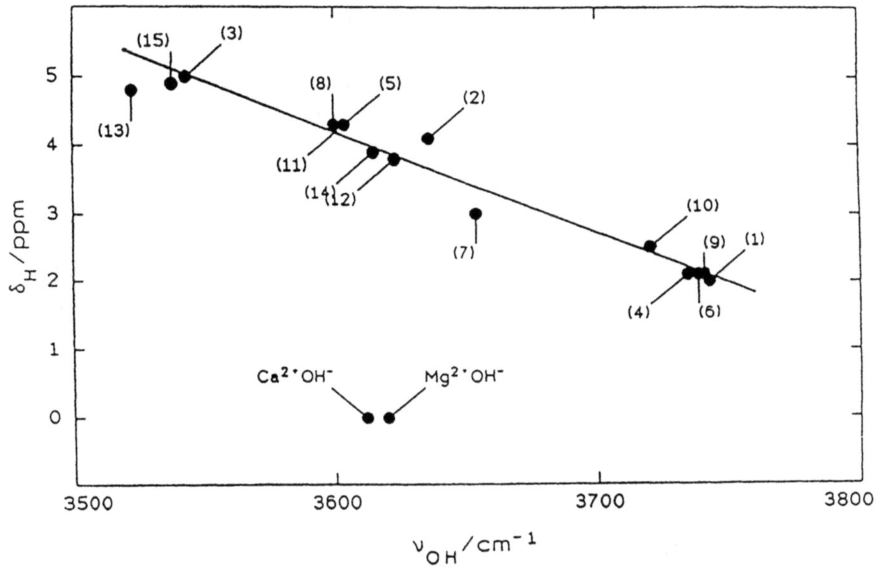

Fig. 9. Experimental values (cf. Table 5) for the wave number ν_{OH} of the IR stretching vibration band and of the chemical shift δ_H of the ^1H MAS NMR signal of OH groups in zeolites [75]

Table 5. Experimental values for the chemical shift δ_H and for the wave number ν_{OH} of the stretching vibration of the various OH groups of zeolites [75]. $^{(*)}$: assignment uncertain, $^{(**)}$: SiOH and SiOHAl as in H-Y, $^{(s)}$: these NMR signals are the sum of two or more components

zeolite	OH group	δ_H/ppm	ν_{OH}/cm^{-1}	data point in Fig. 9
H-Y	SiOH	2.0	3742	(1)
	SiOHAl (HF)	4.1	3636	(2)
	SiOHAl (LF)	5.0	3543	(3)
H-ZSM-5	SiOH	2.1	3734	(4)
	SiOHAl	4.3	3604	(5)
H-ZSM-5	SiOH	2.1	3738	(6)
(dealuminated)	AlOH (extra-framework)	3.0	3654	(7)
	SiOHAl	4.3		
H-SABO	SiOH	2.1	3740	(9)
	SiOHB	2.5	3720	(10)
	SiOHAl	4.3	3604	(11)
SAPO-5	SiOH$^{(*)}$	1.7$^{(s)}$	3739	
	POH$^{(*)}$		3674	
	SiOHAl(I)	3.8	3623	(12)
	SiOHAl(II)	4.8	3523	(13)
SAPO-11	SiOH$^{(*)}$	1.9$^{(s)}$	3741	
	POH$^{(*)}$		3671	
	SiOHAl(I)	3.9	3615	(14)
	SiOHAl(II)	4.9	3538	(15)
Mg-Y$^{(**)}$	$Mg^{2+}OH^-$	0.0	3620	(16)
Ca-Y$^{(**)}$	$Ca^{2+}OH^-$	0.0	3612	(17)

that value, say $\leqq 0.4$ ppm. On the other hand, the position of a resonance line can be determined with an accuracy of ca. 1/5 of its line width so that the resolution of the ^1H MAS NMR method must be less than 0.08 ppm which corresponds, according to Eq. (60), to a resolution of less than ca. $5\,cm^{-1}$ of the IR stretching vibration spectra. This value has to be compared with the resolution of ca. $2\,cm^{-1}$ achieved in [77] which leads to the remarkable statement that for both methods the ultimate resolution is of the same order of magnitude.

As shown in Sect. 3.1, the resonance position of the ^1H MAS NMR line which is given by the isotropic value of the chemical shift can be used as a reliable and sensitive measure for the strength of acidity of isolated Brönsted acid sites of surfaces. On the other hand when relatively small spinning rates of the sample are used, rotational sidebands appear which contain information

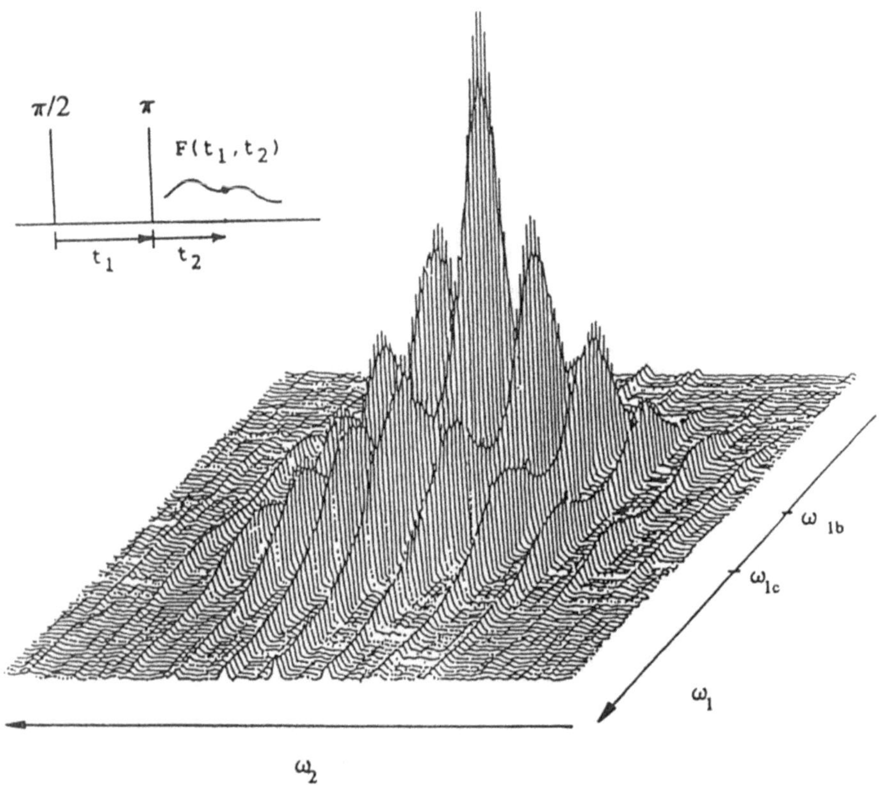

Fig. 10. Two-dimensional ^1H MAS NMR spectrum of a zeolite SAPO-5. The pulse sequence applied is shown in the inlet. The measurements were performed at room temperature, at a resonance frequency of 300 MHz with a spinning rate of 2 kHz, with 64 consecutive values of t_1 and 100 accumulations per free induction decay $F(t_1, t_2)$. The resonance frequencies ω_{1b} and ω_{1c} are related to that of the reference TMS by $(\omega_{1TMS} - \omega_{1b})/\omega_{1TMS} = 3.8$ ppm and $(\omega_{1TMS} - \omega_{1c})/\omega_{1TMS} = 4.8$ ppm [20]

about the anisotropy of the chemical shift and the magnetic dipole–dipole interaction between the H and the neighbouring Al nuclei, or with other words, about their distance. In order to analyse these sideband patterns for each of the various signals, it is necessary to separate them carefully. This can be accomplished by the following experiment which is a typical example of two-dimensional NMR spectroscopy [78]: At first a $\pi/2$ pulse and then after a time interval t_1 a π pulse is applied to the sample. Under the condition that the following equation is fulfilled

$$t_1 \cdot \Omega / 2\pi = \text{integer} \tag{61}$$

where Ω denotes the spinning rate, the nuclear magnetic resonance signal is measured at time t_2 after the π pulse. Hence, this signal which we denote by $F(t_1, t_2)$ depends both on t_1 and t_2. A twofold Fourier transformation of $F(t_1, t_2)$ then gives the two-dimensional ^{1}H MAS NMR spectrum $S(\omega_1, \omega_2)$. A typical example is shown in Fig. 10. The important fact is that $S(\omega_1, \omega_2)$ yields for $\omega_1 =$ const. the ordinary ^{1}H MAS NMR spectrum, i.e. the central lines together with all sideband patterns, while $S(\omega_1, \omega_2)$ gives for $\omega_2 =$ const. the ^{1}H MAS NMR spectrum without the sidebands. Therefore, the sidebands for each line can be analysed separately. In [79] a method is described which allows a determination of both the distance r_{HAl} between the H and the neighbouring Al nuclei and of the chemical-shift anisotropy $\Delta\sigma_H$ for the bridging OH groups (lines *b* and *c* in the ^{1}H MAS NMR spectra) of zeolites. In Table 6 results [79] for the distance r_{HAl} of bridging OH groups which give rise to line *b* and line *c* of zeolites H-Y

Table 6. Results for the distance r_{HAl} of bridging OH groups which give rise to line *b* and lince *c* of zeolites H-Y derived from an analysis of the ^{1}H MAS NMR spinning sideband pattern [79] compared with theoretical data found by a computer simulation [80] and with results of neutron powder diffraction measurements [81]

	line *b* HF band O_1H	line *c* LF band O_3H
r_{HAl}		
Ref. [79]	248 ± 4 pm	237 ± 4 pm
Ref. [80]	238.6 pm	233.2 pm
Ref. [81]	213.2 pm	219 pm
r_{SiO}		
Ref. [79]	—	—
Ref. [80]	169.4 pm	169.7 pm
Ref. [81]	167.4 pm	165.4 pm
r_{AlO}		
Ref. [79]	—	—
Ref. [80]	191.0 pm	193.0 pm
Ref. [81]	167.7 pm	165.4 pm

Table 7. Experimental results for the isotropic value δ_H and for the anisotropy $\Delta\sigma_H$ of the chemical shift, as well as for the distance r_{HAl} between the ^1H and ^{27}Al nuclei of bridging OH groups in zeolites H-Y and SAPO-5 [79]

	line *b*	line *c*
Zeolite H-Y		
δ_H	4.0 ± 1 ppm	5.0 ± 0.1 ppm
$\Delta\sigma_H$	18.3 ± 1.5 ppm	20.2 ± 1.5 ppm
r_{HAl}	248 ± 4 pm	237 ± 4 pm
Zeolite SAPO-5		
δ_H	3.8 ± 0.1 ppm	4.8 ± 0.1 ppm
$\Delta\sigma_H$	14.5 ± 1.5 ppm	19.5 ± 1.5 ppm
r_{HAl}	248 ± 4 pm	234 ± 4 pm

are collected and compared with theoretical data found by a computer simulation [80] and with results of neutron powder diffraction measurements [81]. It can be seen that the values for the distances between the hydrogen and aluminium nuclei derived from the neutron powder diffraction measurements are much smaller and moreover that the ratio of the distances for the two sorts of bridging OH groups is at variance with the result of the two other papers. Presumably this discrepancy is due to the fact that in [81] the Al-O and Si-O distances have been assumed to be equal. Table 7 contains experimental results for r_{HAl} and the anisotropy $\Delta\sigma_H$ of the chemical shift which have been found by the ^1H MAS NMR spinning-sideband analysis [79] for zeolites of type H-Y and SAPO-5. For the bridging OH groups which point into the small cages (line *c*) the H-Al distance has been found to be 237 ± 4 pm for the zeolite H-Y and 234 ± 4 pm for the zeolite SAPO-5. In contrast, for the bridging OH groups in the large cavities the corresponding distances are equal and distinctly larger, viz. 248 ± 4 pm. Within the limits of error, the values for the anisotropy of the chemical shift are equal (19 ± 2 ppm) except for line *b* of the zeolite SAPO-5 which exhibits a smaller value (14.5 ± 2 ppm). In a recent paper [82], the same method was used to study systematically a larger variety of zeolites, viz. zeolites *A*, *X*, *Y*, erionite, mordenite, ZSM-5, SAPO-5, SAPO-17, SAPO-34 and SAPO-37. The measurements were performed at two different magnetic field strengths ($B_0 = 7$ T and 9.3 T) in order to prove the consistency of the method. In Table 8 experimental results are collected for the isotropic value δ_H and for the anisotropy $\Delta\sigma_H$ of the chemical shift, as well as for the distance r_{HAl} between the H and Al nuclei of bridging OH groups attached to the 6-, 8-, 10-, and 12-membered oxygen rings which are constituents of the various zeolites. The structure type [23] and values for the inner diameter d of the oxygen rings are given in Table 9. Using the value for the mean diameter $\langle d \rangle$ of each type of oxygen ring and the mean value of r_{HAl} for the hydroxyl protons attached to these oxygen rings (Tables 8 and 9), the following equation has been derived by linear regression [82]

$$r_{HAl}/\text{pm} = 0.035\langle d \rangle/\text{pm} + 277 \tag{62}$$

Table 8. Experimental results for the isotropic value δ_H and for the anisotropy $\Delta\sigma_H$ of the chemical shift, as well as for the distance r_{HAl} between the ^1H and ^{27}Al nuclei of bridging OH groups attached to the 6-, 8-, 10-, and 12-membered rings of various zeolites. The mean errors for r_{HAl} and $\Delta\sigma_H$ are ±2 pm and ±2.0 ppm, respectively [82]

zeolite	Si/Al(*) Si/(Al + P + Si)	number of oxygen atoms per ring	δ_H/ppm	r_{HAl}/pm	$\Delta\sigma_H$/ppm
25 HNa-A	1.0(*)	8	4.0	238	14.5
50 HNa-X	1.4(*)	12	3.8	246	14.0
		6	—	—	—
52 HNa-Y	2.6(*)	12	3.9	251	14.5
		6	—	—	—
90 HNa-Y	3.5(*)	12	3.9	250	16.0
		6	5.0	236	16.5
37 HNaK-ERI	2.9(*)	8	3.9	246	14.0
H-M	7.1(*)	12	4.2	248	14.0
		8	—	—	—
H-ZSM-5/1	15.0(*)	10	4.2	250	14.5
H-ZSM-5/2	26.0(*)	10	4.2	246	15.5
SAPO-5	0.036	12	3.8	252	15.0
		6	4.8	236	17.5
SAPO-17	0.021	8	3.7	242	14.5
			4.0	240	16.0
SAPO-34	0.114	8	3.8	244	14.5
				242	13.5
SAPO-37	0.121	12	3.8	250	14.0
		6	4.3	234	15.0

Table 9. Denotation of the various zeolites studied in [82], their structure type, number of oxygen atoms per ring, and values for the inner diameters *d* of the rings which are constituents of the structure

zeolite	structure type [23]	number of oxygen atoms per ring	d/pm
A	LTA	8	410
X, Y, SAPO-37	FAU	12	740
		6	250
Erionite, SAPO-17	ERI	8	360 × 510
Mordenite	MOR	12	650 × 700
		8	260 × 570
ZSM-5	MFI	10	530 × 560
			510 × 550
SAPO-5	AFI	12	730
		6	250
SAPO-34	CHA	8	380

with a correlation coefficient of 0.94. Hence, a determination of the H-Al distance by an analysis of the sideband pattern for a bridging OH group as described in [79] allows an estimate of the mean inner diameter $\langle d \rangle$ of the oxygen ring to which the hydroxyl proton is attached by the inverse of Eq. (62):

$$\langle d \rangle/\text{pm} = 28.6\, r_{HAl}/\text{pm} - 6500. \qquad (63)$$

Therefore, an experimental determination of r_{HAl} may be of practical importance since it describes, by the use of Eq. (63), quantitatively the steric accessibility of the bridging OH groups (Brönsted acid sites).

Comparing the results obtained for r_{HAl} with the values found for δ_H and $\Delta\sigma_H$ in these studies [79, 82] one must state that there is no apparent relation between the latter quantities and the H-Al distance r_{HAl} of the bridging OH groups. It should be the goal of forthcoming quantum chemical calculations to throw some light on the missing interdependence between the geometrical (r_{HAl}) and electronic (chemical shift) parameters of the bridging OH groups in zeolites and related catalysts.

By a measurement of the ^{29}Si CP MAS NMR spectra (^{1}H – ^{29}Si cross-polarization in evacuated zeolites) it could be shown [82] that for the original zeolites (cf. Sect. 2.1) the bridging OH groups prefer a location at those silicon atoms for which the number z of aluminium in the second coordination sphere is as large as possible ($0 \leqq z \leqq 4$). For zeolites of SAPO-type with a ratio $Si/(Al + P + Si) \leqq 0.12$ (cf. Table 8), the second coordination sphere of the T positions with respect to the bridging OH group is always occupied by the same number of aluminium and phosphorous atoms.

Solid-state deuterium NMR has been used by Gluszak et al. [83] to study hydroxyl groups in deuterium-exchanged H-Y zeolites. Using a line-shape analysis, it was possible to separate the signals from Brönsted acid sites and silanols, and to determine the quadrupole coupling constant C_Q (cf. [86] or Sect. 4.1) of the former sites as 234 ± 2 kHz. However, the resolution of this method is not sufficient to separate the lines from the two sorts of Brönsted acid sites (lines *b* and *c* in the ^{1}H MAS NMR spectra), of residual ammonium ions (line *d*), and of OH groups at extra-framework aluminium species (line *e*), so that at present ^{2}H NMR cannot compete with ^{1}H MAS NMR in characterizing Brönsted acidity of zeolites and related catalysts.

3.4 Loaded Zeolites

The accessibility of hydroxyl groups can easily be determined through a study of the ^{1}H MAS NMR spectra after loading the adsorbent with a suitable molecule which, however, must be fully deuterated in order to avoid an unwanted additional ^{1}H NMR signal. Using deuterated pyridine the concentrations of accessible and non-accessible silanol groups of silica has been determined [84] since the formation of a hydrogen bond between pyridine and the silanol group shifts the ^{1}H MAS NMR signal of the latter by ca. 8 ppm to higher values of δ_H. This effect is demonstrated in Fig. 11A where a zeolite H-ZSM-5 with a high concentration of silanols (3.7×10^{20} SiOH/g) has been loaded with about the same concentration (4×10^{20} C_5D_5N/g) of deuterated pyridine molecules. In the case of acidic OH groups (Brönsted acid sites) however, the adsorption of pyridine leads to a protonation, i.e., to a formation of pyridinium ions with a larger shift of ca. 12 ppm to higher values of δ_H. The spectra of a zeolite

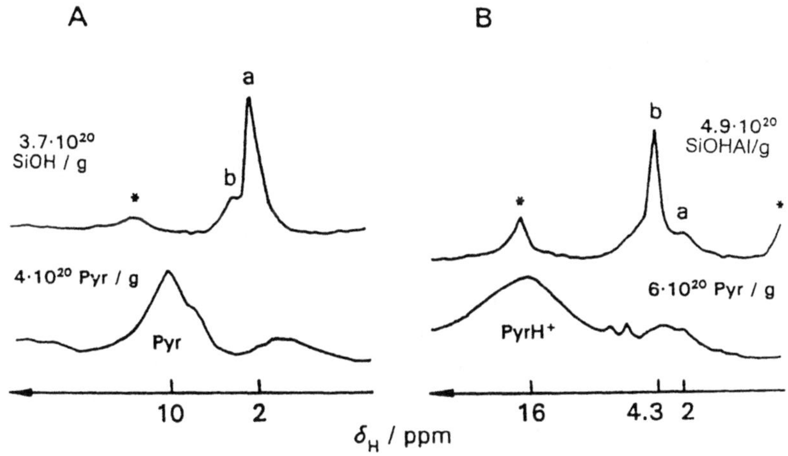

Fig. 11A, B. ¹H MAS NMR spectra of zeolites H-ZSM-5 evacuated and after loading the samples with deuterated pyridine. **A** Specimen with a prevailing concentration of non-acidic OH groups (formation of a hydrogen bond between the OH group and the adsorbed pyridine molecule). **B** Specimen with a prevailing concentration of acidic OH groups (formation of pyridinium ions) [20]

H-ZSM-5 with 4.9×10^{20} Brönsted acid sites per g zeolite and only a negligible concentration of silanols (this can be achieved, e.g., by a synthesis of the zeolite ZSM-5 without template [85]) unloaded and after loading with a concentration of 6×10^{20} deuterated pyridine molecules per g zeolite is shown in Fig. 11B. With the same probe molecule it was also possible to show unambiguously that line *b* in the ^{1}H MAS NMR spectra of zeolites H-Y is due to bridging. OH groups which are easily accessible by pyridine [46] and that line *c* is caused by bridging OH groups pointing into the small cavities (LF band in infrared spectroscopy). In Fig. 12 results [51] are shown for a zeolite SAPO-5 unloaded (A), and after keeping it loaded with deuterated *n*-hexane for 1 hour at 50 °C (B).

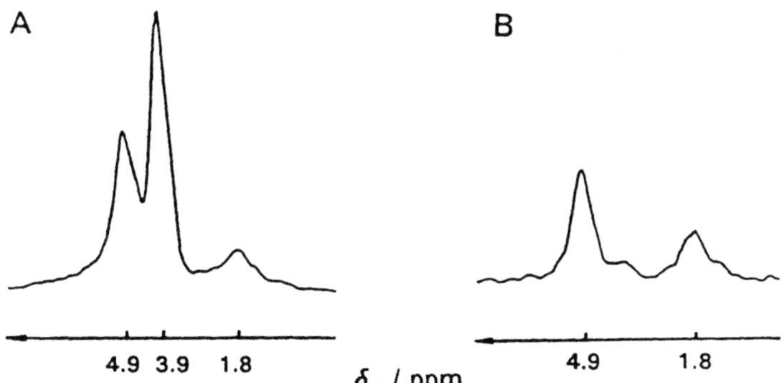

Fig. 12A, B. ¹H MAS NMR spectra of a zeolite SAPO-5. **A** Evacuated sample. **B** After keeping the sample loaded with deuterated *n*-hexane for 1 h at 50 °C [51]

There is no doubt that the 3.9 ppm signal is caused by bridging OH groups which are easily accessible to *n*-hexane in contrast to the 4.9 ppm signal.

^1H MAS NMR studies of hydrated zeolites may be complicated by a superposition of three effects:

(1) Hirschler-Plank-mechanism, i.e. the adsorption and dissociation of water molecules on extra-framework multivalent cations like Ca^{2+} with a formation of cationic and bridging OH groups [88, 89]. The ^1H MAS NMR signal of the former hydroxyls appears at δ_H values of ca. 0 ppm and 2.8 ppm for calcium ions in the large and small cavities of zeolites *Y*, respectively [90], and an inspection of Fig. 9 shows that Eq. (60) is not fulfilled for these OH groups. Up till now a simple interpretation of this experimental result cannot be given.

(2) Formation of hydroxonium ions (chemical shift δ_{h+}) at Brönsted acid sites which, however, take part in a fast proton exchange with physically adsorbed water molecules (chemical shift δ_p) and with accessible bridging OH groups (chemical shift δ_{OH}). Physically adsorbed water molecules may include water molecules adsorbed on the exchangeable cations (adsorption energy: 335 kJ mol^{-1} and 117 kJ mol^{-1} for Mg^{2+} and Na^+, respectively [91], [92]) and water molecules hydrogen bonded to bridging OH groups, to other water molecules and to silanol groups (58.4 kJ mol^{-1}, 20.1 kJ mol^{-1}, and 16.4 kJ mol^{-1}, respectively [91, 92]). Hence, the resulting ^1H MAS NMR shift δ can be described quantitatively by

$$\delta = [(c_{OH} - c_{h+}/3)\delta_{OH} + (c_w - 2c_{h+}/3)\delta_p + c_{h+}\delta_{h+}]/(c_{OH} + c_w) \quad (64)$$

where c_{OH}, c_{h+}, and c_w denote the concentration of the ^1H nuclei in the accessible bridging OH groups, in the hydroxonium ions, and in the physically adsorbed water molecules, respectively. With $\delta_{OH} = 4.3$ ppm, $\delta_p = 3.2 - 4.8$ ppm (the upper limit is for hydrogen bonded water molecules) and $\delta_{h+} = 13$ ppm [93] it is possible to determine quantitatively the concentration of hydroxonium ions from the position δ of the ^1H MAS NMR signal. In shallow-bed (400 °C) treated zeolites H-Y the probability of finding a water molecule in the state of a hydroxonium ion is ca. 0.2 – 0.3 for a rehydration corresponding to one water molecule per accessible bridging OH group (line *b*) [93].

(3) Adsorption of water molecules on Lewis acid sites giving rise to a narrow line at $\delta_H = 6.5$ ppm. In the case of hydrothermally pretreated zeolites H-Y (540 °C; 20 h; 4 kPa water vapour pressure) a concentration of 2 ± 0.5 Lewis acid sites of this type per unit cell could be found. Surprisingly, the MAS sideband pattern of the signal at 6.5 ppm could be explained quantitatively only if these Lewis acid sites are not connected with aluminium. Therefore it was concluded that the signal at 6.5 ppm is caused by water molecules adsorbed on threefold coordinated and positively charged silicon atoms of the zeolite framework [93], i.e., on sites which were proposed in order to explain an infrared band at 4035 cm^{-1} for hydrogen adsorbed on a zeolite H-Y activated at 400 °C under deep-bed conditions, see Refs. [94, 95]. Further experimental and theoretical work seems necessary to prove whether threefold coordinated and positively charged silicon atoms do exist in the framework of zeolites. If the zeolites contain

larger amounts of extra-framework aluminium, the situation becomes even more complicated. Harris et al. [87] were able to show by a thorough investigation of the ^{1}H MAS NMR spectra of steamed zeolites H-Y that the extra-framework aluminium material changes in concentration level by reacting with water: In the dehydrated zeolite the material may be so highly condensed that it bears only few terminating OH groups; during sorption of water, however, this 'alumina' is hydrolyzed to form material which is less concentrated and richer in hydroxyl groups.

Another method of studying the formation of hydroxonium ions in hydrated zeolites was introduced by J. Fraissard et al. [96]. Through a measurement at very low temperatures (4 K) the fast proton exchange among the various species can be excluded and a line-shape analysis should yield quantitative information about the concentration of the various species. Since however, the line widths of the respective ^{1}H NMR signals are so large that they cannot be distinguished by their chemical shift it is necessary to make use of the different line shapes for the ^{1}H NMR signals of rigid one-, two-, and three-spin systems, respectively. Assuming that there are only isolated hydroxyl groups (one-spin systems), isolated water molecules (two-spin systems), water molecules hydrogen bonded to OH groups (three-spin systems with the configuration of an isosceles triangle) and hydroxonium ions (three-spin systems with the configuration of an equilateral triangle) the shape of the observed ^{1}H NMR broad line can be decomposed into contributions of these species by a fitting procedure where additionally the magnetic dipole–dipole interaction among these species must be considered by introducing a further ("line-broadening") parameter. Since the concentration of hydroxonium ions c_{h+} thus determined will depend both on the concentration of the acidic OH groups c_a and on their strength of acidity S_a the ratio c_{h+}/c_a should be a reasonable measure for S_a. Unfortunately, this method fails if the zeolite catalyst contains various sorts of OH groups differing in their concentration and strength of acidity. With respect to the ^{1}H MAS NMR technique it should be of interest whether the sensitivity of this broad line method is sufficient to confirm, e.g., the dependence of the strength of acidity on the electronegativity of zeolite catalysts (Fig. 1).

4 Extra-Framework Aluminium

Apart from relatively weak signals in the IR spectra of adsorbed hydrogen [94] and carbon monoxide [95] which are ascribed to an interaction of these molecules with threefold aluminium and silicon atoms of the zeolite framework affected by, e.g., a hydrothermal treatment, it is generally accepted [99] that Lewis activity of zeolites and related catalysts (silica-alumina, γ-alumina) is connected with the presence of aluminium species on the surface. Therefore two possibilities exist to study Lewis acidity of zeolites by NMR methods: (1) an analysis of highly resolved ^{27}Al NMR spectra similar to the ^{1}H MAS NMR spectroscopy described in Sect. 3, and (2) the use of probe molecules as, e.g., pyridine [100] or the above-mentioned molecules in IR spectroscopy.

4.1 ^{27}Al NMR Spectroscopy

In well-crystallized zeolites exhibiting no Lewis acidity, aluminium is tetrahedrally coordinated with an isotropical chemical shift between 55 and 65 ppm [7] with respect to an aqueous dilute solution of Al^{3+}. After a deep-bed or a hydrothermal treatment, however, extra-framework aluminium species are formed (Lewis acid sites) and the ^{27}Al MAS NMR spectrum becomes more complicated. The intensity of the initial signal at ca. 60 ppm decreases due to ejection of aluminium from the framework into the intracrystalline space and increases due to ^{27}Al signals from tetrahedrally coordinated extra-framework aluminium. This latter aluminium must exist at least partly as $Al(OH)_n$ species as shown by ^{1}H-^{27}Al cross-polarization (CP MAS NMR) experiments [31]. The other extra-framework aluminium species give rise to distinct resonance lines in the interval between −15 and +4 ppm, which could be assigned to polymeric aluminium species [101], and to a line near 30 ppm which is interpreted as being due to either penta-coordinated [102] or tetrahedrally coordinated [103, 104, 105] extra-framework aluminium. In addition, there appears a very broad hump extending from ca. −180 ppm to ca. +230 ppm (for an intensity of the constant magnetic field $B_0 = 7.1$ T) below these more distinct resonances. As has been shown recently [31], this broad hump merges into the lines near 0, 30, and 60 ppm for ultrahigh ($\Omega/2\pi \geqq 18$ kHz) MAS speeds. In hydrated samples a narrow line at 0 ppm is often observed, suggesting the presence of some Al^{3+} cations balancing lattice charges. A survey on these experimental results is given in Table 10. In recent ^{27}Al NMR experiments performed at high fields ($B_0 = 11.7$ T) with a non-spinning sample [106] the following values for the ^{27}Al quadrupole frequency ν_Q defined by Eq. (67) have been found: 0.83 ± 0.05 MHz, 1.9 ± 0.5 MHz and 2.4 ± 0.3 MHz for the framework aluminium

Table 10. Experimental results for the chemical shift δ_{Al} of ^{27}Al NMR signals observed on zeolites

δ_{Al}/ppm	species
0	octah. coord. extra-framework aluminium (Al^{3+}) [7]
4 ··· −15	extra-framework polymeric aluminium species [101]
30	penta-coordinated [102] extra-framework aluminium
30 (50?, ref. [104])	tetrahedrally coordinated extra-framework aluminium: AlOOH associated with two framework oxygens [103], aluminium in amorphous silica-alumina [104]
−180 ··· +230 ($B_0 = 7.1$ T)	extra-framework aluminium of low symmetry ("NMR invisible"); the very broad hump merges into the lines at 0, 30, and 60 ppm for ultrahigh spinning speeds ($\Omega/2\pi \geqq 18$ kHz) [31]
60	framework aluminium [7] and tetrah. coord. extra-framework aluminium (CP MAS NMR [31])

nuclei of zeolites Na-Y, H-Y and H-ZSM-5, respectively. Hence, it must be assumed that for extra-framework aluminium species with a lower symmetry the values of ν_Q may exceed even 3 MHz. Large values of ν_Q give rise to a large second-order quadrupolar shift [86]

$$\nu_{CG} - \nu_L = -\nu_Q^2[(I(I+1) - 3/4](1 + \eta^2/3)(30\nu_L)^{-1} \quad (65)$$

where ν_{CG} and ν_L denote the centre of gravity of the signal and the Larmor frequency (signal for $\nu_Q = 0$) in Hz, respectively, I is the spin quantum number ($I = 5/2$ for ^{27}Al) and η the asymmetry parameter ($0 \leqq \eta \leqq 1$). The quantity ν_Q denoted as the quadrupole frequency, is related to the quadrupole coupling constant

$$C_Q = e^2qQ/h \quad (66)$$

by the equation

$$\nu_Q = 3C_Q/[2I(2I-1)] \quad (67)$$

with eq, eQ and h denoting the electric field gradient, the electric quadrupole moment and Planck's constant, respectively [86]. In addition to this effect (Eq. (65)) which must be taken into consideration if one wants to determine the true chemical shift of a ^{27}Al NMR signal, a strong quadrupole interaction leads to a dramatic broadening of the lines: The widths of the spectra in ppm for a static sample (Δ), for a sample under conditions of magic angle spinning (Δ^{MAS}) and under conditions of double rotation (Δ^{DOR}) are given by [86, 107, 108]:

$$\Delta = (\nu_Q/\nu_L)^2(25 + 22\eta + \eta^2)10^6/18 \quad (68)$$

$$\Delta^{MAS} = (\nu_Q/\nu_L)^2(36 + 12\eta + \eta^2)10^6/63 \quad \text{if } \Omega \geqq \Omega_{crit}^{MAS} \quad (69)$$

$$\Delta^{DOR} = \Delta_C \quad \text{if } \Omega_i = 5\Omega_0 \geqq \Omega_{crit}^{DOR} \quad (70)$$

where Ω is the spinning rate of the rotor in the case of MAS, and Ω_i and Ω_o denote the corresponding rates for the inner and outer rotor of DOR, respectively. The critical values for the rates are given by

$$\Omega_{crit}^{MAS} = 2\pi\Delta^{MAS}\nu_L\,10^{-6} \quad (71)$$

$$\Omega_{crit}^{DOR} = 10\pi\Delta_D\nu_L\,10^{-6}. \quad (72)$$

Δ_C denotes the line width in ppm due to a distribution of the chemical and/or quadrupole shift and Δ_D the line width in ppm due to homonuclear magnetic dipole–dipole interaction. It is the influence of these two latter quantities which determines the ultimate resolution of ^{27}Al MAS and of ^{27}Al DOR NMR spectra. For a constant magnetic field of intensity $B_0 = 11.74$ T, $\nu_Q = 3$ MHz (aluminium species of low symmetry), assuming $\eta = 0$ for simplicity, and an estimated mean distance of ca. 330 ppm between neighbouring aluminium nuclei (condensed extra-framework material) which corresponds to $\Delta_D\nu_L\,10^{-6} \approx 5$ kHz, the following values can be derived: $\nu_L = 130.3$ MHz (Table 1), $\Delta = 736$ ppm (Eq. (68)), $\Omega_{crit}^{MAS}/2\pi = 39.5$ kHz (Eqs. (71) and (69)) and $\Omega_{crit}^{DOR}/2\pi = 25$ kHz (Eq. (72)). An inspection of these values shows that the conditions of Eqs. (69) and (70) cannot be fulfilled experimentally so that we must state that neither by application of ^{27}Al MAS NMR nor of ^{27}Al DOR NMR spectroscopy, will it be possible to detect extra-framework condensed aluminium species of low symmetry.

4.2 *Use of Probe Molecules*

In analogy to infrared spectroscopy, probe molecules can be used to study Lewis acid sites. In the case of NMR spectroscopy however, due to the much smaller resonance frequencies which correspond to a much longer time scale, exchange effects may lead to an average line instead of the series of separate lines caused by the molecules adsorbed on the various adsorption sites: From Heisenberg's uncertainty relation

$$\Delta E \cdot \Delta t \geqq \hbar \tag{73}$$

together with Planck's equation

$$\Delta E = \hbar \Delta \omega \tag{74}$$

it follows that the critical value $\tau_{crit} = \Delta t$ of the mean residence time τ between two succeeding exchanges is given by

$$\tau_{crit} = 1/\Delta\omega. \tag{75}$$

For $\tau \ll \tau_{crit}$ the two lines merge into a single line while for $\tau \gg \tau_{crit}$ they can be observed separately. In infrared spectroscopy, the typical bands of pyridine adsorbed on Brönsted and Lewis acid sites appear at ca. 1540 cm^{-1} and 1450 cm^{-1}, respectively [100], so that the value of τ_{crit} is ca. 0.1 ps. For the ^{15}N NMR of the same molecule at a magnetic field strength of 7 T the frequency difference between the two lines is (cf. Tables 1 and 11) ca. 2 kHz corresponding to $\tau_{crit} \approx 0.1$ ms, which is nine orders of magnitude larger than the value for the IR bands.

If one denotes by ν_k, p_k and τ_k the resonance frequency, the relative concentration and the mean residence time of a nucleus at an adsorption site k, the NMR spectrum for a system with two different sorts of adsorption sites $k = a, b$ with $p_a + p_b = 1$, is given by (see, e.g., [109])

$$M_x + iM_y = -C\tau[2p_a p_b - \tau(p_a\alpha_b + p_b\alpha_a)]/(p_a p_b - \tau^2\alpha_a\alpha_b) \tag{76}$$

where

$$\alpha_a = -2i\pi(\nu_a - \nu) + 1/T_{2a} + p_b/\tau. \tag{77}$$

The equation for α_b is obtained from Eq. (77) by simply exchanging a and b. M_x and M_y denote the x and y components, respectively of the nuclear magnetism perpendicular to the constant magnetic field B_0 which is along the z direction. $(\pi T_{2a})^{-1}$ and $(\pi T_{2b})^{-1}$ are the line widths of lines a and b, respectively, and C

is a constant. The time constant τ is defined by

$$\tau = (1/\tau_a + 1/\tau_b)^{-1} = \tau_a p_b = \tau_b p_a \tag{78}$$

with τ_a and τ_b as the mean residence time of the molecule adsorbed on site *a* and *b*, respectively. For the limiting case of slow exchange

$$2\pi\tau|\nu_a - \nu_b| > 1 \tag{79}$$

the spectrum consists of well separated lines at ν_a and ν_b while for

$$2\pi\tau|\nu_a - \nu_b| < 1 \tag{80}$$

the lines merge to a single line (limiting case of fast exchange):

$$\nu = p_a\nu_a + p_b\nu_b \tag{81}$$

In Fig. 13, numerical results are plotted [118] for the apparent resonance frequencies ν_a' and ν_b' of the two lines into which the NMR spectrum given by Eq. (76) can be decomposed. The sites were denoted such as to fulfill $\nu_a > \nu_b$. If the NMR measurements are performed at a single temperature only, no estimate can be made about the exact value of τ and it must be concluded that an analysis of NMR spectra may be accompanied with large errors since the apparent quantities (ν_k' and p_k') derived from the spectra may be distinctly different from the real values (ν_k and p_K). Therefore, we shall discuss in the following separately experiments where the exchange of the probe molecules among the adsorption sites has been assumed to be slow (Sect. 4.2.1) or fast (Sect. 4.2.2).

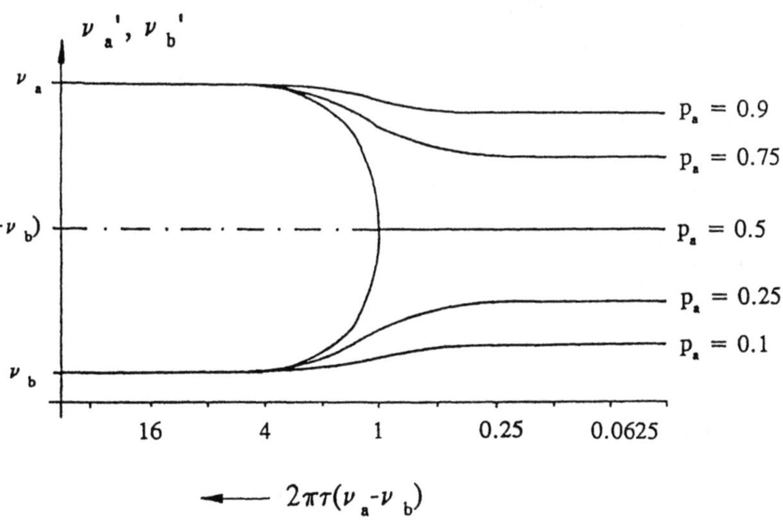

Fig. 13. Values for the apparent resonance frequencies ν_a' and ν_b' of the NMR signals of molecules exchanging between two sorts of adsorption sites where their resonance frequencies are ν_a and ν_b, respectively, [118]. The time constant τ describing the rate of exchange is defined by Eq. (78)

4.2.1 Slow Exchange

In the experiments to be described in this section, the NMR spectra exhibit various lines so that, at least approximately, the limiting case of slow exchange can be assumed for the analysis of the spectra.

^{15}N CP MAS NMR spectra of adsorbed pyridine have been studied by Maciel et al. [110] (pyridine on silica-alumina), Ripmeester [111] (pyridine on γ-alumina, mordenite), and by Majors and Ellis [112] (pyridine on γ-alumina). Values for the resonance shifts relative to solid pyridine are collected in Table 11. The major drawbacks of these experiments are that (1) an absolute determination of concentrations is connected with large errors due to the strong dependence of the line intensities on quantities controlling the efficiency of cross-polarization, see, e.g., Ref. [113], and (2) it is not sure that the exchange rate of the molecules among the various adsorption sites is sufficiently small so that the resonance positions and intensities derived from the spectra may be only the apparent quantities (ν'_k and p'_k, see above).

Compared with pyridine, phosphines are roughly three orders of magnitude stronger bases. ^{31}P CP MAS NMR spectra of trimethylphosphine adsorbed on zeolite H-Y and on γ-alumina were investigated by Lunsford et al. [114], and of various trialkylphosphines adsorbed on silica-alumina and on γ-alumina by Maciel et al. [115]. Values for the resonance shift relative to liquid trimethyl-

Table 11. Values for the ^{15}N NMR shift of pyridine [110–112] in the liquid state, physisorbed, and adsorbed on Brönsted and Lewis acid sites relative to the resonance of solid pyridine

species	δ_N/ppm
solid pyridine (−105 °C)	0
liquid pyridine	−26
physisorbed pyridine	-10 ± 10
pyridinium ion (Brönsted acid sites)	88 ± 2
pyridine on tetrahedral Al^{3+} (Lewis acid sites)	22.5 ± 3.5
pyridine on octahedral Al^{3+} (Lewis acid sites)	46.5 ± 7.7

Table 12. Values for the ^{31}P NMR shift of trimethylphosphine physisorbed and adsorbed on Brönsted and Lewis acid sites, relative to the liquid state [114, 115]

species	δ_P/ppm
liquid trimethylphosphine	0
physisorbed trimethylphosphine	0.7 ± 6
protonated trimethylphosphine (Brönsted acid sites)	59 ± 2
trimethylphosphine on Lewis acid sites	12 ± 10

phosphine are shown in Table 12. For this probe molecule the same drawbacks hold as mentioned above for pyridine, although due to the larger signal-to-noise ratio of the ^{31}P NMR signal compared with ^{15}N, cross-polarization must not be applied. The sensitivity of trimethylphosphine to distinguish between different sorts of Lewis acid sites seems to be less than that of pyridine since, for γ-alumina, only one sort of Lewis acid site could be found (see Table 12).

In princple, water can also be used as a probe molecule since the chemical shift of water molecules adsorbed or Lewis acid sites (6.5 ppm) is different from that of physisorbed water (less than 4.8 ppm) and of water molecules adsorbed on Brönsted acid sites (hydroxonium ions: ca. 13 ppm) [93]. As shown, however, in Sect. 3.4, the ^{1}H MAS NMR spectrum of hydrated zeolites is complicated by a superposition of various effects so that water is not very suitable for probing Lewis acidity even if one realizes the limiting case of slow exchange by a cooling of the sample to low temperature (see, e.g., [96]).

4.2.2 Rapid Exchange

Another way to study Lewis acidity quantitatively results from the fact that probe molecules could be found for which the resonance shift caused by Lewis acid sites is much larger than that caused by Brönsted acid sites or physisorption so that the system can be treated as if an exchange takes place only between molecules adsorbed on sites of type *a* (Lewis acid sites) and type *b* (all other sites including physically adsorbed molecules). Typical examples are carbon monoxide [116], [117], [118] and dinitrogen oxide [119]. Since at room temperature and above, these molecules exchange relatively rapid between the various adsorption sites, only a single line for the ^{13}C NMR of carbon monoxide and for the ^{15}N NMR of the terminal nitrogen of dinitrogen oxide appears. The position of this single line is given by Eq. (81) and it becomes necessary to perform a series of experiments in order to determine the two quantities of interest, namely the resonance shift of a molecule adsorbed on a Lewis acid site and the concentration of these Lewis sites. An interesting method which allows such a determination to be made has been described by Borovkov et al. [120]. If c_L, c_{LM} and c_M denote the concentration of the Lewis acid sites, of the Lewis acid sites occupied by an adsorbed probe molecule, and of the probe molecules, respectively, one may write

$$c_{LM} = K(c_M - c_{LM})(c_L - c_{LM}) \tag{82}$$

where K is the equilibrium constant of the exchange reaction

$$LM \rightleftarrows L + M. \tag{83}$$

On the other hand, Eq. (81) may be rewritten as

$$\nu = (c_{LM}/c_M)\nu_{LM} + [(c_M - c_{LM})/c_M]\nu_M \tag{84}$$

or if one introduces the resonance shift δ with regard to the resonance frequency

ν_M of the probe molecules which are not adsorbed on Lewis acid sites

$$\delta = (\nu_M - \nu)/\nu_M \tag{85}$$

one may write

$$\delta = (c_{LM}/c_M)\delta_{LM} \tag{86}$$

with

$$\delta_{LM} = (\nu_M - \nu_{LM})/\nu_M. \tag{87}$$

A combination of Eqs. (82) and (86) leads to

$$\delta/\delta_{LM} = 0.5(1 + 1/Kc_M + c_L/c_M) \pm [0.25(1 + 1/Kc_M + c_L/c_M)^2 - c_L/c_M]^{1/2}. \tag{88}$$

Inserting the value of δ extrapolated to $c_M = 0$ into this equation:

$$\delta_0 = Kc_L(1 + Kc_L)^{-1}\delta_{LM} \tag{89}$$

Eq. (88) may be rewritten as

$$y = (\delta_{LM}/\delta_0)x - c_L(\delta_{LM}/\delta_0)^2 \tag{90}$$

with

$$y = c_M(\delta/\delta_0)^2(1 - \delta/\delta_0)^{-1} \tag{91}$$

$$x = c_M(\delta/\delta_0)(1 - \delta/\delta_0)^{-1}. \tag{92}$$

The benefit of this representation (Eqs. (90)–(92)) is that x and y are quantities which can be derived from the experimental data, so that a plot of y versus x should yield a straight line from which one can directly determine the resonance shift δ_{LM} of the probe molecule adsorbed on a Lewis acid site and the concentration c_L of these sites. The main disadvantage of this method results from the fact that with decreasing loading of the zeolites ($c_M \to 0$), the signal-to-noise ratio decreases so that δ_0 cannot be determined exactly, and that a given error of δ_0 may lead to relatively large errors for the quantities of interest (δ_{LM}, c_L). Especially in those cases where the absolute value of the slope for the plot of δ versus c_M increases strongly at the lowest loadings, the value of δ_0 and hence to a much larger extent the values of δ_{LM} and c_L are connected with large errors, and the method yields values which are too small in general. Therefore the numerical results for δ_{LM} which have been found in [117] and which are between 300 ppm and 400 ppm (see Table 13) should be taken as a lower limit. In a subsequent study [118], the method of selective saturation introduced by Forsen and Hoffmann [121] has been applied to carbon monoxide adsorbed on various zeolites containing Lewis acid sites: Through a measurement of the ^{13}C NMR signal intensity as a function of the frequency of a selective ^{13}C saturation pulse, the accuracy for the value of δ_{LM} could be enhanced. Depending on the zeolites studied and on their pretreatment, values for δ_{LM} have been found between 440 ppm and 590 ppm with an error of ca. 60 ppm (cf. Table 13).

Table 13. Values for the ^{13}C NMR shift of carbon monoxide physisorbed and adsorbed on Brönsted and Lewis acid sites, relative to the gaseous state [117, 118]

species	δ_C/ppm
carbon monoxide gas	0
physisorbed carbon monoxide	-3 ± 2
carbon monoxide on Brönsted acid sites	-3 ± 2
carbon monoxide on Lewis acid sites	$300 - 400$ [117]
	$(400 - 590) \pm 60$ [118]

Table 14. Values for the ^{15}N NMR shift of dinitrogen oxide physisorbed and adsorbed on Brönsted and Lewis acid sites relative to the gaseous state [119]

species	δ_N/ppm
dinitrogen oxide gas	0
physisorbed dinitrogen oxide	4.5 ± 3
dinitrogen oxide on Brönsted acid sites	4.5 ± 3
dinitrogen oxide on Lewis acid sites	50 ± 20

In Table 14, results are given for the resonance shift of dinitrogen oxide physisorbed and adsorbed on Brönsted and Lewis acid sites. The latter result was derived by the authors using the method of Borovkov described above. The error given in Table 14 was added after a consideration of the experimental values, of the extrapolation procedure and application of Eq. (90)

Summarizing the results obtained up till now with probe molecules (Sects. 4.2.1 and 4.2.2), the following statements can be made:

1. The accuracy is not sufficient to take the values of the resonance shift for the molecules adsorbed on Lewis acid sites as a measure of their strength of acidity.

2. In those cases, where due to a poor signal-to-noise ratio, cross-polarization has been applied (pyridine, trimethylphosphine) only a semiquantitative determination of the concentration of Lewis acid sites is possible.

3. Steric effects connected with larger probe molecules may lead to large errors. As an example the result presented in a paper of Lunsford et al. [122] should be mentioned: With trimethylphosphine as a probe molecule, 37 protonated species per unit cell were found for a zeolite H-Y in contrast to the real value of 54 Brönsted acid sites (framework aluminium atoms) per unit cell.

5 Molecular Transport

Among the various methods of NMR spectroscopy which have been applied to the study of solid surfaces and their interaction with adsorbed molecules, the pulsed field gradient (PFG) method has revealed the most exciting results

Table 15. Comparison of the intracrystalline self-diffusion coefficients D_{intra} obtained by the PFG NMR method with the corresponding values derived from classical uptake experiments [152]

	$D_{intra}/m^2 s^{-1}$	
	PFG NMR	Class. Uptake
methane/NaCa-A (23 °C)	$2 \cdot 10^{-9}$	$5 \cdot 10^{-14}$
ethane/NaCa-A (23 °C)	$2 \cdot 10^{-10}$	$1 \cdot 10^{-14}$
propane/NaCa-A (23 °C)	$5 \cdot 10^{-12}$	$3 \cdot 10^{-15}$
n-heptane/Na-X (164 °C)	$5 \cdot 10^{-9}$	$3 \cdot 10^{-13}$
c-hexane/Na-X (164 °C)	$4 \cdot 10^{-9}$	$4 \cdot 10^{-13}$
benzene/Na-X (164 °C)	$2 \cdot 10^{-10}$	$1 \cdot 10^{-14}$

both from the standpoint of applied and basic research. As an example, in Table 15, values for the intracrystalline self-diffusion coefficient D_{intra} obtained by the PFG NMR method [152] and by classical uptake experiments are compared, and one observes that the NMR values are larger by up to five orders of magnitude than the 'classical' results which were generally accepted and in some cases also founded theoretically at this time. While in these early years, all over the world, doubts arose about the reliability of the NMR method, the situation has now completely changed and besides quasielastic neutron scattering experiments [126] which yield results in complete agreement with the PFG NMR method, the latter has been proven to be the method of choice. Since during the last years several reviews have been published, describing the results of systematic studies on the self-diffusion of adsorbed molecules [20, 51, 123, 125, 153, 154, 155], in the following mainly typical aspects of the method will be treated.

5.1 *Pulsed Field Gradient (PFG) NMR*

In general, zeolites are only available as crystallites with a size of the order of a few micrometres. Therefore, measurements of molecular diffusion have to be carried out with assemblages of zeolite crystallites (powder samples). The conventional way of measuring molecular diffusion in such systems is to follow the rate of mass change of the sample after changing the pressure of the surrounding atmosphere [123]. However, the interpretation of such sorption or uptake experiments, which are apparently of macroscopic nature, is not always unambiguous and may lead to large errors (cf. Table 15). A microscopic method to study molecular migration is provided by the pulsed field gradient NMR spectroscopy (PFG NMR) [124, 125]. Although this method is mostly combined with the formation of a so-called spin echo in order to eliminate the influence of an inhomogeneity of the constant magnetic field B_0, its essential features can be explained by a consideration of the free induction decay, i.e., the NMR signal following a short intense rf pulse. During the decay time of this signal ($t \geqq 0$), two succeeding magnetic field gradient pulses of opposite direction are applied.

Because these pulses cause additional fields much smaller than the constant magnetic field $\mathbf{B}_0$ which is along the z direction, the Larmor frequency is affected only by the z component of these fields. Orthogonal components have only the effect of slightly tilting the net field direction. If we define **g** in the usual manner as the **grad** of the z component of the magnetic field, the effective external magnetic field is given by

$$\begin{array}{lll} B_0 + \mathbf{g}\cdot\mathbf{r} & \text{for} & 0 \leqq t \leqq \delta \\ B_0 & \text{for} & \delta < t < \delta + \Delta \\ B_0 - \mathbf{g}\cdot\mathbf{r} & \text{for} & \delta + \Delta \leqq t \leqq 2\delta + \Delta. \end{array} \tag{93}$$

g denotes the intensity and δ the duration of the magnetic field gradient pulses. Assuming that δ is much smaller than the time interval Δ between the two magnetic field gradient pulses, the ratio ψ of the amplitudes of the free induction decay at time $t = 2\delta + \Delta$ and $t = 0$ can be shown to be [125]

$$\psi = \exp(-\Delta/T_2) \int p(\mathbf{r}' - \mathbf{r}, \Delta) \exp\{i\gamma\mathbf{g}(\mathbf{r}' - \mathbf{r})\} \mathrm{d}(\mathbf{r}' - \mathbf{r}). \tag{94}$$

Here T_2 denotes the transverse nuclear magnetic relaxation time of the adsorbed molecules and γ the gyromagnetic ratio of the resonating nuclei. $p(\mathbf{r}' - \mathbf{r}, \Delta)$ is the mean propagator defined by

$$p(\mathbf{r}' - \mathbf{r}, \Delta) = \int p(\mathbf{r}) p(\mathbf{r}, \mathbf{r}', \Delta) \, \mathrm{d}\mathbf{r} \tag{95}$$

with the a-priori probability density $p(\mathbf{r})$ to find a molecule at **r**, and the conditional probability density, the 'propagator', $p(\mathbf{r}, \mathbf{r}', \Delta)$ to find a molecule at time Δ at $\mathbf{r}'$ if it has been at **r** at time 0. Assuming

$$T_2 \gg \Delta \tag{96}$$

and introducing the vector

$$\mathbf{q} = \mathbf{g}\gamma\delta/2\pi \tag{97}$$

Eq. (94) can be rewritten as

$$\psi = \int p(\mathbf{r}' - \mathbf{r}, \Delta) \exp\{i2\pi\mathbf{q}(\mathbf{r}' - \mathbf{r})\} \, \mathrm{d}(\mathbf{r}' - \mathbf{r}). \tag{98}$$

This equation is very similar to the expression for the incoherent fraction of the quasielastic neutron scattering function (see, e.g., [126])

$$S_{\text{inc}} = \langle N^{-1} \Sigma_i \exp\{i2\pi\mathbf{q}(\mathbf{r}_i(t) - \mathbf{r}_i(0))\} \rangle \tag{99}$$

where **q** denotes the scattering wavevector and the sum is taken over all scattering centres ($i = 1$ to $i = N$). By contrast, the coherent fraction is described by

$$S_{\text{coh}} = \langle N^{-2} \Sigma_i \Sigma_j \exp\{i2\pi\mathbf{q}(\mathbf{r}_i(t) - \mathbf{r}_j(0))\} \rangle. \tag{100}$$

This sensitivity to the relative motion of different centres $(\mathbf{r}_i(t) - \mathbf{r}_j(0))$ makes an interpretation of coherent neutron scattering and of quasielastic light scattering considerably more difficult: The direct measurement of self-motion is a major advantage of the PFG NMR method. With the gradient **g** directed parallel to the constant magnetic field $\mathbf{B}_0$ which is along the z direction, Eq. (94) reduces to

$$\psi = \exp(-\Delta/T_2) \int p(z'-z, \Delta) \cos[g\gamma\delta(z'-z)] \mathrm{d}(z'-z) \tag{101}$$

and it can be easily shown that for $T_2 \gg \Delta$ the Fourier transform of the signal decay ψ directly yields the mean propagator with regard to the z component of the diffusing molecule:

$$p(z'-z, \Delta) = (2\pi)^{-1} \int \psi \cos[(z'-z)g\gamma\delta] \mathrm{d}(g\gamma\delta). \tag{102}$$

In [127], where the concept of the mean propagator has been used for the first time to interpret PFG NMR signals, examples for a determination of the mean propagator for molecular diffusion in assemblages of zeolites can be found. In homogeneous systems the mean propagator (solution to the Fickian equation) is given by a Gaussian function

$$p(z'-z, \Delta) = (4\pi\Delta D)^{-1/2} \exp\{-(z'-z)^2/4D\Delta\} \tag{103}$$

and it follows from Eq. (101)

$$\psi = \exp(-\Delta/T_2 - (g\gamma\delta)^2 D\Delta). \tag{104}$$

The important fact to be mentioned here is that a single parameter, viz. the self-diffusion coefficient D determines completely the diffusional process. In contrast, for heterogeneous systems, due to substantial difficultics in the mathematical analysis [128], only in some limiting cases an analytical expression can be found for the mean propagator. In order to describe the diffusional process also in these cases by a single parameter, the initial slope of a plot of $\ln\psi$ versus $(g\gamma\delta)^2$ may be used since this quantity is directly related to the mean square displacement $\langle(z'-z)^2\rangle$ of the adsorbed molecules:

$$(\mathrm{d}\ln\psi/\mathrm{d}(g\gamma\delta)^2)_{g\gamma\delta=0} = \langle(z'-z)^2\rangle/2 \tag{105}$$

as can be easily shown by the use of Eq. (101). On the other hand, Eq. (103) yields, for homogeneous systems,

$$\langle(z'-z)^2\rangle_{\mathrm{homog}} = 2D\Delta \tag{106}$$

so that we may introduce an apparent self-diffusion coefficient D_{app} by the equation

$$\begin{aligned} D_{\mathrm{app}} &= \langle(z'-z)^2\rangle/2\Delta \\ &= (\mathrm{d}\ln\psi/\mathrm{d}(g\gamma\delta)^2{}_{g\gamma\delta=0}/\Delta. \end{aligned} \tag{107}$$

5.2 *Limitations of the PFG NMR Method*

The limits of the PFG NMR method are determined by the gyromagnetic ratio γ of the resonating nuclei and the maximum achievable values of Δ, g and δ. An inspection of Eq. (104) shows that the maximum value of Δ should not exceed T_2 significantly, and we write

$$\Delta_{max} = T_2. \tag{108}$$

The maximum value of g depends on the design of the electronic equipment and of the probe head and will be not much larger than $20\,T\,m^{-1}$. δ is limited by the condition $\delta \ll \Delta$ and we assume $\delta_{max} \approx 10^{-3}\,s$. From Eq. (104) the condition for the minimum value of the self-diffusion coefficient D_{min} which is accessible by the PFG NMR method follows as

$$D_{min} \cdot (g^2\gamma^2\delta^2\Delta)_{mas} = 1. \tag{109}$$

With the value of the gyromagnetic ratio of the proton $\gamma_{prot} \approx 2.675 \times 10^8\,T^{-1}\,s^{-1}$ and the maximum values for Δ, g, and δ given above, this condition may be rewritten as

$$D_{min}/m^2s^{-1} = 3 \cdot 10^{-14}(\gamma_{prot}/\gamma)^2(T_2/s)^{-1}. \tag{110}$$

For hydrocarbons adsorbed on zeolites Na-X, the transverse ^1H magnetic relaxation time T_2 is of the order of $10^{-2}\,s$, so that the minimum value of the intracrystalline self-diffusion coefficient which can be measured is ca. $3 \times 10^{-12}\,m^2\,s^{-1}$. Similarly from the condition

$$(2D\Delta)_{min} \cdot \tfrac{1}{2}(g\gamma\delta)^2_{min} = 1 \tag{111}$$

which is equivalent to Eq. (109), the minimum value of the rms displacement (cf. Eq. (106)) follows as

$$\sqrt{\langle (z'-z)^2 \rangle}_{min} = 300 \cdot (\gamma_{prot}/\gamma)\,nm \tag{112}$$

i.e. a value of 300 nm for the ^1H PFG NMR method.

The estimation presented here depends on the NMR signal attenuation given by Eq. (104). It is valid for the free induction decay as well as for the amplitude of the so-called primary spin echo which is the NMR signal at $t = 2\tau$ if a $\pi/2$ pulse applied at $t = 0$ is followed by a π pulse at $t = \tau$ (see, e.g., [125]). In contrast, for the so-called stimulated spin echo which appears at $t = \tau_1 + \tau_2$ if three $\pi/2$ pulses are applied at $t = 0$, τ_1 and τ_2, the attenuation due to longitudinal (T_1) and transverse (T_2) relaxation and self-diffusion ($\delta \ll \Delta$) is given by [125]

$$\psi = \tfrac{1}{2}\exp\{-2\tau_1/T_2 - (\tau_2 - \tau_1)/T_1 - (g\gamma\delta)^2 D\Delta\}. \tag{113}$$

Therefore, the minimum value of the self-diffusion coefficient D_{min} which is accessible in this case will depend not only on the value of T_2 but also on the ratio T_1/T_2. For the stimulated echo the two field gradient pulses must be applied between the first and second $\pi/2$ pulse, and the third $\pi/2$ pulse and the echo, respectively. Therefore τ_2 is equal to Δ and a comparison of Eqs. (113) and (104) shows that the relaxation attenuation for the stimulated echo is less than that for the primary echo if

$$\Delta/T_2 > 2\tau_1/T_2 + (\Delta - \tau_1)/T_1 + \ln 2. \tag{114}$$

With Eq. (108) and $2\tau_1 \ll \Delta$ this condition can be rewritten as

$$T_1/T_2 > (1 - \ln 2)^{-1}. \quad (115)$$

Hence, if the ratio T_1/T_2 is larger than ca. 3, the stimulated echo should be preferred and the minimum value of D which can be measured is given approximately by Eq. (110) with T_1 instead of T_2. A more detailed analysis can be found in [129]. If the transverse relaxation time T_2 decreases, the time interval τ_1 between the first and the second $\pi/2$ pulse should be as small as possible. The 'natural' limit is given by the width δ of the (first) field gradient pulse which must be applied during the time interval between the first and the second $\pi/2$ pulse:

$$\tau_{1\,\mathrm{min}} = \delta. \quad (116)$$

However, with decreasing differences between τ_1 and δ, the second rf pulse no longer acts as a $\pi/2$ pulse since the tail of the field gradient pulse causes a resonance offset which leads to an additional attenuation of the stimulated echo. Hence, the lower limit of τ_1 given by Eq. (116) cannot be achieved. A much better approach to the lower limit given by Eq. (116) is possible if one applies instead of the second and third $\pi/2$ pulses a spin-locking field during the time interval $\tau_2 - \tau_1$ which leads to the formation of a spin-locking echo at the same time $(\tau_1 + \tau_2)$ as the stimulated echo. The amplitude of this echo is less sensitive to the tails of the field gradient pulses, thus allowing a measurement of even smaller self-diffusion coefficients than with the stimulated echo [130].

5.3 *Internal Magnetic Fields, Multi-Phase Relaxation*

In heterogeneous systems like assemblages of zeolite crystallites, a study of molecular self-diffusion by the PFG NMR method may be complicated by internal magnetic field gradients and/or multi-phase relaxation. The general formula for the attenuation of the primary echo caused by the self-diffusion of molecules describable by a self-diffusion coefficient D can be shown to be [124], [125]:

$$\psi = \exp\{-(2/3)\gamma^2 g_0^2 \tau^3 D - (g\gamma\delta)^2(\Delta - \delta/3)D + \delta\gamma^2 \mathbf{g}_0 \cdot \mathbf{g}[(t_1^2 + t_2^2) + \delta(t_1 + t_2) + (2/3)\delta^2 - 2\tau^2]D\} \quad (117)$$

where τ is the time interval between the $\pi/2$ and the π pulse. Δ denotes the time interval between the first and the second field gradient pulse, δ the width of each of the field gradient pulses, t_1 the time interval between the $\pi/2$ pulse and the first field gradient pulse, and t_2 the time interval between the end of the second field gradient pulse and the echo maximum (cf. Fig. 14). $\mathbf{g}_0$ is a time independent field gradient caused either by the external magnetic field or by the inhomogeneity of the sample (internal magnetic fields) and $\mathbf{g}$ the amplitude of the field gradient pulses. For the special case $\delta = \tau = \Delta$, i.e. $t_1 = t_2 = 0$, Eq. (117)

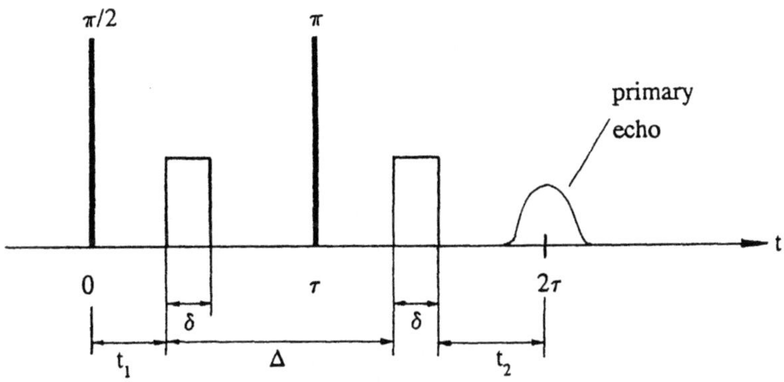

Fig. 14. Time dependence of the field gradient and rf pulses in a PFG NMR experiment (primary spin echo method)

simplifies to

$$\psi = \exp\{-\gamma^2\delta^2(\Delta - \delta/3)D(\mathbf{g}_0 + \mathbf{g})^2\} \tag{118}$$

which has been used in [131]. Concerning the internal field gradient $\mathbf{g}_0$ the following two assumptions have been made [132]:

(1) During the whole time interval 2τ, the internal field gradient $\mathbf{g}_0$ experienced by an individual molecule is constant, but different molecules in different parts of the sample experience a different internal field gradient.

(2) For the projection g_0 of $\mathbf{g}_0$ along the pulsed field gradient $\mathbf{g}$, a Gaussian distribution is assumed:

$$f(g_0) = (2\pi\sigma^2)^{-1/2}\exp(-g_0^2/2\sigma^2). \tag{119}$$

Then it follows [133] from the general formula (Eq. (117)) if one assumes a symmetric pulse sequence ($t_1 = t_2 = \tau - \Delta/2 - \delta/2$)

$$\psi = (1 + 4D\gamma^2\tau^3\sigma^2/3)^{-3/2}\exp\{-(g\gamma\delta)^2(\Delta - \delta/3)D + \tfrac{1}{2}(\gamma^2 g\sigma\delta)^2 \cdot[2\tau\Delta - \Delta^2/2 - \delta^2/6]^2 D^2(1 + 4D\gamma^2\tau^3\sigma^2/3)^{-1}\}. \tag{120}$$

For

$$\Delta, \tau \gg \delta \quad \text{and} \quad 4D\gamma^2\tau^3\sigma^2/3 \ll 1 \tag{121}$$

Eq. (120) simplifies to the result of Ref. [132]

$$\ln\psi = (g\gamma\delta)^2\Delta D\{1 - \tfrac{1}{2}\Delta D(\gamma\sigma)^2[2\tau - \Delta/2]^2\} \tag{122}$$

or

$$D_{\mathrm{app}} = -D\{1 - \tfrac{1}{2}\Delta D(\gamma\sigma)^2[2\tau - \Delta/2]^2\}. \tag{123}$$

From Eqs. (122) and (123) three important conclusions can be drawn:

(1) Also in those cases where the internal field gradient $\mathbf{g}_0$ affects the attenuation of the primary echo, $\ln\psi$ increases with the square of the 'intensity' $(g\gamma\delta)$ of the field gradient pulses. This result revises a statement in [131].

(2) The apparent self-diffusion coefficient D_{app} derived from a PFG NMR experiment is a function of the time interval Δ between the two field gradient pulses, and

(3) D_{app} is smaller than the real self-diffusion coefficient D.

The other problem which must be taken into consideration if PFG NMR measurements are performed with heterogeneous samples occurs if the various phases of the heterogeneous sample exhibit different nuclear magnetic relaxation times: Since the contribution of the various phases to the spin echo is controlled by their nuclear magnetic relaxation rates, with increasing values of the time interval Δ, the contribution of the phases with the largest relaxation times will become more and more dominant leading to a dependence of the apparent self-diffusion coefficient on Δ. While investigating samples containing a small amount of molecules with an extremely large relaxation time, this effect may lead to completely erroneous results: For sufficiently large values of Δ, it is the diffusivity of these molecules rather than that of the bulk phase that is obtained in PFG NMR experiments. Hence, in diffusion studies of zeolites a formal analysis of the attenuation ψ of the spin echo would yield the diffusivity of a small amount of extra-crystalline molecules. In Refs. [134] and [135] a simple method has been described which allows a variation of the influence of relaxation upon the spin echo without changing the field gradient pulse sequence. In this way any change in the spin echo attenuation can be attributed unambiguously to the influence of multi-phase relaxation. The pulse sequence is shown in Fig. 15. It may be considered to be a conventional primary spin echo pulsed field gradient experiment, in which instead of the starting $\pi/2$ pulse one has applied a preparatory spin echo sequence. Evidently, the function of the second half of the spin echo brought about by the preparatory sequence is equivalent to that of the free induction decay following the $\pi/2$ pulse in the original version of the pulsed field gradient experiment. However, in contrast to this original experiment, by varying the time interval τ_p between the preparatory rf pulses, the proposed sequence allows a variation of the influence of relaxation on the

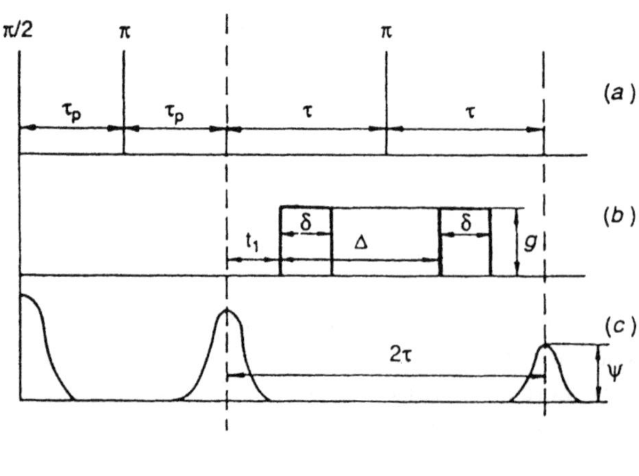

Fig. 15. (*a*) rf pulses, (*b*) magnetic field gradient pulses, (*c*) NMR signals in a modified PFG NMR experiment. By the use of the preparatory rf pulses it is possible to prove whether the measured self-diffusion coefficient is falsified by the influence of multi-phase relaxation [134]

observed NMR signal without the necessity for any change of the pulsed field gradient sequence. This procedure may be carried out with both the primary and the stimulated spin echo sequences. Examples for an application of this method can be found in [135] (*n*-hexane adsorbed on Na-X), [136] (*n*-butane adsorbed on ZSM-5), and [137] (benzene adsorbed on ZSM-5). Considering these experiments it must be assumed that the high values of the self-diffusion coefficients which were obtained for benzene adsorbed on silicalite by an application of the simple PFG NMR method [138] and which are far above the mobilities determined by other authors [137], [139], [140], [141] are caused by the artifacts described above.

5.4 Anisotropic Diffusion

In zeolites of non-cubic symmetry as, e.g., silicalite, molecular diffusion should be anisotropic. Assuming that condition (96) is fulfilled it follows from Eq. (101) for molecules adsorbed in a single crystal

$$\psi = \exp\{-\gamma^2 g^2 \delta^2 \cdot (D_x \cos^2\varphi \sin^2\vartheta + D_y \sin^2\varphi \sin^2\vartheta + D_z \cos^2\vartheta)\Delta\} \tag{124}$$

where D_x, D_y, D_z are the principal values of the self-diffusion tensor and φ and ϑ determine the direction of the magnetic field gradients in the principal axes system of the crystal. For an assemblage of zeolite crystallites (powder sample) one has to average over all orientations which yields

$$\psi = (4\pi)^{-1} \int \exp\{-\gamma^2 g^2 \delta^2 (D_z \cos^2\vartheta + D_y \sin^2\vartheta)\} I(\vartheta) \mathrm{d}(\cos\vartheta) \qquad (126)$$

with

$$I(\vartheta) = \int \exp\{-\gamma^2 g^2 \delta^2 (\sin\vartheta \cos\varphi)^2 (D_x - D_y)\Delta\} \mathrm{d}\varphi. \qquad (126)$$

In the case of silicate the intracrystalline space consists of intersecting linear and zig-zag channels. Assuming that, on passing a certain channel intersection, a molecule will take its subsequent direction independent of the direction it has come from, the principal values of the self-diffusion tensor are related by [142]

$$c^2/D_z = a^2/D_x + b^2/D_y \tag{127}$$

with a, b and c denoting the length of the unit cell in x, y and z direction. Inserting this relation into Eq. (125) and introducing the mean self-diffusion coefficient

$$\langle D \rangle = (D_x + D_y + D_z)/3 \tag{128}$$

it is possible to calculate ψ numerically as a function of $\gamma^2 g^2 \delta^2 \langle D \rangle$ for given values of the ratio D_y/D_x. Due to the structure of the channel system in silicate, $D_y < D_x$ can be excluded so that in Fig. 16 only curves are plotted for $D_y/D_x \geqq 1$. The experimental data presented also in Fig. 16 were measured at various

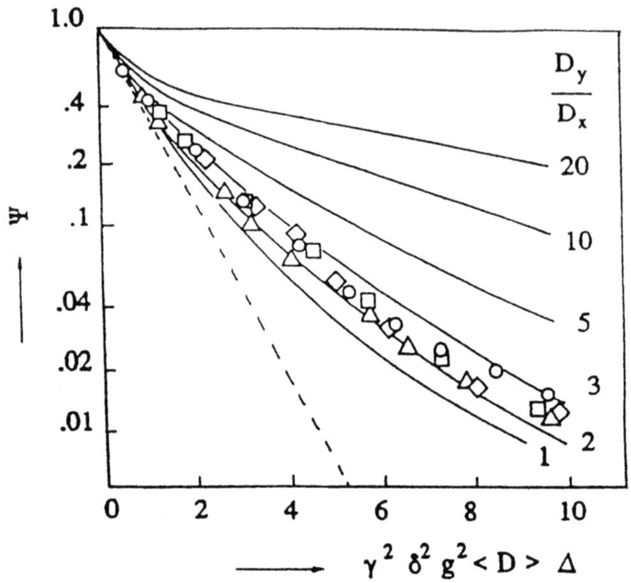

Fig. 16. Theoretical dependence of the signal decay ψ for anisotropic molecular diffusion in silicalite for various values of the ratio D_y/D_x. The *broken line* represents the result for an isotropic self-diffusion. $\langle D \rangle$ denotes the mean self-diffusion coefficient defined by Eq. (128). The experimental data were measured at 193 K (○), 223 K (□), 273 K (◇), and 298 K (△) for methane adsorbed on silicalite with a concentration of 3 molecules per channel intersection [143]

temperatures for methane adsorbed in silicalite at a sorbate concentration of 3 molecules per channel intersection [143].

Another method of studying the anisotropy of self-diffusion has been achieved by introducing the crystallites into a system of parallel capillaries. Since, in the case of silicalite, the longer axis of the crystallites is along the z-direction, the z-axis of the channel system is preferentially oriented along the direction of the capillaries [144]. Therefore, with the magnetic field gradients applied in the direction of the capillary axes, the measured self-diffusion coefficient will be

$$D_{\text{parallel}} = \langle D \rangle + (2/3)S\{D_z - \tfrac{1}{2}(D_x + D_y)\} \qquad (129)$$

and for the orientation perpendicular to the axes

$$D_{\text{perpendicular}} = \langle D \rangle + (1/3)S\{\tfrac{1}{2}(D_x + D_y) - D_z\} \qquad (130)$$

where S denotes the order parameter

$$S = \tfrac{1}{2}\langle 3\cos^2\theta - 1 \rangle \qquad (131)$$

with θ as the angle between the z-axis of the crystallites and the direction of the capillaries. By the use of capillaries with an internal diameter of $200 \pm 20\ \mu m$ and crystallites of typical size $300 \cdot 60 \cdot 60\ \mu m^3$, values between 0.63 and 0.73 were

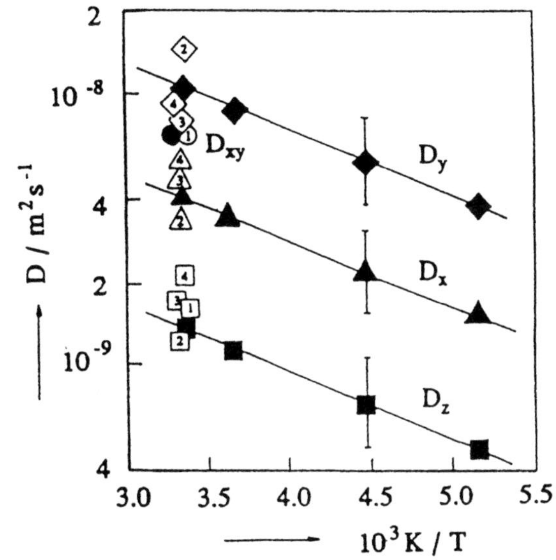

Fig. 17. Temperature dependence of the principal values of the self-diffusion tensor for methane adsorbed on silicalite as determined from PFG NMR measurements on powder samples (*full symbols*) [143], and comparison with the results of oriented samples (*1*) [144] and of molecular dynamics calculations (*2*) [145], (*3*) [146], and (*4*) [147]

realized for S. For a concentration of 3 methane molecules per channel intersection, the measurements performed at room temperature yielded a value of $1.6 \times 10^{-9} m^2 s^{-1}$ for D_z and of $7.2 \times 10^{-9} m^2 s^{-1}$ for the quantity $D_{xy} = \frac{1}{2}(D_x + D_y)$. It can be seen from Fig. 17 that these values are in good agreement with the results of the measurements on powder samples (full symbols). For comparison, in Fig. 17 also the results of molecular dynamics calculations are included which were published by three independent groups [145], [146], [147]. While the averages of these results are in excellent agreement with the principal values of the self-diffusion tensor determined experimentally by the PFG NMR method, it is apparent that in Ref. [145] the anisotropy viz. the ratio D_y/D_x is overestimated.

5.5 NMR Tracer Desorption Technique

If one approximates the assemblage of zeolite crystallites loaded with one sort of molecule by a 'two-phase diffusion model' which means that the behaviour of the system can be described by only four quantities, viz. the relative number p_a of molecules of phase a, their mean residence time τ_a in this phase, and the self-diffusion coefficients D_a and D_b in phases a and b, respectively, an analytical expression for ψ can be derived (see, e.g., [125]). Neglecting the influence of the transverse nuclear magnetic relaxation ($T_2 \gg \Delta$) and assuming

$$p_b = 1 - p_a \ll 1 \quad \text{and} \quad D_a \ll D_b \tag{132}$$

the general formula [125] simplifies to

$$\psi = \exp\{-\gamma^2 g^2 \delta^2 [D_a + p_b D_b (1 + \gamma^2 g^2 \delta^2 \tau_a p_b D_b)^{-1}] \Delta\}. \tag{133}$$

The practical value of this equation lies in the fact, that the suppositions (132) are met if phase a is related to molecules adsorbed in the intracrystalline space and b to molecules outside (intercrystalline space), so that we may write:

$$D_a \equiv D_{intra}; \quad D_b \equiv D_{inter}; \quad p_a \equiv 1 - p_{inter}; \quad \tau_a \equiv \tau_{intra} \tag{134}$$

where D_{intra} is the intracrystalline self-diffusion coefficient and τ_{intra} the mean residence time of a molecule in a zeolite crystallite. A plot of $\ln\psi$ as a function of $\gamma^2 g^2 \delta^2$ yields then for large values of $\gamma^2 g^2 \delta^2$ a straight line

$$\ln\psi = -\gamma^2 g^2 \delta^2 \Delta D_{intra} - \Delta/\tau_{intra} \tag{135}$$

so that the intracrystalline self-diffusion coefficient (D_{intra}), and the mean residence time in a crystallite (τ_{intra}) can be determined independently. The latter information is equivalent to that obtained in conventional tracer exchange experiments, and therefore this method of analysing PFG NMR data has been called NMR tracer desorption technique [148]. For spherical crystallites of radius R the mean intracrystalline residence time τ_{intra} is determined only by the intracrystalline self-diffusion coefficient and the radius if there exists no additional transport resistance (diffusion barrier) at the outer surface. It can be shown easily that in this case τ_{intra} must be equal to

$$\tau_{D,\,intra} = R^2/(15 D_{intra}). \tag{136}$$

In those cases where the shape of the crystallites cannot be approximated by a sphere, $\tau_{D,\,intra}$ must be calculated numerically. Results can be found in [141]. If one finds that the experimentally determined quantity τ_{intra} is larger than $\tau_{D,\,intra}$ there must be an additional transport resistance at the external surface, and the ratio $\tau_{intra}/\tau_{D,\,intra}$ can be taken as measure of the strength of this resistance.

In Table 16 experimental results are collected for a NaCa-A zeolite pretreated at 400 °C under shallow bed and deep bed conditions respectively, for the powder compacted with a pressure of 20 MPa and for the powder granulated with a

Table 16. Experimental results for the time constant $\tau_{D,intra}$ (see Eq. (136)) and the corresponding real residence time τ_{intra} of a methane molecule in zeolite crystallites NaCa-A with a mean diameter of 5 μm. The measurements were performed at 20 °C and with a concentration of 6 methane molecules per large cavity. SB and DB denotes the shallow bed and deep bed pretreatment, respectively, described in Sect. 3.3

Sample	$\tau_{D,intra}$/ms	τ_{intra}/ms	$\tau_{intra}/\tau_{D,intra}$
NaCa-A SB	0.3 ± 0.1	0.3 ± 0.1	1
NaCl-A DB	0.3 ± 0.1	2.5 ± 0.5	8.3
NaCa-A DB compacted	0.3 ± 0.1	2.3 ± 0.5	7.7
NaCa-A DB granulated	0.3 ± 0.1	5.6 ± 1	19

binder, as usual, if zeolites are used as commercial catalysts. It can be seen that neither the deep bed pretreatment nor the compaction or the granulation leads to a change of the interacrystalline self-diffusion coefficient which means that the intracrystalline pore system is unaffected. In contrast, the external surface of the crystallites is deteriorated both by the deep-bed procedure and the granulation while the compaction is of no influence within the limits of accuracy.

With this method it is also possible to answer the question whether the deactivation of a zeolite catalyst by coking is due to a blocking of the entrance of molecules through the external surface or by a blocking of the active sites in the intracrystalline space. As an example the coking of ZSM-5 by *n*-hexane and mesitylene will be discussed [149]. In Fig. 18, values for the mean intercrystalline residence time τ_{intra} and the time constant $\tau_{D,\,intra}$ defined by Eq. (136) are plotted as a function of the coking time. Apparently, the two compounds used lead to completely different dependences: For *n*-hexane τ_{intra} and $\tau_{D,\,intra}$ coincide over a large range of coking times, whereas for mesitylene only τ_{intra} is found to increase, while the intracrystalline self-diffusion coefficient represented by $\tau_{D,\,intra}$ remains unaffected. Hence, one must conclude that during mesitylene coking, the carbonaceous compounds are deposited exclusively on the external surfaces of the crystallites which is reasonable since the mesitylene molecules are too large to penetrate the intracrystalline channel system. For *n*-hexane, two stages of the coke deposition become visible: At shorter coking times, *n*-hexane is mainly deposited in the intracrystalline space, thus simultaneously affecting a retardation of intracrystalline diffusion and of the mean intracrystalline residence time. In a second stage, similar to the behaviour observed with

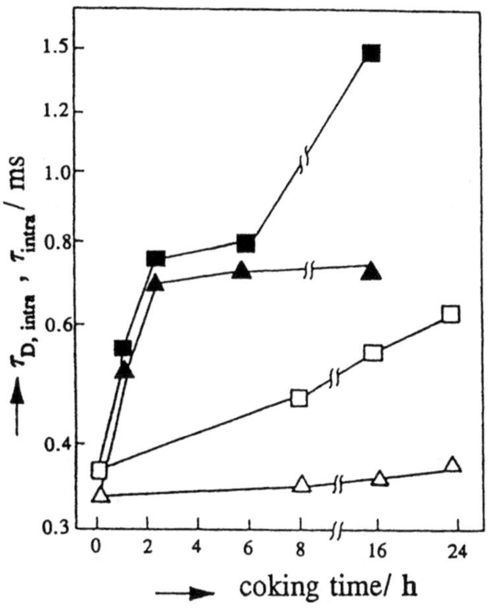

Fig. 18. Mean intracrystalline residence time τ_{intra} (□) and the time constant $\tau_{D,intra}$ (△) which is related to the intracrystalline self-diffusion coefficient (see Eq. (136)) for methane at 296 K with a concentration of 3 molecules per channel intersection in ZSM-5, coked by *n*-hexane (*full symbols*) and mesitylene (*open symbols*) in dependence on the time on stream [149]

mesitylene, coke is predominantly deposited at the external surfaces of the crystallites.

It is one of the main experimental problems of the PFG NMR method that any mismatch ($\delta_1 g_1 \neq \delta_2 g_2$) between the intensities of the two field gradient pulses which is inevitable in general, leads to an additional damping of the NMR signal which may be interpreted erroneously as an enhanced diffusivity. By the application of a constant magnetic field gradient $\mathbf{g}_0$ along the direction of the pulsed field gradients, however, the influence of such mismatches can be drastically reduced (see, e.g., [125]) so that most PFG NMR experiments are performed with such an additional field gradient. As has been shown in [150] instead of Eq. (133), the attenuation of the primary spin echo is then given by

$$\psi = \exp\{-\gamma^2 A[D_a + p_b D_b(1 + \gamma^2 \tau_a p_b D_b A/\Delta)^{-1}]\} \tag{137}$$

with

$$A = (2/3)\tau^3 g_0 + \delta^2 g^2 \Delta - \delta\{t_1^2 + t_2^2 + \delta(t_1 + t_2) + (2/3)\delta^2 - 2\tau^2\} g g_0. \tag{138}$$

Here t_1 denotes the time interval between the $\pi/2$ pulse and the onset of the first field gradient pulse and t_2 the time interval between the end of the second field gradient pulse and the maximum of the primary spin echo. The corresponding formula for the stimulated spin echo can be found in [150]. For simplicity we assume $\Delta = \tau$, a condition which is often fulfilled in PFG NMR experiments using the primary spin echo. Moreover we introduce the quantities defined by Eq. (134) so that a plot of $\ln\psi$ as a function of $(g\gamma\delta)^2$ yields the same straight line as given by Eq. (135) for large values of $(g\gamma\delta)^2$. In contrast however to the $g_0 = 0$ experiment, $\ln\psi$ does not vanish for $(g\gamma\delta)^2 = 0$, instead it is given by

$$(\ln\psi)_{(g\gamma\delta)=0} = -(2/3)\gamma^2 g_0 \Delta^3 \cdot\{D_{\text{intra}} + p_{\text{inter}} D_{\text{inter}}(1 + 2\gamma^2\Delta^2 g_0 \tau_{\text{intra}} p_{\text{inter}} D_{\text{inter}}/3)^{-1}\}. \tag{139}$$

If one now determines analogously to Eq. (135) a time constant from the distance between $(\ln\psi)_{g(\gamma\delta)=0}$ and the intersection of the straight line, which approximates $\ln\psi$ at large values of $(g\gamma\delta)^2$, with the ordinate, it follows an apparent mean residence time τ'_{intra} given by

$$1/\tau'_{\text{intra}} = 1/\tau_{\text{intra}} - (2/3)\gamma^2 g_0^2 \Delta^2 \cdot\{D_{\text{intra}} + p_{\text{inter}} D_{\text{inter}}(1 + (2/3)\gamma^2 g_0^2 \Delta^2 p_{\text{inter}} D_{\text{inter}} \tau_{\text{intra}})^{-1}\}. \tag{140}$$

The important fact is that this apparent mean residence time is larger than τ_{intra} so that the ratio $\tau'_{\text{intra}}/\tau_{\text{D, intra}}$ becomes larger than $\tau_{\text{intra}}/\tau_{\text{D, intra}}$ which is tantamount to an increased sensitivity of the method with regard to the formation of a transport resistance at the outer surface of the zeolite crystallites. This method has been applied to study the deterioration of granulated zeolites

NaCa-A by hydrothermal treatment and during their industrial use as molecular sieves [150], [151].

Acknowledgements. I wish to thank Prof. Dieter Freude, Prof. Jörg Kärger and Dr. Eike Brunner for many helpful comments and Mrs. Cordula Krause for her kind assistance in preparing the manuscript.

6 References

1. Ertl G (1987) Langmuir 3: 4
2. Ertl G (1988) Z. phys. Chemie, Leipzig 269: 306
3. Ertl G (1990) Angew. Chemie 102: 1258
4. Pfeifer H (1976) Physics Reports 26C: 293
5. Pfeifer H, Meiler W, Deininger D (1983) Annual Reports on NMR Spectroscopy 15: 291
6. Wang P-K, Ansermet J-P, Rudaz SL, Wang Z, Shore S, Slichter CP, Sinfeld JH (1986) Science 234: 35
7. Engelhardt G, Michel D (1987) High-resolution solid-state NMR of silicates and zeolites. J. Wiley, Chichester
8. Corbin DR, Burgess BF, Vega AJ, Farlee RL (1987) Analytical Chem 59: 2722
9. Nagy JB, Derouane EG (1988) ACS Symposium Series 368: 2
10. Fraissard J, Ito T (1988) Zeolites 8: 350
11. Engelhardt G (1989) Trends Analyt. Chem. 8: 343
12. Bell AT (1989) ACS Symposium Series 398: 66
13. Thomas JM, Vaughan DEW (1989) J. Phys. Chem. Sol. 50: 449
14. Clague ADH, Alma NCM (1989) Analytical NMR. J. Wiley, New York, p 115
15. Pfeifer H (1991) Nato ASI Series B 265: 151
16. Klinowski J (1991) Chem. Rev. 91: 1459
17. Fyfe CA, Feng Y, Grondey H, Kokotailo GT, Gies H (1991) Chem. Rev. 1525
18. Chen R, Springuel-Huet MA, Fraissard J (1991) Stud. Surf. Sci. Catalysis 65: 219
19. Mastikhin VM, Mudrakowsky JL, Nosov AV (1991) Progress in Nuclear Magnetic Resonance Spectroscopy 23: 211
20. Pfeifer H, Ernst H (1993) Annual Reports on NMR Spectroscopy 28: (in press)
21. Breck DW (1974) Zeolite molecular sieves, structure, chemistry and use. J. Wiley, London
22. Barrer RM (1978) Zeolite and clay minerals as sorbents and molecular sieves. Academic, London
23. Meier WM, Olson DH (1987) Atlas of Zeolite Structure Types, 2nd rev. edn. Butterworths, London
24. van Bekkum H, Flanigen EM, Jansen JC (1987) Studies in surface science and catalysis. Elsevier, Amsterdam (vol. 58)
25. Öhlmann G, Pfeifer H, Fricke R (1987) Studies in surface science and catalysis. Elsevier, Amsterdam (vol. 65)
26. Jacobs PA, Jaeger NI, Kubelkova L, Wichterlova B (1987) Studies in surface science and catalysis. Elsevier, Amsterdam (vol. 69)
27. Mortier WJ (1982) Compilation of extra framework sites in zeolites. Butterworths, London
28. Lide DR (1990) Handbook of chemistry and physics. CRC Press, Boston (71st Edition)
29. Freude D, Brunner E, Pfeifer H, Prager D, Jerschkewitz H-G, Lohse U, Öhlmann G (1987) Chem. Phys. Lett. 139: 325
30. Haase J, Pfeifer H, Oehme W, Klinowski J (1988) Chem. Phys. Lett. 150: 189
31. Kellberg L, Linnen M, Jakobson HJ (1991) Chem. Phys. Lett. 182: 120
32. Bonardet JL, Fraissard J (1976) J. Magn. Res 22: 543
33. Jackowski K, Raynes WT (1977) Mol. Phys. 34: 465
34. Gay JD, Kriz JF (1978) J. Phys. Chem. 82: 319
35. Duthaler RO, Roberts JD (1978) J. Am. Chem. Soc. 100: 4969
36. Witanowski M, Stefaniak L, Szymanski S, Januszewski H (1977) J. Magn. Res 28: 217

37. Michel D, Meiler W, Gutsze A, Wronkowski A (1980) Z. phys. Chemie, Leipzig 261: 953
38. Pfeifer H, Gutsze A, Shdanov SP (1976) Z. phys. Chemie, Leipzig 257: 721
39. Sauer J (1987) Mol. Cat. 54: 312
40. Yesinowski JP, Eckert H (1987) J. Am. Chem. Soc. 109: 6274
41. Berglund B, Vaughan RW (1980) J. Chem. Phys. 73: 2037
42. Chauvel JP, True NS (1985) Chem. Phys. 95: 435
43. Bartmess JE, Scott JA, McIver RT (1979) J. Am. Chem. Soc. 101: 6046
44. Taft RW, Koppel IA, Topsom RD, Anvia F (1990) J. Am. Chem. Soc. 112: 2047
45. Sanderson RT (1976) Chemical bonds and bond energy. Academic, New York
46. Freude D, Hunger M, Pfeifer H, Schwieger W (1986) Chem. Phys. Lett. 128: 62
47. Pfeifer H (1989) Colloids Surf. 36: 312
48. Fleischer U, Kutzelnigg W, Bleiber A, Sauer J (1993) J. Am. Chem. Soc. (in press).
49. Akitt JW (1973) J. Chem. Soc., Dalton Trans. 49
50. Mastikhin VM, Mudrakowsky IL, Schmachkova VP, Kotsarenko NS (1987) Chem. Phys. Lett. 139: 93
51. Pfeifer H, Freude D, Kärger J (1991) Studies in surface science and catalysis. Elsevier, Amsterdam (vol 65) p 89
52. Brunner E, Freude D, Hunger M, Pfeifer H, Staudte B (1991) Studies in surface science and catalysis. Elsevier, Amsterdam (vol 69) p 453
53. Alla M, Lippmaa E (1982) Chem. Phys. Lett. 87: 30
54. Brunner E (1990) J. Chem. Soc., Faraday Trans. 86: 3957
55. Freude D, Pfeifer H (1980) In: Proc. 5th Intern. Conf. on Zeolites, Heyden, London, p 732
56. Fenzke D, Gerstein BC, Pfeifer H (1992) J. Magn. Res. 98: 469
57. Anderson PW, Weiss PR (1953) Rev. Mod. Phys. 25: 269
58. Schneider B, Doskocilova D, Jakes J (1979) In: Proceedings 20th Congress Ampere, Tallinn. Springer, Berlin
59. Andrew ER, Jasinski A (1971) J. Phys. Chem. Solid State Phys. 4: 391
60. Hunger M (1991) Thesis (habil.), University of Leipzig
61. Brunner E (1993) J. Chem. Soc., Faraday Trans. 89: 165
62. Opella SJ, Frey M, Gross TA (1979) J. Am. Chem. Soc. 101: 5856
63. Böhm J, Fenzke D, Pfeifer H (1983) J. Magn. Res. 55: 197
64. Olivieri AC (1989) J. Magn. Res. 81: 201
65. Stevenson RL (1971) J. Catal. 21: 113
66. Freude D, Klinowski J, Hamdan H (1988) Chem. Phys. Lett. 149: 355
67. Brunner E, Freude D, Gerstein BC, Pfeifer H (1990) J. Magn. Res. 99: 90
68. Maricq MM, Waugh JS (1979) J. Phys. Chem. 70: 3300
69. Brunner E, Fenzke D, Freude D, Pfeifer H (1990) Chem. Phys. Lett. 69: 591
70. Wind R (1991) In: Modern NMR techniques and their application in chemistry. M. Dekker, New York
71. Fenzke D, Hunger M, Pfeifer H (1991) J. Magn. Res. 95: 477
72. Pfeifer H, Freude D, Hunger M (1985) Zeolites 5: 274
73. Lippmaa E, Samoson AV, Brey VV, Gorlov YI (1981) Doklad. Akad. Nauk 259: 403
74. Halik S, Lercher JA, Mayer H (1988) J. Chem. Soc., Faraday Trans. I 84: 4457
75. Brunner E, Karge H-G, Pfeifer H (1993) Z. Phys. Chemie 176: 173
76. Fyfe CA (1983) Solid State NMR for Chemists. C.F.C. Press, Guelph
77. Dwyer J, Dewing J, Thompson NE, O'Malley PJ, Karim K (1989) J. Chem. Soc., Chem. Comm. 843
78 Ernst RR, Bodenhausen G, Wokaun A (1986) Principles of nuclear magnetic resonance in one and two dimensions. Clarendon, Oxford
79. Fenzke D, Hunger M, Pfeifer H (1991) J. Magn. Res. 95: 477
80. Schröder K-P, Sauer J, Leslie M, Catlow CRA, Thomas JM (1992) Chem. Phys. Lett. 188: 320
81. Czjzek M, Jobic H, Fitch AN, Vogt T (1992) J. Phys. Chem. 96: 1535
82. Hunger M, Anderson MW, Ojo A, Pfeifer H (1993) Microporous Materials 1: 17
83. Gluszak TJ, Chen DT, Sharma SB, Dumesic JA, Root TW (1992) Chem. Phys. Lett. 190: 36
84. Rosenberger H, Ernst H, Scheler G, Jünger I, Sonnenberger R (1982) Z. phys. Chemie, Leipzig 263: 864
85. Hunger M, Freude D, Fröhlich T, Pfeifer H, Schwieger W (1987) Zeolites 7: 108
86. Freude D, Haase J (1993) NMR – Basic Principles and Progress 29: 1
87. Clarke PG, Gosling K, Harris RK, Smith EG (1993) Zeolites 13: 388
88. Hirschler AE (1963) J. Catal. 2: 428

89. Plank CJ (1964) Proceedings of the Third International Congress on Catalysis, Amsterdam (vol 1) p 568
90. Hunger M, Freude D, Pfeifer H, Prager D, Reschetilowski W (1989) Chem. Phys. Lett. 163: 221
91. Sauer J (1985) Proceedings Int. Symp. Zeolite Catal., Siofok/Hungary, p 19
92. Sauer J, Hobza P (1984) Theor. Chim. Acta (Berlin) 65: 279 and 291
93. Hunger M, Freude D, Pfeifer H (1991) J. Chem. Soc., Faraday Trans. 87: 657
94. Kustov LM, Kazanski VB (1991) J. Chem. Soc., Faraday Trans. 87: 2675
95. Kustov LM, Kazanski VB, Beran S, Kubelkova L, Jiru P (1987) J. Phys. Chem. 91: 5247
96. Batamack P, Doremieux-Morin C, Fraissard J, Freude D (1991) J. Phys. Chem. 95: 3790
97. Bronnimann CE, Zeigler RC, Maciel GE (1988) J. Am. Chem. Soc. 110: 2023
98. Maciel GE, Sindorf DW (1980) J. Am. Chem. Soc. 102: 7606
99. Dwyer J (1987/1988) Innovation in zeolites materials science. Elsevier, Amsterdam, p 333
100. Karge HG (1991) Studies in surface science and catalysis. Elsevier, Amsterdam (vol 65), p 133
101. Müller D, Gessner W, Behrens HJ, Scheler G (1981) Chem. Phys. Lett. 79: 59
102. Gilson JP, Edwards GC, Peters AW, Rajagopalan K, Wormbecher RF, Roberie TG, Shatlock MP (1987) J. Chem. Soc., Chem. Comm. 91.
103. Freude D, Hunger M, Pfeifer H (1987) Z. phys. Chemie (NF) 152: 171
104. Grobet PJ, Geerts H, Martens JA, Jacobs PA (1989) Stud. Surf. Sci. Catal. 46: 721
105. Klinowski J, Fyfe CA, Gobbi GC (1985) J. Chem. Soc., Faraday Trans. I 81: 3003
106. Ernst H, Freude D, Wolf I (1993) Chem. Phys. Lett. (in press)
107. Samoson A, Pines A (1989) Rev. Sci. Instr. 60: 3239
108. Woessner DE, Timken HKC (1990) J. Magn. Res. 90: 411
109. Kalinowski H-O, Berger S, Braun S (1984) ^{13}C-NMR-Spektroskopie. G. Thieme, Stuttgart
110. Maciel GE, Haw JF, Chuang I, Hawkins BL, Early TA, McKay DR, Petrakis L (1983) J. Am. Chem. Soc. 105: 5529
111. Ripmeester JA (1983) J. Am. Chem. Soc. 105: 2925
112. Majors PD, Ellis PD (1987) J. Am. Chem. Soc. 109: 1648
113. Schulze D, Ernst H, Fenzke D, Meiler W, Pfeifer H (1990) J. Phys. Chem. 94: 3499
114. Lunsford JH, Rothwell WP, Shen V (1985) J. Am. Chem. Soc. 107: 1540
115. Baltusis L, Frye JS, Maciel GE (1987) J. Am. Chem. Soc. 109: 40
116. Michael A, Meiler W, Michel D, Pfeifer H (1981) Chem. Phys. Lett. 84: 30
117. Michael A, Meiler W, Michel D, Pfeifer H, Hoppach D, Delmau J (1986) J. Chem. Soc., Faraday Trans. I 82: 3053
118. Wutscherk T (1990) Thesis, University of Leipzig
119. Mastikhin VM, Mudrakowsky JL, Filimonova SV (1988) Chem. Phys. Lett. 149: 175
120. Borovkov VJ, Zhidomirov GM, Kazanski VB (1975) Zhurn. Strukt. Chim. 16: 308
121. Forsen S, Hoffmann RA (1963) J. Chem. Phys. 39: 2892, (1964) 40: 1189
122. Lunsford JH, Tutunjian PN, Chu P, Yeh EB, Zalewski DJ (1989) J. Phys. Chem. 93: 2590
123. Kärger J, Ruthven DM (1992) Diffusion in zeolites and other microporous solids. J. Wiley, New York
124. Stejskal EO, Tanner JE (1965) J. Chem. Phys. 42: 288
125. Kärger J, Pfeifer H, Heink W (1988) Adv. Magn. Resonance 12: 1
126. Jobic H (1991) Nato ASI Series B 265: 255
127. Kärger J, Heink W (1983) J. Magn. Res. 51: 1
128. Crank J (1975) The mathematics of diffusion. Oxford Univ. Press
129. Kärger J, Heink W (1971) Exp. Technik d. Physik 19: 453
130. Germanus A, Pfeifer H, Heink W, Kärger J (1983) Annalen Physik 40: 161
131. Kärger J, Pfeifer H, Rudtsch S (1989) J. Magn. Res. 85: 381
132. Zhong J, Kennan RP, Gore JC (1991) J. Magn. Res. 95: 267
133. Fenzke, private commun.
134. Heink W, Kärger J, Pfeifer H (1990) J. Chem. Soc., Chem. Comm. 1454
135. Heink W, Kärger J, Pfeifer H (1991) Z. phys. Chemie 170: 199
136. Heink W, Kärger J, Pfeifer H, Datema KP, Nowak AK (1992) J. Chem. Soc., Faraday Trans. 88: 3505
137. Förste C, Heink W, Kärger J, Pfeifer H, Feokistova NN, Zhdanov SP (1989) Zeolites 9: 299
138. Morgan M, Cosgrove T, Richardson R (1989) Coll. Surf. 36: 209
139. Zikanova A, Bülow M, Schlodder H (1987) Zeolites 7: 115
140. Beschmann K, Kokotailo GT, Riekert L (1987) Chem. Eng. Proc. 22: 233
141. Förste C, Kärger J, Pfeifer H, Riekert L, Bülow M, Zikanova A (1990) J. Chem. Soc., Faraday Trans. 86: 881

142. Kärger J, Pfeifer H (1992) Zeolites 12: 872
143. Hong U, Kärger J, Pfeifer H, Müller U, Unger KK (1991) Z. phys. Chemie 173: 225
144. Hong U, Kärger J, Kramer R, Pfeifer H, Seiffert G, Müller U, Unger KK, Lücke H-B, Ito T (1991) Zeolites 11: 816
145. Demontis P, Fois ES, Sufritti G-B, Quartieri S (1990) J. Phys. Chem. 94: 4329
146. Goodbody SJ, Watanabe K, MacGowan D, Walton JPB, Quirke N (1991) J. Chem. Soc., Faraday Trans. 87: 1951
147. June RL, Bell AT, Theodorou DN (1990) J. Phys. Chem. 94: 8232
148. Kärger J (1982) AIChE J. 28: 417
149. Kärger J, Pfeifer H (1988) ACS Symposium Series 368: 376
150. Richter R, Seidel R, Kärger J, Heink W, Pfeifer H, Fürtig H, Höse W, Roscher W (1986) Z. phys. Chemie, Leipzig 267: 841
151. Richter R, Seidel R, Kärger J, Heink W, Pfeifer H, Fürtig H, Höse W, Roscher W (1986) Z. phys. Chemie, Leipzig 267: 1145
152. Kärger J, Caro J (1977) J. Chem. Soc., Faraday Trans. I 73: 1363
153. Kärger J, Pfeifer H (1993) In: NMR and Catalysis. M. Dekker, New York (in press)
154. Kärger J, Pfeifer H (1987) Zeolites 7: 90
155. Kärger J, Pfeifer H (1992) In: Magnetic resonance spectroscopy. VCH Verlagsgesellschaft, Weinheim, FRG

MAS and CP/MAS NMR of Less Common Spin-1/2 Nuclei

Angelika Sebald

Bayerisches Geo-Institut, Universität Bayreuth, 95440 Bayreuth, FRG

Table of Contents

Magic angle spinning (MAS) and cross polarisation (CP/MAS) NMR applications involving the less commonly used spin-1/2 nuclei ^{77}Se, ^{89}Y, ^{109}Ag, ^{113}Cd, ^{119}Sn, ^{125}Te, ^{183}W, ^{195}Pt, ^{199}Hg and ^{207}Pb will be reviewed. Areas of application include inorganic and organometallic chemistry as well as some related aspects from materials research-related activities. In addition to this literature survey, some practical aspects of MAS and CP/MAS methodology concerned with these nuclei will be discussed.

NMR Basic Principles and Progress, Vol. 31

1 Introduction

The following chapter started from a working title "NMR of rare nuclei" and thus required some further definition of the area to be covered. "Rare" was defined as rarely, or at least, less commonly, used nuclei. Considering the periodic table of the elements, a fair number of nuclei with $I = 1/2$ and $I > 1/2$ qualify for this categogy. In the next step, quadrupolar nuclei were eliminated from this chapter: a whole volume in this series (Vol. 30) is devoted to $I > 1/2$ nuclei, and chapters 4 and 5 in this volume as well as the review in Vol. 29 by D. Freude and J. Haase cover experimental aspects concerning quadrupolar nuclei.

This selection process leaves us with less commonly used $I = 1/2$ nuclei under high-resolution solid-state NMR conditions. Certainly, ^{13}C, ^{29}Si and ^{31}P do not belong to this category and will, therefore, not be considered. The nucleus ^{129}Xe would be a candidate for this chapter, if not a whole chapter (Vol. 30, Chapter 3) in this series would be dealing with ^{129}Xe NMR. Similarly, also ^{1}H and ^{19}F are to some extent less commonly used $I = 1/2$ nuclei in high-resolution solid-state NMR, but aspects of multiple pulse line narrowing experiments are covered in Vol. 32. After elimination, the remaining nuclei for this chapter are ^{77}Se, ^{89}Y, ^{109}Ag, ^{113}Cd, ^{119}Sn, ^{125}Te, ^{183}W, ^{195}Pt, ^{199}Hg and ^{207}Pb.

It is clear that the following will heavily lean on MAS and CP/MAS applications and its seems justified to first consider some general practical aspects, followed by an overview of what has been reported in the literature. It is equally obvious that such MAS and CP/MAS applications will cover a wide range of chemistry.

To come straight to the point, what can be reviewed concerning these less commonly used spin-1/2 nuclei, is—with a few exceptions—mostly qualitative usage of isotropic chemical shifts and/or chemical-shift anisotropies. In many branches of chemistry this qualitative approach is (and will remain) an important concept of using NMR as an analytical tool, both in solution and in the solid state. On the other hand, this approach propagates tendencies to "under-use" what high-resolution solid-state NMR has to offer. As a consequence, the level of sophisticated use of MAS and CP/MAS methodology of these exotic nuclei is by far less well-developed than applications involving common nuclei, like ^{13}C or ^{31}P. Still, it is/was probably necessary and reasonable that the first years of these exotic nuclei in high-resolution solid-state NMR were mostly dedicated to exploration and to attempts to gain more practical experience. Now that for several of these nuclei, like, e.g., ^{113}Cd or ^{119}Sn, a considerable bulk of such practical experience is available, it becomes straightforward to predict what the next few years will see with respect to exotic spin-1/2 nuclei in high-resolution solid-state NMR. First, more branches in chemistry will "discover" MAS and CP/MAS of these nuclei for their purposes. Second, and maybe even more importantly, there are first signs of a—spectroscopically speaking—more refined, "second round" with nuclei like ^{119}Sn; taking full advantage of the favourable

NMR properties (wide range of relative magnitudes of dipolar vs scalar coupling interactions, etc.) of these nuclei which are rarely provided by nuclei like ^{13}C or ^{29}Si. Clearly, also a more vigorous consideration of anisotropic interactions, both with respect to chemical shielding and scalar coupling, should be expected to be high on the list of priorities: heavy spin-1/2 nuclei offer exceptional ease of accessibility of such experimental data and it seems rather wasteful not to fully exploit the considerable amount of theoretical and computational background available.

2 Some Practical Aspects

When considering the whole lot of less commonly used spin-1/2 nuclei, applications of MAS and CP/MAS NMR in every conceivable field of chemistry cross one's mind. Not always will it, therefore, be perfectly obvious what the best, i.e., the most efficient and convenient way to set up the spectrometer for a given nucleus will be. In many cases, even if the desired mode of operation were single-pulse experiments, a Hartmann-Hahn CP experiment using a suitable standard compound is the method of choice for the set-up procedure. This approach requires a selection of suitable set-up samples for the various nuclei, whereby the term "suitable" is defined by considerations of convenience: such samples should be chemically stable, preferably commercially available, should yield a satisfactory CP response with a single transient and/or should display moderate ^{1}H T_1 relaxation times along with small chemical-shift anisotropies for the observed X-nucleus. Table 1 compiles such a selection of compounds for the $^{1}H \rightarrow X$ CP case. The contents of Table 1 are the most up-to-date results of a continued search for the most convenient options and could certainly be improved in some cases, but all compounds listed have a proven record as set-up samples.

In order to further illustrate set-up CP/MAS spectra of these standard compounds, Fig. 1 shows typical CP/MAS spectra obtained on a 300 MHz instrument under conditions of routine operation.

Of course, some general trends can be obtained from Table 1, like, for instance, that contact times of approximately 1 ms will be suitable for $^{1}H \rightarrow {}^{119}Sn$ CP/MAS experiments on many organotin compounds. However, the information in Table 1 should not be misread: the quoted optimised CP parameters refer to the respective compounds and will not in general apply for a given nucleus in a variety of chemical environments. Considering the wide range of chemistry circumstances in which these nuclei might be involved, we cannot a priori predict suitable experimental CP parameters for a given ^{1}H/X-nucleus combination. In fact, what we can predict, is that variation in chemistry will be accompanied by highly variable CP/MAS $^{1}H \rightarrow$ X-nucleus spin dynamics. In the familiar $^{1}H/^{13}C$ case in most solid organic compounds we can rely on strong homonuclear

Table 1. Some standard CP/MAS compounds and appropriate CP parameters[a]

nucleus	compound	isotropic chemical shift [ppm]	relative to external: (=0 ppm)	contact time [ms]	recycle delay [s]	Ref.
^{13}C	adamantane	38.5; 29.5	TMS	5	6	
^{15}N	NH_4NO_3, 20% ^{15}N enriched	0.0; −353.2	aqueous NO_3^- solution	8	10	
^{29}Si	Q_8M_8	+11.7; −108.4; −108.6 +11.5; −109.4; −109.7	TMS	2	15	1
^{31}P	$CaHPO_4 \cdot 2H_2O$	+1.2	$H_3PO_4/CDCl_3$	1	5	
^{77}Se	$(NH_4)_2SeO_4$	+1040.2	Me_2Se	3	8	2
^{89}Y	$Y(NO_3)_3 \cdot 6H_2O$	−53.2	aqueous YCl_3^- solution	10	15	3
^{109}Ag	$Ag[CH_3CHC(OH)CO_2]$, silver lactate	+345.9; +320.2 +219.7; +210.7	aqueous $AgNO_3^-$ solution	50	10	4
^{113}Cd	$Cd(NO_3)_2 \cdot 4H_2O$	−100.0	1 m $Cd(ClO_4)_2$ in H_2O	15	10	5
^{119}Sn	$Sn(C_6H_{11})_4$	−97.35	Me_4Sn	1	8	6
^{125}Te	$Te(OH)_6$	+693.3; +686.6	Me_2Te	5	5	7
^{129}Xe	Hydrochinone-Xe clathrate	+220.0	Xe-gas	30	5	8
^{183}W	$(NH_4)_2WS_4$	+3639.6	solid $W(CO)_6$ = −3470 ppm	100	10	9
^{195}Pt	$K_2Pt(OH)_6$	+8024	$\Xi\,^{195}Pt$	2	4	6
^{199}Hg	$Hg(O_2CCH_3)_2$	−2490	Me_2Hg	5	10	6
^{207}Pb	$Pb(C_7H_7)_4$, tetrakis-*p*-tolyl-lead	−148.8	Me_4Pb	5	30	6

[a] with the exception of the hydrochinone-Xe clathrate and $Pb(p\text{-tolyl})_4$ all compounds are commercially available; with the exception of Ag-lactate (8 transients), $(NH_4)_2WS_4$ (16 transients) and $Hg(OAc)_2$ (8 transients) all other compounds in Table 1 allow optimisation of the Hartmann-Hahn match from a single transient response

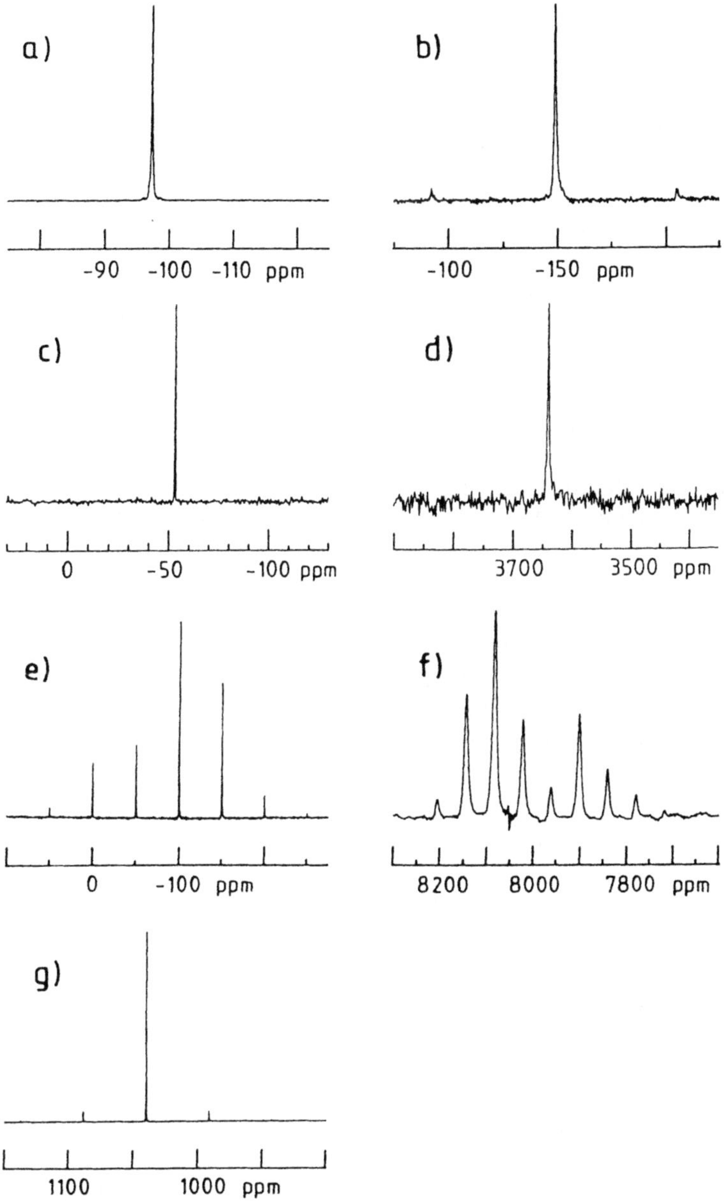

Fig. 1a–g. Typical CP/MAS spectra of various $^{1}H \rightarrow X$ CP set-up compounds, all obtained on a 300 MHz instrument, parameters as in Table 1. **a** ^{119}Sn CP/MAS $Sn(C_6H_{11})_4$, 8 transients, $\nu_{rot} = 3$ kHz; **b** ^{207}Pb CP/MAS Pb(p-tolyl)$_4$, 8 transients, $\nu_{rot} = 3.6$ kHz; **c** ^{89}Y CP/MAS $Y(NO_3)_3 \cdot 6H_2O$, 32 transients, $\nu_{rot} = 2.7$ kHz; **d** ^{183}W CP/MAS $(NH_4)_2WS_4$, 400 transients, $\nu_{rot} = 2.6$ kHz; **e** ^{113}Cd CP/MAS $Cd(NO_3)_2 \cdot 4H_2O$, 8 transients, $\nu_{rot} = 3.3$ kHz; **f** ^{195}Pt CP/MAS $K_2Pt(OH)_6$, 160 transients, $\nu_{rot} = 4$ kHz; **g** ^{77}Se CP/MAS $(NH_4)_2SeO_4$, 8 transients, $\nu_{rot} = 2.8$ kHz CP/MAS. Spectra for ^{109}Ag and ^{119}Hg CP/MAS standard compounds Ag-lactate and $Hg(OAc)_2$ are shown in Figs. 5 and 6, respectively.

^1H-^1H and heteronuclear ^1H-^{13}C dipolar interactions, leading to favourable $^1H/^{13}C$ CP spin dynamics, and some educated guess concerning suitable experimental CP parameters is usually granted. Very rarely will the heteronuclei in Table 1 confront us with short ^1H-X distances, so that we usually have to deal with much smaller ^1H-X dipolar interactions and longer cross-relaxation times $T_{^1HX}$. Especially in many organometallic compounds with small and mobile organic ligands, or in inorganic compounds where crystal lattice-H_2O is the only source of proton magnetisation, the homonuclear ^1H-^1H dipolar interaction will also often be scaled by motion. Also proton $T_{1\rho}$s may tend to be short. In other words: CP/MAS experiments involving heteronuclei will quite regularly represent an evil regime where the ^1H-X CP spin dynamics severely interfer with MAS frequencies. This condition requires conservative MAS frequencies, which in turn can be rather counterproductive if the chemical shift anisotropy of the observed X nucleus happens to be large. Unfavourable interplay of relatively long cross relaxation times $T_{^1HX}$ versus relatively short proton $T_{1\rho}$s can prevent a satisfactory CP response altogether. In fact, when dealing with CP/MAS experiments of low-γ nuclei this turns into a severely limiting constraint.

Despite these pessimistic-sounding remarks, the majority of chemical circumstances involving nuclei from Table 1 *will* allow the use of $^1H \rightarrow X$ CP/MAS experiments. Even under less promising circumstances due to unfavourable $T_{^1HX}/T_{1\rho}$ interplay, the gain in sensitivity in such CP/MAS spectra as compared to single-pulse operation is usually still enormous. However, for the application of CP/MAS experiments involving heteronuclei for a new class of compounds, prior knowledge of proton T_1s and $T_{1\rho}$s is recommended, if not a necessary prerequisite: simply for the reason that obtaining the desired CP/MAS spectrum might critically depend on the choice of truly optimised experimental parameters.

Meanwhile the exploitation of the Hartmann-Hahn condition for $^1H \rightarrow X$ CP/MAS spectroscopy of less commonly used spin-1/2 X-nuclei can look back on a history of approximately 10–15 years. As will also be obvious from the following literature survey, this period has seen a strong emphasis on the sensitivity-aspect of $^1H \rightarrow X$-nucleus CP/MAS spectroscopy, while alternative usage of CP/MAS methodology for less commonly used spin-1/2 nuclei in terms of selectivity has not (yet) been established.

More recently, CP/MAS spectroscopy of spin-1/2 nuclei has "gone multinuclear" in yet another sense. Not only have the observed X nuclei, under conditions of Hartmann-Hahn matching $\gamma_{^1H}B_{^1H} = \gamma_X B_X$, conquered the periodic table (with a few noticeable exceptions like, e.g., ^{57}Fe or ^{103}Rh), but nuclei other than ^1H have begun to attract attention as alternative magnetisation sources for CP/MAS spectroscopy. In many ways this development parallels the evolution of more widespread applications of solution-state NMR experiments in "heteronuclear" chemistry which we have seen in the past.

For reasons of sensitivity (but not only for such reasons), the obvious choice for such $\gamma_Y B_Y = \gamma_X B_X$ CP experiments are high natural abundance, high γ-nuclei such as ^{19}F or ^{31}P. What has been documented in terms of useful standard compounds for $^{19}F \rightarrow$ X-nucleus CP/MAS spectroscopy so far is shown in Table 2 and Figure 2; $^{19}F \rightarrow {}^{13}C$ CP/MAS spectroscopy has been used for the characterisation of polymer blends ([12]).

Of course, when using ^{19}F as the magnetisation reservoir for $^{19}F \rightarrow X$ CP/MAS experiments the build-up and decay of X-magnetisation as a function of contact time is governed by the same principles as for the $^1H/X$ case, but the operative time-scale with regard to cross relaxation times $T_{^{19}FX}$ and $T_{1\rho}$ (^{19}F) can be quite different. From the little which is known there seems to be a trend to fairly long cross relaxation times $T_{^{19}FX}$ and ^{19}F $T_{1\rho}$s (see Table 2,

Table 2. Suitable reference compounds for $^{19}F \rightarrow$ X-nucleus CP/MAS NMR

nucleus	compound	isotropic chemical shift [ppm]	contact time [ms]	recycle delay [s]	Ref.
^{29}Si	topaz, $[Al_2(F,OH)_2(SiO_4)]$	−85.9	20	20	10
	Na_2SiF_6	−189.5	10	15	10
^{31}P	KPF_6	−145.6	6	20	11
^{113}Cd	CdF_2	−206.5	5	10	
^{119}Sn	Na_2SnF_6	−823	5	20	10

[a] isotropic chemical shifts are quoted with respect to the same references as in Table 1

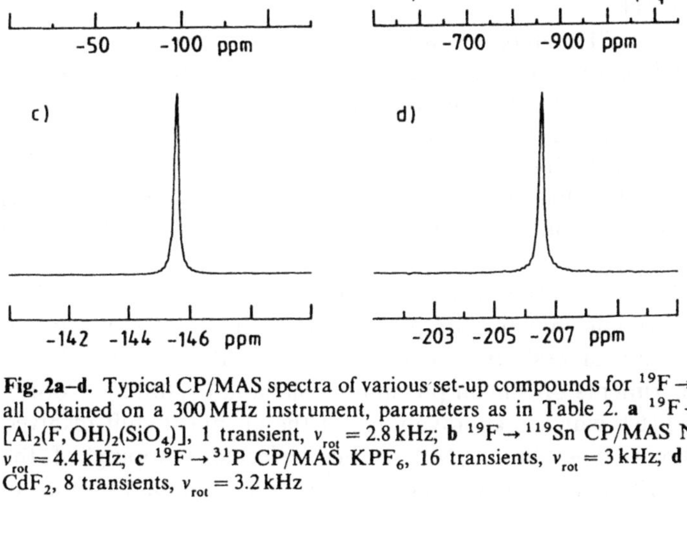

Fig. 2a–d. Typical CP/MAS spectra of various set-up compounds for $^{19}F \rightarrow X$ CP/MAS operation, all obtained on a 300 MHz instrument, parameters as in Table 2. **a** $^{19}F \rightarrow {}^{29}Si$ CP/MAS topaz, $[Al_2(F,OH)_2(SiO_4)]$, 1 transient, $\nu_{rot} = 2.8$ kHz; **b** $^{19}F \rightarrow {}^{119}Sn$ CP/MAS Na_2SnF_6, 16 transients, $\nu_{rot} = 4.4$ kHz; **c** $^{19}F \rightarrow {}^{31}P$ CP/MAS KPF_6, 16 transients, $\nu_{rot} = 3$ kHz; **d** $^{19}F \rightarrow {}^{113}Cd$ CP/MAS CdF_2, 8 transients, $\nu_{rot} = 3.2$ kHz

[13]). Apart form such CP-timing considerations, $^{19}F \rightarrow X$ CP/MAS spectroscopy requires consideration of further experimental parameters. First of all, it needs to be remembered that ^{19}F spin-lattice relaxation times T_1 are usually longer than what we are used to from ^1H operation, leading to, accordingly, longer relaxation delays for $^{19}F \rightarrow$ X-nucleus CP/MAS experiments. Most importantly (and different from the familiar ^1H/X CP-situation), both the chemical shift range and possible chemical shift differences for different sites of the ^{19}F magnetisation reservoir need to be taken fully into account in order to be able to realise proper on-resonance Hartmann-Hahn ^{19}F/X matching conditions. These two points clearly indicate that ^{19}F MAS and/or multiple pulse experiments are an unrenouncable preparation-step for $^{19}F \rightarrow X$ CP/MAS studies. The much greater ^{19}F chemical shift range as compared to ^1H might be considered a blessing or a curse. In the presence of more than one ^{19}F resonance it may become difficult to spin-lock the entire ^{19}F magnetisation, substantial ^{19}F shielding anisotropies and large scalar coupling constants $J(^{19}FX)$ can add to this problem. These problems can be partially cured by the use of moderate to low external magnetic field strengths B_0, while often fast MAS is of only limited use. On the other hand, when using a magnetisation reservoir other than ^1H for cross-polarisation purposes, we gain ease of access to further experimental freedom. If there is more than one resonance for different sites of the spin reservoir and if the chemical shift differences for such inequivalent sites of the reservoir are larger than the respective homonuclear dipolar interactions between inequivalent sites, it then becomes a rather straightforward task to selectively excite (and spin-lock) a particular resonance from the magnetisation reservoir. Either DANTE-type pulse schemes [14] or selective excitation by the SELDOM pulse sequence [15] can be used for the purpose of preparation, followed by cross-polarisation. Clearly, such selective CP experiments are not normally provided by ^1H as the CP spin reservoir, and much is to be gained in terms of spectral editing techniques from the exploitation of magnetisation reservoirs other than ^1H.

Whether a particular ^{19}F spin reservoir belongs to this chemical shift-related category of additional degrees of experimental freedom, or whether one is confronted with a more proton-like, ^{19}F homonuclear dipolar coupling-governed situation, will, of course, be largely determined by chemistry and by the external magnetic field strength used. Given a ^{19}F chemical-shift range of approximately 400 ppm, it is to be expected that for the more commonly used B_0 field strengths of 4–7 T, the ^{19}F magnetisation reservoir will mostly fall in between these two categories.

If we consider ^{31}P as yet another magnetisation source for $^{31}P \rightarrow X$ CP/MAS experiments, it then becomes much more probable that this spin reservoir will provide "selective circumstances". In fact, very recently $^{31}P \rightarrow {}^{29}Si$ and $^{31}P \rightarrow {}^{113}Cd$ CP/MAS experiments on silicon phosphide, SiP and crystalline $CdSiP_2$ have been reported [16], which take advantage of the chemical shift-resolved ^{31}P-situation for the $^{31}P \rightarrow {}^{29}Si$ CP/MAS-edited selected mapping of the complete Si-P network in SiP. Also very recently, a preliminary report of

$^{27}Al \leftrightarrow ^{31}P$ CP/MAS experiments on the molecular sieve VPI-5 has appeared [17], again highlighting the possibility of establishing $^{27}Al/^{31}P$-connectivities by magnetisation transfer to or from ^{31}P, to or from ^{27}Al. Quadrupolar nuclei like ^{27}Al carry substantial potential, owing to their fast T_1-relaxation characteristics, as spin reservoirs for CP/MAS-detected rare spin-1/2 nuclei in inorganic solids devoid of protons. However, cross-polarisation to or from half-integer quadrupolar nuclei under MAS conditions is a non-trivial task because the efficiency of the spin locking and the Hartmann-Hahn condition will depend on the relative magnitudes of the quadrupole frequency, the rotation frequency and the *rf* amplitude [18, 19]. Whether or not this condition precludes reliable routine applications of CP/MAS experiments involving quadrupolar nuclei, remains to be seen.

3 Applications

3.1 Nuclei with a Low Magnetogyric Ratio: ^{89}Y, ^{109}Ag, ^{183}W, (^{103}Rh, ^{57}Fe)

All of the following elements—yttrium, silver, tungsten, rhodium and iron—play an important role in various branches of chemistry, in particular in solid-state chemistry. To name just a few important areas, superconductors, ceramics and ceramic precursors (Y), heterogeneous and hybrid-catalysis (W, Rh, Fe), homo- and heteropolyanions (W) need to be mentioned, let alone the wide field of molecular compounds of these elements. Despite this prominent role in chemistry, none of these elements have, so far, played an equally prominent part in high-resolution solid-state NMR. This apparent discrepancy is caused by the NMR properties of these nuclei: they have low resonance frequencies and often prohibitively long solid-state relaxation times T_1 in common. Such low resonance frequencies cause acoustic ringing problems for single pulse excitation experiments, extremely long relaxation times T_1 (of the order of several minutes to hours) lead to unacceptable sensitivity/time problems with respect to the acquisition of spectra with a reasonable signal-to-noise ratio, even for 100% naturally abundant nuclei, like e.g. ^{89}Y. An obvious way to avoid sensitivity problems caused by long relaxation times T_1 is to use cross-polarisation methods. But in contrast to the situation for more commonly used nuclei like ^{31}P or ^{13}C, where fulfilling of the Hartmann-Hahn matching condition $\gamma_{^1H} B_{^1H} = \gamma_X B_X$ is a straightforward task, the large difference in $B_{^1H}$ and B_X amplitudes for Hartmann-Hahn matching for such low-γ nuclei X is technically much more demanding as far as power requirements and power handling of the CP/MAS probe are concerned. Most commercially available standard CP/MAS probes will, in fact, require optimisation of the tuning circuit for such purposes [3].

However, strategies to scale down the X-channel power requirements for CP experiments by using time-averaged precession frequencies [20] or by utilising off-resonance effects [21] have been proposed and demonstrated for ^{13}C and ^{15}N, respectively.

One can say that the current low-γ nuclei MAS and CP/MAS state of affairs is one of combatting these technical problems. Accordingly, to date, only a few and mostly preliminary reports, involving low-γ nuclei MAS and CP/MAS results, have appeared in the literature, and isotropic chemical shift $\delta^{89}Y$, $\delta^{109}Ag$, $\delta^{183}W$ have so far been the prevalent focus of attention.

^{89}Y

^{89}Y is the most "user-friendly" of these low-γ nuclei: ^{89}Y has the highest resonance frequency within that group, is 100% naturally abundant, displays very sharp resonances for crystalline yttrium compounds, the moderate chemical shift range of only a few hundred ppm is still more than adequate for the use of isotropic ^{89}Y chemical shifts for diagnostic purposes, and ^{89}Y shielding anisotropies are moderate to very small. Accordingly, the majority of low-γ nuclei MAS and CP/MAS studies deal with this nucleus. As for all other low-γ nuclei, ^{89}Y single-pulse MAS experiments are also often severely handicapped by long relaxation times T_1. A recent study [3] demonstrates that, for proton-bearing yttrium compounds, cross-polarisation offers a practical and time-saving alternative. Figure 3 illustrates the enormous gain in terms of spectrometer time, comparing the ^{89}Y CP/MAS with the SP/MAS spectrum of $Y(NO_3)_3 \cdot 6H_2O$. A similar, perhaps more realistic comparison is shown in Fig. 4 where the ^{89}Y CP/MAS spectrum of $Y(OH)_3$ (2.4 hours) is depicted in comparison with the ^{89}Y SP/MAS spectra of $Y_2SiBe_2O_7$ and Y_2O_3 (ca. 22 hours) [22]. Y_2O_3 displays two ^{89}Y resonances in a 3:1 intensity ratio, at 314 and 272 ppm

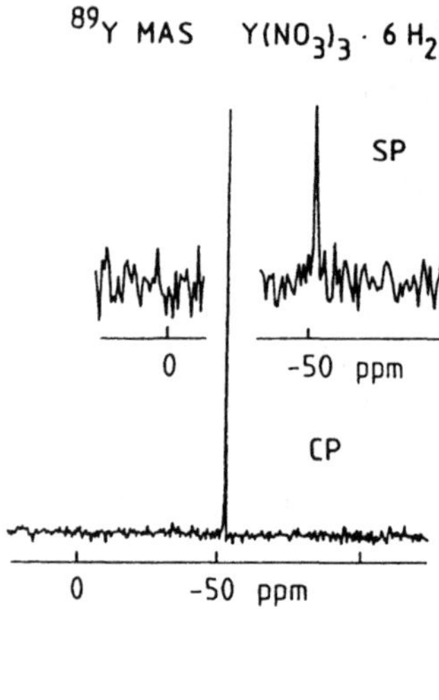

Fig. 3. Comparison ^{89}Y MAS vs ^{89}Y CP/MAS: $Y(NO_3)_3 \cdot 6H_2O$. *Top*: ^{89}Y MAS, 4 transients, 45 min recycle delay; total time 180 min. *Bottom*: ^{89}Y CP/MAS, 32 transients, 10 s recycle delay; total time 320 s

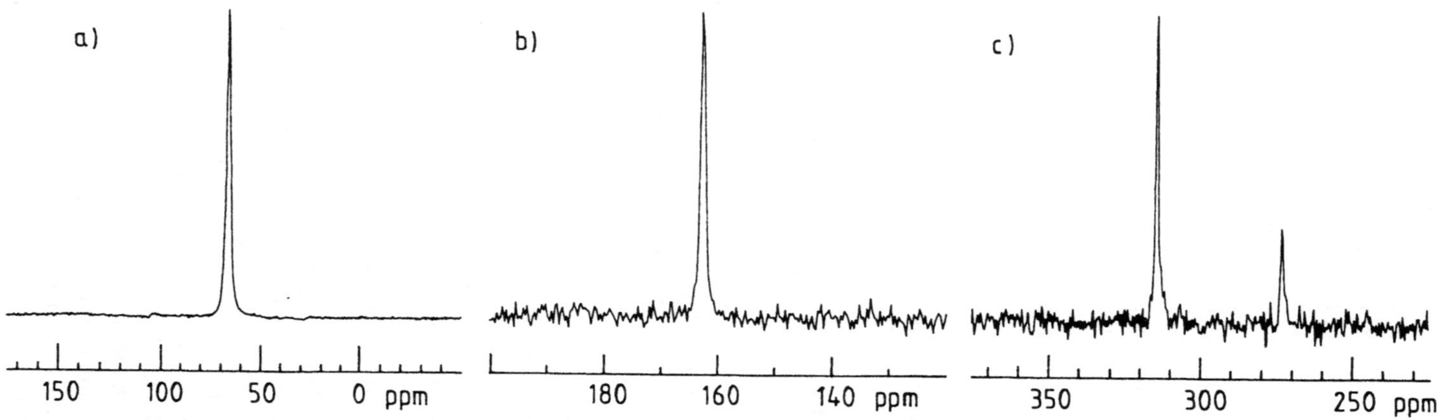

Fig. 4a–c. ^{89}Y CP/MAS spectrum of **a** $Y(OH)_3$ and ^{89}Y MAS spectra 14.75 MHz of **b** $Y_2SiBe_2O_7$ and **c** Y_2O_3. **a** $Y(OH)_3$, 870 transients, 15 ms contact time, 10 s recycle delay, total time 2.4 hours; **b** $Y_2SiBe_2O_7$, 44 transients, 20° pulse, 30 min recycle delay, total time 22 hours; **c** Y_2O_3, 64 transients, 40° pulse, 20 min recycle delay, total time 21 hours

[23–25], in agreement with two different YO_6 oxygen-nearest-neighbours environments as shown by the crystal structure. This approximately 40 ppm chemical shift difference for pure Y_2O_3 has been used to characterise the oxygen-vacancy ordering about yttrium in the solid solution $(Bi_2O_3)_{0.6}(Y_2O_3)_{0.4}$ [24]: a broad isotropic resonance, centered at 275 ppm, leads to the conclusion that the Y^{3+} ions in this solid solution prefer centrosymmetric sites created by $\langle 111 \rangle$ oxygen-vacancy pairs (similar to the less abundant Y site in pure Y_2O_3), while the Bi^{3+} ions will then be located at less symmetric sites created by $\langle 110 \rangle$ oxygen-vacancy pairs.

An attempt has been made to rationalise structural influences on $\delta^{89}Y$ in the ^{89}Y SP/MAS spectra of various yttrium-aluminate and silicate phases [25]. Similarly as for other nuclei such as ^{27}Al and ^{29}Si, $\delta^{89}Y$ depends on the number of nearest oxygen-neighbours and on the nature of the next-nearest neighbours. Trends like an increase in ^{89}Y-shielding as a function of increasing local coordination number n, YO_n, or as a function of next-nearest neighbours group-electronegativity, are obvious. However, in contrast to ^{29}Si or ^{27}Al, the ^{89}Y shift ranges for different nearest neighbour environments show severe overlap. Arguments based on isotropic ^{89}Y chemical shifts should, therefore, be used with caution.

Considerably deshielded ^{89}Y resonances have been observed in the ^{89}Y SP/MAS spectrum of Y_2Cl_3 (762 and 506 ppm) and have been attributed to the presence of additional metal–metal d-bonding states in this sesquihalide [26]. The ^{89}Y resonance for anhydrous YCl_3 is much broader and more strongly shielded (− 230 ppm) but other than this broadening, no well-defined ^{89}Y-$^{35/37}Cl$ coupling is detectable from these room temperature ^{89}Y SP/MAS spectra.

^{89}Y SP/MAS spectroscopy, in conjunction with variable temperature measurements and ^{89}Y spin lattice relaxation rate determinations, has been used for the characterisation of $YBa_2Cu_3O_7$ and of the effects of substitution in $YBa_2(Cu_{1-x}M_x)_3O_7$ with M = V, Fe, Co, Ni, Zn, Al, Ga [27]. Both ^{89}Y shifts and T_1 relaxation rates, reflect Cu-chain-site substitution by Fe, Co, Al and Ga. By comparison, these data suggest that V also substitutes on the chain sites, while Ni and Zn seem to substitute plane sites. The ^{89}Y shift of undoped $YBa_2Cu_3O_7$ is − 103 ppm, but extrapolation of the ^{89}Y shifts for Fe, Co, Al or Ga-doped $YBa_2(Cu_{1-x}M_x)_3O_7$ to $x = 0$ yields a shift of -90 ± 2 ppm for the undoped material. This 10 ppm difference has been attributed to the variation in ^{89}Y chemical shift as the structure changes from orthorhombic (pure $YBa_2Cu_3O_7$) to tetragonal in the doped material. The measured room-temperature ^{89}Y shifts result from a chemical shift and a Knight shift contribution. It seems generally accepted that the isotropic ^{89}Y chemical shift of orthorhombic $YBa_2Cu_3O_7$ is close to + 200 ppm, that the Knight shift is − 303 ppm, and that shift changes in the doped phases as a function of substitution correspond to a reduction of the Knight shift contribution.

Including the diamagnetic end-member compounds $Y_2Sn_2O_7$ and $Y_2Ti_2O_7$, a series of paramagnetic yttrium pyrochlores $Y_{2-y}Ln_yM_2O_7$ (Ln = Ce, Pr, Nd,

Sm, Eu, Yb; M = Sn, Ti) has been investigated by means of ^{89}Y SP/MAS spectroscopy [28]. Incorporation of these paramagnetic ions into the pyrochlore lattice causes an enormous reduction in the respective ^{89}Y spin-lattice relaxation times and gives rise to additional ^{89}Y resonances. The ^{89}Y shifts of these additional resonances were found to be proportional to the number of lanthanide ions in the first yttrium coordination sphere; their relative intensities were used to determine the concentration of paramagnetic dopant in the diamagnetic end-member phase and to explore the respective solid-solution limits. Additional ^{89}Y resonances due to $Y(OY)_{6-n}(OLn)_n$ for $n = 1$–3 were observed, for $n > 3$ signal-to-noise limitations prevented the identification of the corresponding ^{89}Y resonances. No ^{89}Y resonance could be obtained from Ho^{3+} and Tm^{3+} doped pyrochlores, probably due to excessive ^{89}Y line broadening.

Table 3. Isotropic ^{89}Y chemical shifts of some diamagnetic solid yttrium compounds, referenced to aqueous YCl_3-solution

Compound	$\delta^{89}Y$ [ppm]	Reference
$Y(NO_3)_3 \cdot 6H_2O$	−53	9
$Y_2(SO_4)_3 \cdot 8H_2O$	−46	9
$Y(OAc)_3 \cdot 4H_2O$	+47	9
$Y(acac)_3 \cdot 3H_2O$	+28; +21	9
$(Bi_2O_3)_{0.6}(Y_2O_3)_{0.4}$	+275	24
Y_2O_3	+315; +270	23
	+314; +272	24, 25
	+314; +273	22
$Y(OH)_3$	+66	22
$Y_2SiBe_2O_7$	+163	22
$Y_3Al_5O_{12}$	+222	25
$YAlO_3$	+214.5	25
Y_2SiO_5	+237; +148	25
α-$Y_2Si_2O_7$	+144	25
β-$Y_2Si_2O_7$	+208	25
γ-$Y_2Si_2O_7$	+198	25
δ-$Y_2Si_2O_7$	+122	25
$Y_2Sn_2O_7$	+149	28
$Y_2Ti_2O_7$	+65	28
Y_2Cl_3	+762; +506	26
YCl_3	−230	26
YF_3	−112	29
$YCl_3 \cdot 6H_2O$	+58	23
$YBr_3 \cdot 6H_2O$	+80	29
$YJ_3(THF)_3$	+663; +547; +529; +356	29
$YBr_3(THF)_3$	+441	29
$YCl_3(THF)_3$	+370; +361; +253; +218	29
$cp_2YBr(THF)$	−90	29
$[cp_2YCl]_2$	−122	29
$cp_2YCl(THF)$	−200	29
$cp_3Y(THF)$	−405	29
$cp_5Y_5(OMe)_8O$	+6; −1	29
$Y_3(O^tBu)_7Br_2(THF)_2$	+227; +157	29
$Y_3(O^tBu)_7Cl_2(THF)_2$	+247; +200	29
$Y_3(O^tBu)_9(HO^tBu)_2$	+161	29
$Y_3(O^tBu)_9(THF)_2$	+173; +152	29
$[(C_5H_4SiMe_3)Y(O^tBu)_2]_2$	+158; −11	29

A very recent study [29] has extended the application of ^{89}Y CP/MAS to include solid yttrium alkoxide complexes and compounds where protons in cyclopentadienyl ligands or in solvate THF-molecules provide the ^{1}H CP magnetisation source.

^{109}*Ag*

After the introduction of polarisation transfer experiments for the solution state (INEPT) in the early 1980s, the following years have been named "the golden age for silver NMR" [30] as far as solution-state NMR spectroscopy is concerned. Even if INEPT (using scalar coupling J(^{109}Ag^{1}H) or J(^{109}Ag^{31}P)) and reverse detection schemes have provided easy access to solution-state ^{109}Ag NMR spectra, solution-state ^{109}Ag NMR spectroscopy is severely hampered by the frequent occurrence of fast chemical exchange for silver compounds in solution. In many cases not even low-temperature solution-state ^{109}Ag NMR spectroscopy can provide information other than exchange-averaged spectra. The more desirable it seems to be able to utilise high-resolution solid-state ^{109}Ag NMR methods (in a quasi-slow-exchange regime) – and the more surprising it may seem that, so far, there are hardly any reports to be found in the literature. This is certainly caused by a lack of suitable solid-state NMR methodology rather than by a lack of interest in the spectroscopic properties of silver compounds. For most applications ^{109}Ag single-pulse MAS spectroscopy will be highly impractical; only very recently have the first ^{109}Ag CP/MAS spectra been reported [4].

Figure 5 shows the ^{109}Ag CP/MAS spectrum of silver lactate as an example. This spectrum illustrates several points: 1) a ^{109}Ag chemical shift range of several hundred ppm allows easy detection of crystallographic inequivalences; 2) optimised ^{109}Ag CP/MAS conditions can be quite efficient and yield ^{109}Ag CP/MAS spectra with an excellent signal-to-noise ratio in reasonable amounts of time, but the CP characteristics are such that fairly long contact times (of the order of 50–80 ms) are necessary for optimum CP results—such circumstances

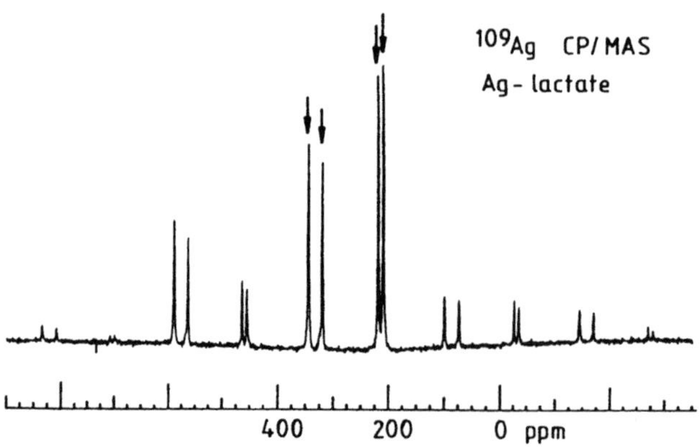

Fig.5. ^{109}Ag CP/MAS of silver lactate, displaying ^{109}Ag resonances of four inequivalent silver sites. 5133 transients, $\nu_{rot} = 3.4$ kHz, 10 s recycle delay, 50 ms contact time

require sufficiently long proton $T_{1\rho}$s, and the determination of the respective ^1H $T_{1\rho}$-data prior to ^{109}Ag CP/MAS experiments is strongly recommended; 3) substantial chemical-shift anisotropies of the order $\Delta\sigma \approx 1000$ ppm will be common as many molecular Ag(I) compounds in the solid state show linear two-fold silver coordination. Due to the low resonance frequency, such ^{109}Ag chemical-shift anisotropies do not create any problems, on the contrary, "convenient" MAS frequencies of 2–3 kHz (i.e., MAS frequencies which will not normally interfer with the CP spin dynamics) will then yield ^{109}Ag CP/MAS spectra suitable for spinning sideband analysis. ^{109}Ag shielding tensor information will be particularly important: it is well documented that the electronic structure of many solid silver compounds is dominated by mutually attractive Ag–Ag interactions (and, accordingly, show unusually short Ag–Ag distances), these bonding properties will also be mirrored by the properties of ^{109}Ag shielding tensors.

The magnetic moments of the isotopes $^{107/109}Ag$ attracted attention at an early stage [31, 32] and isotropic ^{109}Ag chemical shifts for the silver halides have been obtained using static ^{109}Ag spectra (AgF −100 ppm, AgCl 370 ppm, AgBr 350 ppm, AgI 710 ppm, with respect to aqueous $AgNO_3$-solution) [33, 34]. To date only a few ^{109}Ag MAS results have been reported. $\delta^{109}Ag = 320$ ppm for $Ag_{10}Si_4O_{13}$ has been quoted [35], ^{109}Ag MAS spectra have been used to demonstrate polymorphism for AgI, in conjunction with ^{109}Ag wideline spectra of silver borate glasses of the $AgI:Ag_2O:B_2O_3$ system [36]. The substantial line width and the irregular shape of the ^{109}Ag MAS resonance of crystalline β-AgI [36] has also been interpreted in terms of polytypes, but incompletely (by MAS) averaged dipolar and scalar ^{109}Ag-^{127}I coupling is more likely to be responsible for this line broadening. Very recently ^{109}Ag MAS spectra have been used to characterise silver particles supported on SiO_2, η-Al_2O_3 and zeolite A [37]. Only Ag-particles larger than 50 nm in diameter gave rise to a single bulk Knight-shifted metallic resonance at 5232 ± 10 ppm, while smaller particles—probably because of excessive line broadening due to surface boundary effects—yielded no observable ^{109}Ag resonance.

^{183}W

If there is not (yet) much to report from the literature concerning ^{89}Y or ^{109}Ag MAS and CP/MAS, there is even less to report about ^{183}W MAS and CP/MAS results. Given the even lower ^{183}W resonance frequency, the lower natural abundance (14.4%) and a much larger ^{183}W chemical shift range of several thousand ppm [38, 39], this finding is not so surprising. Only two publications dealing with ^{183}W high-resolution solid-state NMR have appeared in the literature, both merely demonstrate that it is possible, in principle, to obtain ^{183}W MAS [40] and ^{183}W CP/MAS [9] spectra. As expected, ^{183}W single-pulse MAS operation turns out "somewhat time consuming" [40], while ^{183}W CP/MAS operation is at the mercy of sufficiently long proton $T_{1\rho}$s: rather long contact times of approximately 100 ms were necessary for optimum cross-polarisation. Tungsten compounds where H_2O-protons are the only, or the predominant, source of magnetisation did not yield any ^{183}W CP response. Using MAS frequencies of ca. 2.5 kHz and external magnetic field strengths of 7 and 8.45 T, only isotropic resonances have been observed for ^{183}W in tetrahedral WO_4^{2-} and WS_4^{2-} environments, isotropic ^{183}W chemical shifts for WO_4^{2-} anions are found in the range -100 to $+30$ ppm, depending on the respective cation. $\delta^{183}W$ for $(NH_4)_2WS_4$ is found at 3640 ppm, while solid $W(CO)_6$ has an isotropic chemical shift of -3470 ppm. The slightly distorted octahedral WO_6-environments in WO_3 ($\delta^{183}W = -414, -438$ ppm) and $H_3[P(W_{12}O_{40})] \cdot nH_2O$ ($\delta^{183}W = -174$ ppm) give rise to spinning sidebands in the ^{183}W MAS spectra, $\Delta\sigma$ for the latter compound is substantial (ca. 1500 ppm). These few ^{183}W high-resolution solid-state NMR data clearly demonstrate the potential usefulness of ^{183}W MAS and CP/MAS spectroscopy, the same conclusion can be reached upon inspection of the wealth of useful ^{183}W data obtained from solution-state NMR [38, 39]. Taking experimental difficulties, both with respect to long T_1s and to unfavourable CP dynamics into account, it would, however, be unrealistic to predict a boom for ^{183}W MAS and CP/MAS spectroscopy in the near future.

Over the last few years, first steps have been taken to utilise the spin properties of low-γ nuclei like ^{89}Y, ^{109}Ag and ^{183}W for high-resolution solid-state NMR spectroscopy; two important isotopes, ^{103}Rh and ^{57}Fe, have not yet been tackled. Much in this area of low-γ nuclei remains to be explored, and probably this will remain an area of practical compromise for some time to come. To some extent, high external magnetic field strengths will help to overcome probe ringing and sensitivity problems when single-pulse excitation is the desired experiment. For the heavier nuclei like ^{103}Rh, ^{109}Ag, ^{183}W the beneficial effects of high magnetic fields will be scaled down, depending on chemical-shift anisotropies and on chemical-shift dispersion in disordered systems. Single-pulse operation for these nuclei will remain a matter of patience (or rather: availability of spectrometer time), unless procedures like doping of the samples with paramagnetic species are employed. This in itself is a fairly expensive price to pay with respect to reliability and quantification of the results, and will only be helpful for some selected applications. In our view, the use of cross-polarisation methods, where applicable, offers more promising perspectives for a wider and more routine use of low-γ nuclei in high-resolution solid-state NMR. Especially at moderate magnetic field strengths, and even if standard Hartmann-Hahn CP experiments for these nuclei represent a brute force approach and are demanding with respect to probes and other hardware aspects, polarisation transfer techniques are certainly the most efficient way to circumvent T_1 problems. The above-mentioned strategies [20, 21] to reduce the power requirements in the X-channel for CP experiments have not yet been exploited for low-γ nuclei. Of course, the power requirements for Hartmann-Hahn cross-polarisation are most demanding if ^1H is used as the magnetisation source. Given that many important inorganic and organometallic compounds of these elements contain phosphorus or fluorine, ^{31}P and ^{19}F could be promising alternative sources for magnetisation transfer experiments involving low-γ nuclei in the solid state.

3.2 Nuclei with Large Chemical-Shift Anisotropies: ^{195}Pt, ^{199}Hg, ^{207}Pb, (^{205}Tl)

If nuclei like ^{89}Y, ^{109}Ag or ^{183}W in the previous section were characterised by experimental problems due to low resonance frequencies and long relaxation times T_1, then the experimental problems encountered with nuclei like ^{195}Pt, ^{199}Hg, ^{207}Pb in this section can be summarised as due to enormous chemical-shift anisotropies $\Delta\sigma$. Spectral widths exceeding 150 kHz cause problems with pulse excitation and digitisation under MAS conditions. Even if MAS frequencies > 10 kHz are no longer difficult to achieve, fast magic-angle spinning will often be of only limited use because of severe interference with cross-polarisation spin dynamics. For nuclei like ^{195}Pt or ^{199}Hg, in most cases the extremely useful (i.e. informative) slow-spinning regime (which we often find for nuclei like ^{31}P or ^{119}Sn at spinning frequencies of a few kHz and at B_0 field strengths of 4.7–9.3 T) deteriorates into a barely manageable ultra-slow-spinning regime when using "normal" MAS frequencies and "normal" magnetic field strengths. Just to mention an example: ^{195}Pt shielding anisotropies for many relevant square-planar Pt(II)-compounds will be of the order of 5000–10000 ppm [41]; assuming $B_0 = 7$ T this corresponds to a chemical shielding anisotropy pattern width of 0.32–0.64 MHz.

Accordingly, MAS and CP/MAS reports in the literature involving nuclei like ^{195}Pt, ^{199}Hg and ^{207}Pb are not particularly numerous—and deal predominantly with highly symmetric species where shielding anisotropies are moderate (on this several thousand ppm csa scale). Such cases with "pathologically" small ^{195}Pt or ^{199}Hg shielding anisotropies should not propagate any illusive views that an extension of these experiments to a routine application of, for example, ^{195}Pt MAS spectroscopy in platinum chemistry was only going to be a "small next step": highly symmetric (tetrahedral or octahedral) coordination is not very common for solid platinum or mercury compounds.

More or less regular octahedral environments will only be found for Pt(IV)-complexes; the vast majority of platinum chemistry deals with Pt(II) compounds and, accordingly, square-planar Pt coordination or, to a lesser extent, with Pt in even lower oxidation states. Given the enormous importance of Pt chemistry within the field of transition metal chemistry (ranging from homogeneous and heterogeneous catalysis to cluster chemistry and anti-cancer drugs), there should definitely be a wide range for useful future ^{195}Pt MAS and CP/MAS applications if large-csa-effects were kept under control by the use of low magnetic field strengths.

Most solid organomercury compounds prefer a linear two-fold coordination, but more or less symmetric 3-or 4-fold coordination, or strongly distorted tetrahedral environments and irregular higher coordination schemes are also found; for most of these cases large ^{199}Hg shielding anisotropies will be the consequence. Similar coordination patterns are found in thallium chemistry, with similar ^{205}Tl csa consequences. Furthermore, the ^{205}Tl resonance frequency falls outside the normal tuning range of most CP/MAS probes, and ^{205}Tl solid-state NMR has so far been limited to wide-line NMR studies of ternary chalcogen phases [42–45], of some Tl complexes of biochemically relevant ligands [46, 47] and of some further inorganic Tl compounds [48–50].

By and large, elements like mercury, thallium or lead play a less prominent role in chemistry than do, for instance, platinum or tin; many organolead compounds are less stable than the corresponding tin compounds. Whether the isotope ^{207}Pb will belong to the problematic, large-csa category of nuclei, or whether ^{207}Pb falls into a much more manageable csa range (like e.g. ^{113}Cd or ^{119}Sn) is in each single case largely governed by the respective chemistry circumstances, and there will be many "borderline" cases, depending on the magnetic field strength used. As a general trend (always assuming more common B_0 field strengths in the 4.7–9.3 T range), most Pb(II) compounds will qualify for the large anisotropy category, with all concomitant experimental difficulties, while the majority of Pb(IV)-compounds (with lead in tetrahedral or distorted tetrahedral coordination) will display somewhat less expressed chemical shift anisotropies. However, even Pb(IV) in more strongly distorted coordination, like in, e.g., trigonal-bipyramidal 5-fold coordination, will give rise to csa patterns exceeding a width of 1000 ppm.

^{195}*Pt*

The known isotropic chemical shift range for ^{195}Pt spans approximately 14 000 ppm; in solution-state ^{195}Pt NMR it is common practice to use Ξ ^{195}Pt = 21.4 MHz as the chemical shift reference [51, 52]. This shift reference should also be adopted for ^{195}Pt MAS and CP/MAS studies, so far different solid-state ^{195}Pt ppm-scales have been used by different authors. Numbers quoted in this paragraph will refer to the Ξ ^{195}Pt = 21.4 MHz scale, some isotropic chemical shifts δ^{195}Pt obtained from MAS and CP/MAS spectra are collected in Table 4.

For solid K_2PtCl_6 symmetry dictates a fully symmetric, i.e. isotropic ^{195}Pt shielding tensor. This has been found in the 43.0 MHz ^{195}Pt MAS spectrum of K_2PtCl_6 [53] and the line width of approximately 24 ppm has been tentatively assigned to residual dipolar and scalar coupling ^{195}Pt-$^{35/37}$Cl. This point has recently been addressed in somewhat more detail for the compounds M_2PtCl_6 ($M = NH_4$, K, Na) using a magnetic field strength of 9.4 T [54]: a slightly asymmetric ^{195}Pt line shape with further splittings is obtained for all $PtCl_6^{2-}$ anions. Assuming axial symmetry for the J-tensor and using known Pt-Cl bond distances, these line shapes can be simulated and yield a scalar coupling constant $^1J(^{195}Pt^{35}Cl) = 140$ Hz, the anisotropy ΔJ is estimated to be approx. 158 Hz

Table 4. Isotropic ^{195}Pt chemical shifts of some solid platinum compounds, referenced to Ξ^{195}Pt = 21.4 MHz[a]

Compound	δ^{195}Pt [ppm]	Reference
Na_2PtCl_6	4640	54
K_2PtCl_6	4705	53, 54
$(NH_4)_2PtCl_6$	4737	54
K_2PtCl_4	2906	53
$K_2Pt(OH)_6$	8024	6
$[Na_2PtCl_6]\cdot 6H_2O$	4607	55, 56
$[Pt(en)_3]Cl_4\cdot 2H_2O$	3579	55, 56
cis-$[PtCl_4(NH_3)_2]$	4285	55, 56
trans-$[ClPt(P^nBu_3)_2$-C≡C-*p*-C_6H_4-C≡C-$(P^nBu_3)_2PtCl]$	102	59
cis-$[Me_2Pt(PEt_3)_2]$	132	58
cis-$[Me_2Pt(PPh_3)_2]$	104	58

[a] where necessary, shifts have been converted using $\delta^{195}PtCl_6^{2-}$ in H_2O = 4522 ppm [74] or δ^{195}Pt solid K_2PtCl_6 = 4705 ppm

for solid K_2PtCl_6. The 43.0 MHz ^{195}Pt MAS spectrum of K_2PtCl_4 [53] could only be obtained as a composite of several overlapping partial spectra. This composite spectrum yields δ^{195}Pt = + 2906 ppm and a chemical-shift anisotropy $\Delta\sigma = 10340 \pm 200$ ppm for K_2PtCl_4. The D_{4h} crystal point group symmetry requires an axially symmetric ^{195}Pt shielding tensor for this compound, and further single-crystal ^{195}Pt studies helped to refine $\Delta\sigma$ for K_2PtCl_4 to 10419 ± 9 ppm.

Both $K_2Pt(OH)_6$ [6, 55] and $(NH_4)_2PtCl_6$ [54] have been proposed as set-up samples for ^{195}Pt CP/MAS experiments. Despite an almost perfect octahedral PtO_6 coordination in $K_2Pt(OH)_6$, a substantial ^{195}Pt shielding anisotropy of approx. 425 ppm is observed [56]. ^{195}Pt shielding anisotropies and isotropic ^{195}Pt chemical shifts have been obtained from ^{195}Pt CP/MAS spectra of several Pt(IV)-complexes with platinum in a more or less symmetric octahedral environment. Shielding anisotropies of ca. 1000 ppm for solid $[Me_3Pt(acac)]_2$ and $[Me_3Pt]_2SO_4\cdot 4H_2O$ are consistent with strongly field-dependent ^{195}Pt T_1-relaxation rates obtained in solution, caused by csa contributions to T_1: measured T_1 rates in solution suggest a csa contribution comparable with the experimentally obtained $\Delta\sigma$ values for the solid state [57]. For $[Pt(en)_3]Cl_4\cdot 2H_2O$ and for *cis*-$[PtCl_4(NH_3)_2]$ moderate shielding anisotropies of only a few hundred ppm have been observed [58], but both compounds exhibit broad line widths (ca. 23 ppm) which is not surprising given that in both cases ^{195}Pt is neighboured to six quadrupolar nuclei (^{14}N, $^{35/37}$Cl).

^{195}Pt CP/MAS spectra of no more than three Pt(II)-complexes have been reported: *cis*-$[Me_2Pt(PEt_3)_2]$ and *cis*-$[Me_2Pt(PPh_3)_2]$ [56, 58], and *trans*-$[ClPt(P^nBu_3)_2$-C≡C-p-C_6H_4-C≡C-$(P^nBu_3)_2PtCl]$ [59]. In all three cases sharp ^{195}Pt resonances, further split due to scalar coupling $^1J(^{195}Pt^{31}P)$, and broad csa patterns have been observed. For the latter compound with Pt(II) in *trans*-configuration each spinning sideband consists of a triplet due to scalar coupling to two (possibly equivalent) ^{31}P-nuclei ($^1J(^{195}Pt^{31}P)$ = 2454 Hz). Even at $B_0 = 4.7$ T and a MAS frequency of 12.5 kHz only part of the ^{195}Pt CP/MAS spectrum could be recorded, the width of the csa pattern must be more than 2000 ppm. The ^{31}P CP/MAS spectra display two ^{31}P resonances for *cis*-$[Me_2Pt(PEt_3)_2]$ as well as for *cis*-$[Me_2Pt(PPh_3)_2]$. Whether this inequivalence is due to intra- or to intermolecular effects can, in the absence of X-ray diffraction results, only be decided on the basis of the respective ^{195}Pt CP/MAS spectra. For both compounds only one ^{195}Pt resonance is observed, either split into a pseudo triplet, or into a double doublet due to $^1J(^{195}Pt^{31}P)$, respectively [58]. Platinum(II)-phosphine complexes are an important class of platinum compounds, the broad general interest in the spectroscopic properties of such compounds is nicely documented by a fairly large number of publications dealing with ^{31}P CP/MAS spectroscopy of such compounds [60–62]. However, there are contradictory reports concerning the number of ^{31}P resonances in some cases, as well as concerning the origin of splittings in some such ^{31}P CP/MAS spectra. Especially for *cis*-P_2Pt configurations the effects of homonuclear ^{31}P dipolar interaction will have pronounced effects on the ^{31}P CP/MAS spectra [63]; detailed analysis of such ^{31}P CP/MAS spectra would be greatly facilitated if the X part, i.e. the ^{195}Pt spectra, of such P_2Pt spin systems were more readily accessible. Clearly, similar arguments also hold for phosphine-adducts of other transition metals such as rhodium, mercury or silver. So far ^{195}Pt CP/MAS spectroscopy of Pt(II)-complexes in general has been grossly underrepresented, due to experimental difficulties.

^{199}*Hg*

The first reported ^{199}Hg CP/MAS spectrum was that of mercury acetate [6]. Obviously, this spectrum suffered from instable and/or slightly off-angle MAS which caused an apparent splitting. As may be seen from the ^{199}Hg CP/MAS spectrum of $Hg(OAc)_2$ in Fig. 6, there is only one sharp ^{199}Hg resonance for $Hg(OAc)_2$ as has also been reported elsewhere (64.) Apart from this singular ^{199}Hg CP/MAS spectrum, only two further ^{199}Hg CP/MAS studies have been published so far.

In analogy to well-established ^{113}Cd solid-state NMR studies (vide infra), ^{199}Hg CP/MAS could be a suitable probe for the investigation of mercury-sulfur coordination in metalloproteins. With this application in mind, two ^{199}Hg CP/MAS studies of mercury (II) thiolate model compounds of known Hg-S geometry have been reported [64, 65]. With possible coordination schemes ranging from regular (or almost regular) tetrahedral (e.g., $Hg(SC_6H_4Cl)_4^{2-}$), to strongly distorted tetrahedral with bridging thiolate groups (e.g., $Hg(S^tBu)_2$), to trigonal-planar with proper C_3-symmetry (e.g., $Hg(S^tBu)_3^-$), to trigonal-planar with Y-shape distortion (e.g., $Hg(SPh)_3^-$) and to linear two-fold coordination (e.g., $Hg(S\text{-}2,4,6\text{-}^iPrC_6H_2)_2$), typical ^{199}Hg csa patterns and asymmetry parameters could be expected for each type of Hg-S coordination. The expected trends, such as small to very small $\Delta\sigma$ for tetrahedral coordination, intermediate

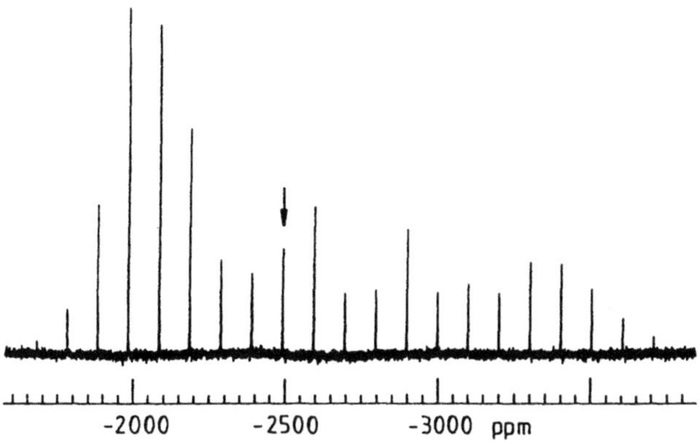

Fig. 6. 35.68 MHz ^{199}Hg CP/MAS of $Hg(OAc)_2$. 5900 transients, 5 s recycle delay, 5 ms contact time, $\nu_{rot} = 3.6$ kHz

$\Delta\sigma$ (1200–1400 ppm) for trigonal coordination (accompanied by a small asymmetry parameter η in the case of C_3-symmetry and large η (>0.8) for Y-shape distortion) and very large $\Delta\sigma$ (>4000 ppm) plus small asymmetry parameter η for two-fold coordination have been established [64].

^{199}Hg shielding anisotropies in solid (preferably two-fold coordinated) organomercury compounds will certainly be of the same order of magnitude, if not larger, than what has been observed for the linear S-Hg-S arrangement. It is, therefore, not surprising that no such data on organomercury compounds have been reported. It seems, however, somewhat more surprising that "hard inorganics" have so far not utilised ^{199}Hg MAS spectroscopy: within that field of application less problematic Hg-coordination patterns are more common than in organomercury chemistry.

^{207}*Pb*

When trying to summarise the currently available ^{207}Pb MAS and CP/MAS results, a certain lack of systematic studies is noticeable. Several "singularities" have to be reported, ranging from the static and MAS ^{207}Pb spectrum of $Pb(NO_3)_2$ which became a text-book example for an axially symmetric shielding tensor [66], to the ^{207}Pb CP/MAS spectrum of lead (II) tetradecanoate [67] which indicates, by similar isotropic ^{207}Pb chemical shifts for the solid state, for the liquid-crystalline and for the liquid phase, that such phase changes in Pb(II) carboxylates are not connected with dramatic changes in the metal coordination. In contrast to the ^{195}Pt MAS spectra of $PtCl_6^{2-}$ anions, the 62.8 MHz ^{207}Pb MAS spectrum of $(NH_4)_2PbCl_6$ (see Fig. 7) displays, within experimental error, a regular line shape as expected for scalar coupling $^1J(^{207}Pb\ ^{35/37}Cl)$ to six equivalent chlorine neighbours ($\delta^{207}Pb = -2058$ ppm, $^1J(^{207}Pb^{35}Cl) = 305 \pm 10$ Hz).

Based on δ^{207}Pb and on an almost axially symmetric ^{207}Pb shielding tensor pattern, a polymeric chain structure with penta-coordinated lead in a trigonal-

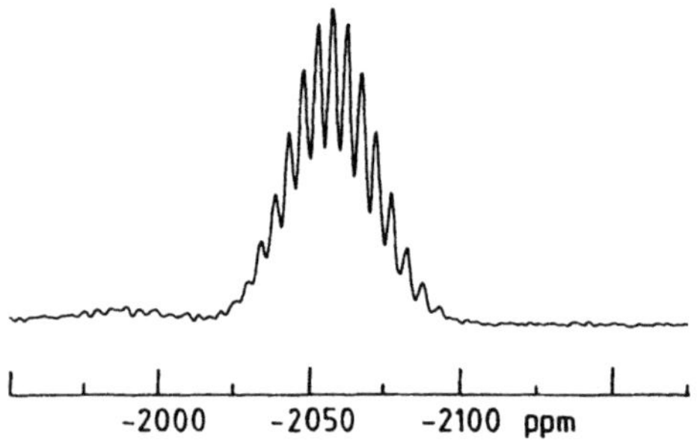

Fig. 7. 62.8 MHz ^{207}Pb MAS of $(NH_4)_2 \cdot PbCl_6$. 4600 transients, 20 s recycle delay, 40° pulse, $\nu_{rot} = 4$ kHz; $\delta^{207}Pb = -2058$ ppm, $^1J(^{207}Pb^{35/37}Cl)$ approximately 305 Hz

bipyramidal Me_3PbO_2 coordination has been postulated for solid Me_3PbOMe [68]. The ^{207}Pb CP/MAS spectrum of Me_3PbOMe has been compared to the ^{207}Pb CP/MAS spectra of two further organolead compounds,

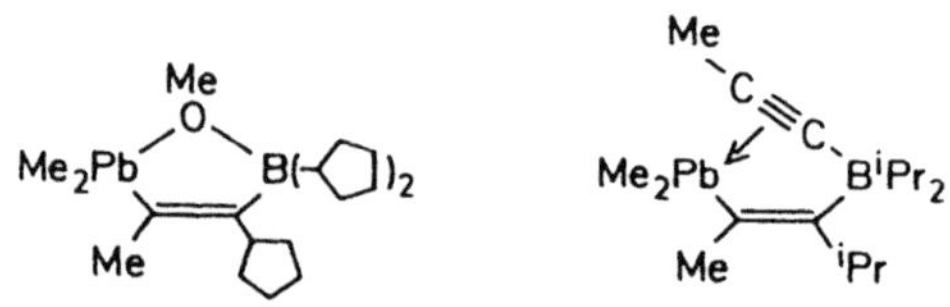

The—formally—lower coordination number for these latter two compounds results in ^{207}Pb resonances ca. 500 ppm less shielded than for Me_3PbOMe. In all three cases the ^{207}Pb csa pattern exceeds a width of 1500 ppm, the asymmetry parameter gradually changes from almost axially symmetric in Me_3PbOMe (*trans*-Me_3PbO_2 geometry) to $\eta = 0.4 \pm 0.2$ for the distorted Me_2RPbO tetrahedral environment to $\eta = 0.7 \pm 0.1$ for the highly irregular Me_2RPb ($\leftarrow C_2$) coordination [68]. Solid $(C_5Me_5)_2Pb$ displays an isotropic ^{207}Pb chemical shift very similar to that found in solution. The bent-sandwich arrangement of the two pentamethylcyclopentadienyl ligands in solid $(C_5Me_5)_2Pb$ may account for an assymetry parameter $\eta = 0.35$, although it should be pointed out that the influence of the presence of lone pair electrons in this Pb(II) compound on the ^{207}Pb shielding parameters is by no means clear [69]. The ^{207}Pb CP/MAS spectrum of $(Ph_3Pb)_2S$ readily reveals $^2J(^{207}Pb^{207}Pb) = 351$ Hz scalar coupling information since, unlike in solution, the two lead nuclei in solid $(Ph_3Pb)_2S$ give rise to an AB spin system [70].

The static and MAS ^{207}Pb spectra of $Pb(NO_3)_2$ have recently been reinvestigated [71]. $Pb(NO_3)_2$ crystallizes in space group Pa3 with the Pb^{2+} ions occupying sites of C_3-symmetry, thus an axially symmetric ^{207}Pb shielding tensor is observed in the static spectrum (see Fig. 8): A substantial temperature dependence of $\delta^{207}Pb$ is found (0.7 ppm/K) in the temperature range 290–340 K for a static sample of $Pb(NO_3)_2$. In addition to this temperature effect, the isotropic ^{207}Pb chemical shift of $Pb(NO_3)_2$ under MAS conditions is reversibly, strongly and apparently instantaneously dependent on the MAS frequency (see

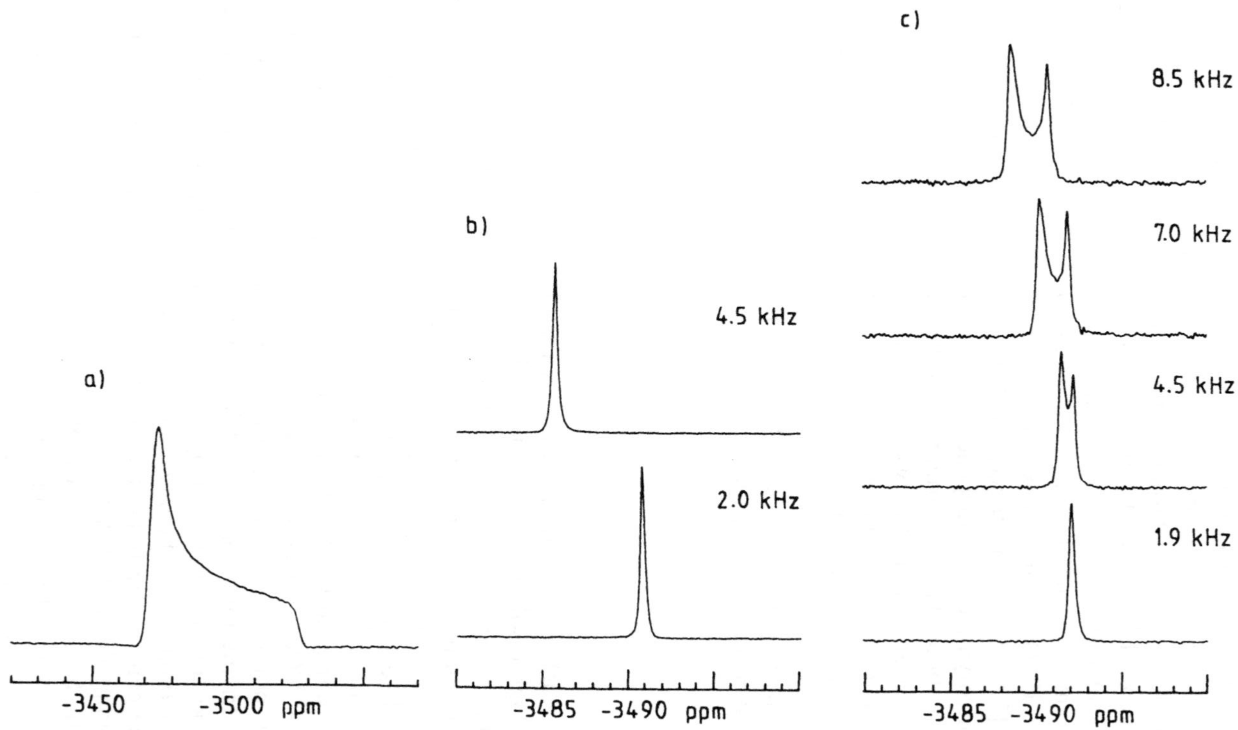

Fig. 8a–c. 62.8 MHz ^{207}Pb MAS and static spectra of $Pb(NO_3)_2$: **a** static; $\sigma_{\parallel} = -3474$ ppm, $\sigma_{\perp} = -3524$ ppm; **b** MAS, using a 7 mm o.d. ZrO_2 rotor; **c** MAS, using a 4 mm o.d. ZrO_2 rotor

Fig. 8): at higher MAS frequencies, a second resonance occurs. This observation may be explained if we assume that we observe the onset of a MAS-induced phase transition. Clearly, the determination of the exact nature of this process requires more detailed ^{207}Pb NMR studies of solid $Pb(NO_3)_2$, as well as further diffraction work and thermodynamic considerations.

Similarly dramatic, symmetry-related effects on the ^{207}Pb shielding tensor components for Pb(IV) in solid hexaorganyl-dilead compounds Pb_2R_6 have been reported [72, 73]. For instance, the X-ray crystal structure of Pb_2Ph_6 shows two centrosymmetric, independent molecules per unit cell, in fairly different conformations. This conformational difference is reflected in the ^{207}Pb shielding tensor parameters of the two respective ^{207}Pb resonances (molecule 1: $\sigma_{iso} = -14.5$ ppm, $\sigma_{11} = +219$ ppm, $\sigma_{22} = +2$ ppm, $\sigma_{33} = -265$ ppm, $\eta = 0.86$; molecule 2: $\sigma_{iso} = +131.8$ ppm, $\sigma_{11} = +467$ ppm, $\sigma_{22} = +291$ ppm, $\sigma_{33} = -363$ ppm, $\eta = 0.35$).

Just like in the case of ^{199}Hg, also for the nucleus ^{207}Pb the application field of inorganic chemistry has so far remained largely unexplored.

3.3 ^{113}Cd

Finally, with the nucleus ^{113}Cd we come across a candidate for which a considerable amount of high-resolution solid-state NMR information has been accumulated. This is only partly attributable to the convenient NMR properties of this nucleus: ^{113}Cd often displays considerable chemical-shift anisotropies $\Delta\sigma$, but very rarely exceeds $\Delta\sigma$ for ^{113}Cd the highly manageable regime where "normal" MAS and CP/MAS experiments (apart from static and single-crystal work) provide easy access to ^{113}Cd csa information from the analysis of MAS spinning sideband patterns. In addition, for the majority of Cd-compounds we find rather favourable $^{1}H \rightarrow {}^{113}Cd$ CP-characteristics, and even ^{113}Cd single-pulse experiments are often not severely hampered by unduly long ^{113}Cd T_1 relaxation times. Another driving force for ^{113}Cd NMR studies is to be found in several well-defined tasks and problem-solving strategies which require ^{113}Cd MAS and CP/MAS methodology. These two circumstances reinforce each other and thus explain the comparatively large bulk of "^{113}Cd experience" with respect to high-resolution solid-state NMR.

One leitmotiv for solid-state ^{113}Cd NMR studies stems from the fact that many Cd-compounds and Cd-complexes undergo rapid chemical exchange in solution, to the extent that often only averaged ^{113}Cd chemical shift information is available from solution-state NMR studies (similar solution-state NMR problems are also common for nuclei like ^{109}Ag or ^{199}Hg) [74]. The solid state then provides a slow-exchange regime with Cd in a defined coordination environment, a regime which may not be accessible by low-temperature solution-state NMR. This dynamic solution-state ^{113}Cd NMR problem applies equally to inorganic and organometallic complexes, and to Cd-complexes with ligands of biochemical relevance. The latter field of application is of particular importance since ^{113}Cd is a well-established probe for the study of metal-binding sites in proteins and other biochemical compounds [75, 76]. Within the framework of inorganic chemistry, questions concerning both short-range and long-range order as well as spin-interactions involving ^{113}Cd, set the scene for the application of ^{113}Cd MAS and CP/MAS NMR.

In an attempt to maintain some kind of order parameter within this chapter, the following ^{113}Cd overview has been subdivided into three major fields of application. Roughly speaking, these three sections could be characterised as:

1. truly inorganic binary and ternary Cd-bearing phases, devoid of protons (glasses and semiconductors)—where mutual spin interactions, also with other NMR-active isotopes, and their exploitation play an important role;
2. inorganic and organometallic Cd-compounds—where the application of ^{113}Cd MAS and $^{1}H \rightarrow {}^{113}Cd$ CP/MAS experiments can be exploited for a variety of problems;
3. Cd-compounds and model compounds of biochemical relevance—where the interpretation of isotropic ^{113}Cd chemical shifts, ^{113}Cd shielding anisotropies and relative orientations of shielding tensor components within a given framework of O-, N-, S-donor atoms is of prime importance.

re 1

An attempt has been made to control the size of CdS clusters by adding variable amounts of thiophenol during the preparation, whereby PhS functions as a covalently bonded surface passivation agent for the cluster particles [77]: ^{113}Cd MAS NMR monitors the structural evolution as a function of the S^{2-}/PhS^{-} ratio, taking advantage of the chemical shift difference between Cd in purely CdS-type coordination in the cluster-core (60 ± 5 ppm) and Cd in a purely PhS environment ($\delta^{113}Cd$ $Cd(SPh)_2 = -78$ ppm). The combination of ^{113}Cd wide-line, MAS and spin-echo experiments on crystalline and glassy compounds in the $CdAs_2$-$CdGeAs_2$ system reveal striking short-range order differences for the glasses and their stoichiometrically analogous crystalline counterparts [78]. For example, crystalline $CdAs_2$ shows $\delta^{113}Cd = +158$ ppm, while the glass with $CdAs_2$ composition has an isotropic ^{113}Cd chemical shift of -67 ± 5 ppm. The combined results indicate the presence of Cd-rich domains of the Cd_3As_2 type in the glasses. The observed faster ^{113}Cd spin-echo decays for the glasses would support this view of shorter Cd-Cd distances if it was possible to safely exclude that ^{75}As may also be responsible for this effect. Similar differences between crystalline model compounds and the analogous glasses have also been found for the CdP_2-$CdGeP_2$ system [79], indicating a reduced average number of Cd-P bonds per Cd in the glasses. For these phosphorus-containing systems ^{31}P MAS, spin-echo and ^{31}P-^{113}Cd SEDOR experiments have been added to the NMR toolbox. Exploiting the spin properties of ^{31}P and ^{113}Cd for

$CdGeAs_{2-x}P_x$ phases, contradicts a phase-segregation model but indicates a non-statistical cation/anion distribution [80]. For the $Zn_xCd_{1-x}GeP_2$ system $\delta^{113}Cd$ changes monotonically as a function of composition [81]. For this system the characterisation of local next-nearest environments did not rely on heteronuclear dipole interactions ^{31}P-^{113}Cd like in $^{31}P/^{113}Cd$ SEDOR experiments but scalar coupling $J(^{113}Cd^{31}P)$ served as the site-discriminating interaction: ^{31}P-^{113}Cd heteronuclear double quantum filtering MAS experiments allow selective detection of Cd-P bonded sites and provide useful constraints for line shape simulations, thus disfavouring e.g. mixed PGe_2CdZn sites. Heteronuclear decoupling and double quantum filtered $^{31}P/^{113}Cd$ MAS experiments have also been reported for the crystalline compounds $CdSiP_2$ and $CdGeP_2$ [82] where scalar coupling constants $^1J(^{113}Cd^{31}P)$ of 432 Hz ($CdSiP_2$) and 330 Hz ($CdGeP_2$) provide convenient circumstances for such editing techniques. The possible extension to 2D hetero-correlation ^{31}P-^{113}Cd spectroscopy has also been illustrated. Editing techniques based on heteronuclear J interactions elegantly complement editing methods which rely on heteronuclear dipole interactions such as SEDOR. In chemistry terms, an even higher degree of selectivity can be expected from J-based experiments but T_2^* relaxation times (in this case ^{31}P) will constrain the applicability. However, optimism is in place given the usually substantial magnitude of $^1J(^{31}PX)$ scalar coupling constants. Yet another possibility to utilise heteronuclear ^{31}P-^{113}Cd dipole interactions is represented by $^{31}P \rightarrow {}^{113}Cd$ cross-polarisation experiments, again exemplified on crystalline $CdSiP_2$ (16, see also paragraph 2). The ease of achieving $^{19}F \rightarrow {}^{113}Cd$ cross-polarisation for crystalline CdF_2 (see paragraph 2) can be taken as an indication that related selective $^{19}F/^{113}Cd$ MAS and CP/MAS experiments should be feasible for a variety of crystalline and amorphous Cd-F systems.

re 2

Some early examples, demonstrating the sensitivity advantages of ^{113}Cd CP operation (MAS and static) over ^{113}Cd single-pulse excitation included simple hydrous inorganic salts like $CdCl_2 \cdot xH_2O$, $Cd(OH)_2 \cdot xH_2O$, $3CdSO_4 \cdot 8H_2O$, $Cd(OAc)_2 \cdot 2H_2O$ and $Cd(NO_3)_2 \cdot 4H_2O$. Cross-relaxation times $T_{^1H^{113}Cd}$ have been discussed, pointing out the danger in quantitation of ^{113}Cd CP results [83]. In fact, since then the ^{113}Cd CP/MAS spectra of $Cd(O_2CH) \cdot 2H_2O$ and $Cd(O_2CH) \cdot 2D_2O$ have been used to demonstrate the unwanted modulation effect of small homonuclear ^1H dipole–dipole interactions on the ^{113}Cd-^1H Hartmann-Hahn matching condition under MAS. A variable-angle spinning experiment (90° during the CP period, 54.7° during data acquisition) has been proposed and was shown to be less susceptible to such spinning frequency-related modulation effects [84]. Three different types of Cd-coordination (CdS_4O, CdS_4 and CdS_3O_3 sites), displaying a fairly wide range of isotropic ^{113}Cd chemical shifts and ^{113}Cd shielding tensor properties, have been identified for the decanuclear complex $[Cd_{10}(SCH_2CH_2OH)_{16}](ClO_4)_4$ [85]. Both MAS and off-angle spinning (OMAS) have been used to obtain the shielding tensor components which then served for assignment purposes in terms of nearest-neighbours coordination. A similar approach, using ^{113}Cd CP/MAS and OMAS, was chosen to probe the nearest neighbours cadmium-oxygen coordination in $3CdSO_4 \cdot 8H_2O$ and $CdSO_4 \cdot H_2O$ [86] and to discuss the ^{113}Cd shielding parameters on the basis of Cd-O bond lengths and bond angles. In general, it seems surprising that off-angle spinning has to date not been more widely used (in combination with on-angle MAS) to obtain reliable shielding tensor parameters for cases with moderate shielding anisotropies—even if one takes into account that calibration of the angle limits the accuracy of this approach.

^{113}Cd CP/MAS spectra of a variety of Cd complexes of the type $(NR_4)_2CdX_4$ (X = Cl, Br, J) have been reported [87] in order to provide "benchmark" isotropic ^{113}Cd chemical shifts to support the assignment of related solution-state ^{113}Cd NMR spectra. Combined solution-state and CP/MAS ^{113}Cd results have been used to establish typical isotropic ^{113}Cd chemical shift ranges for representative Cd-halogen coordination patterns Cd^{2+}, CdX^-, CdX_2, CdX_6^{4-}, CdX_5^{3-}, CdX_3^-, CdX_4^{2-} (X = Cl, Br, J) [88]. Some of these coordination patterns cannot be distinguished by considering δ_{iso} ^{113}Cd alone.

Questions concerning the ^{113}Cd shielding tensor and its orientation in the respective crystal axes frame have been re-considered for $Cd(NO_3)_2 \cdot 4H_2O$ and for $3CdSO_4 \cdot 8H_2O$, employing single-crystal studies [89]. Indeed, as is to be expected from point group symmetry considerations, $Cd(NO_3)_2 \cdot 4H_2O$ gives rise to a non-axially symmetric ^{113}Cd shielding tensor. More ^{113}Cd single crystal studies have been carried out on several Cd(II) complexes which represent a variety of cadmium-oxygen coordination patterns: $CdCa(OAc)_4 \cdot 6H_2O$, Cd(maleate)$\cdot 2H_2O$, Cd(formate)$\cdot 2H_2O$, $(NH_4)_2Cd(SO_4)_2 \cdot 6H_2O$, $Cd(OAc)_2 \cdot 2H_2O$ [90]. The driving force for these efforts results from the failure of δ_{iso} ^{113}Cd to unambiguously discriminate 6-, 7- and 8-coordinated cadmium-oxygen environments. These single-crystal studies revealed four symmetry-related ^{113}Cd shielding tensors for $Cd(OAc)_2 \cdot 2H_2O$, a strategy of how to assign these symmetry-related shielding tensors to specific lattice sites has been outlined. Later on, this assignment was confirmed by single-crystal studies on ^{13}C-enriched $Cd(OAc)_2 \cdot 2H_2O$ [91], where the heteronuclear ^{113}Cd-^{13}C dipolar interaction gives rise to orientation-dependent ^{113}Cd line widths for the four distinguishable symmetry-related ^{113}Cd shielding tensors and thus allows unambiguous assignment to specific lattice positions. A single-crystal ^{113}Cd CP study of the $CdCl_2 \cdot$18-crown-6 complex revealed an almost axially symmetric ^{113}Cd shielding tensor [92], the most shielded component only deviates by 2.4° from the threefold rotation axis defined by the axial chlorine ligands and the CdO_6 pseudoplane created by the crown ether ligand. This Cd-crown ether complex is an instructive example that fairly remote ligands (in this case: Cl) can dominate shielding patterns for metal nuclei like ^{113}Cd. Similarly unexpected ^{113}Cd shielding effects have been observed in the CP/MAS spectra of a related complex with a CdN_6O_2 nearest neighbours coordination [93], in this case an unsaturated macrocycle with N donor sites provides the CdN_6 plane, the oxygens occupying axial positions are due to molecular H_2O. The unusual shielding for this compound has been ascribed to mostly electrostatic Cd-N interactions and predominantly covalent Cd-O bonding properties; by comparison with other ^{113}Cd shielding tensor components of known orientation, it has been suggested that the most shielded component of the non-axially symmetric shielding tensor should be oriented approximately orthogonal to the CdN_6-plane.

With a view of environmental hazards, Cd-exchanged montmorillonite samples have been investigated by ^{113}Cd MAS and spin-echo techniques [94]. The substantial ^{113}Cd line widths had to be ascribed mostly to dynamic processes, and conclusions concerning macroscopic Cd diffusion in/on natural clay minerals could be drawn, allowing an estimating of the spread of contamination from highly Cd-polluted natural sites.

The X-ray crystal structure of $(N^nPr_4)[Cd(SCN)_3]$ shows a trigonal bipyramidal S_2CdN_3-coordination with three inequivalent nitrogen atoms, two of which occupy the axial positions of the trigonal bipyramid. Despite the lack of a threefold axis of symmetry, an axially symmetric ^{113}Cd shielding tensor has been reported for this compound [95]. In addition, the ^{113}Cd CP/MAS spectrum shows further fine structure which has been explained as due to scalar coupling $^1J(^{113}Cd^{14}N)$ and incompletely averaged dipolar interactions ^{113}Cd-^{14}N; the largest coupling constant $^1J(^{113}Cd^{14}N) = 178$ Hz was assigned to the shortest Cd-N bond. The ^{113}Cd MAS and CP/MAS spectra of $Cd(CN)_2$ and $Cd(CN)_2 \cdot C_6H_{12}$, respectively, show several well-resolved ^{113}Cd resonances of variable intensity. If one assumes a static disorder of the CN groups, then these resonances can be attributed to the various statistically possible CdC_4 to CdN_4 coordinations (a slight deviation from the expected intensity distribution for a completely statistical distribution has been reported for $Cd(CN)_2$). A gradual increase of the ^{113}Cd line widths with an increasing number of directly neighboured nitrogen atoms has also been reported [96] but no resolved splittings due to scalar coupling $J(^{113}Cd^{14}N)$ have been observed. The ^{113}Cd CP/MAS spectrum of the host-guest complex $Cd(CN)_2 \cdot 2/3H_2O \cdot {}^tBuOH$ shows two types of Cd-sites for the host-lattice in a 2:1 intensity ratio: tetrahedrally coordinated $Cd(CN)_4$-type sites and octahedral Cd in $Cd(CN)_4(H_2O)_2$ coordination [97]. The tetrahedral Cd-sites give rise to several resonances, accounting for the different CdC_4 to CdN_4 possibilities of coordination. The intensity distribution of these various resonances of the tetrahedral Cd-sites in the ^{113}Cd CP/MAS spectra have been used to suggest a partially ordered Cd-CN coordination model in this complex. The local order around Cd in another similar host-guest complex, $[Cd_5(CN)_{10}(H_2O)_4 \cdot 4C_6H_{11}OH]$ has also been determined by ^{113}Cd CP/MAS spectroscopy [98]. The $C_6H_{11}OH$ guest molecules were shown to be mobile and to interact with the host lattice via hydrogen bonding. A slight dependence of the isotropic ^{113}Cd chemical shifts of this compound on the spinning frequency was noted and ascribed to pressure effects caused by MAS.

Several CdX_2/PR_3 complexes (X = Cl, Br; R = Ph, Bu, Et) with Cd:P ratios of 1:1, 1:2 and 2:3 have been studied by solution- and solid-state NMR [99]. The latter complexes $(R_3P)_3Cd_2X_4$ exist in solution only as equilibrium mixtures of the 1:1 and 1:2 complexes, while ^{113}Cd CP/MAS spectra show that

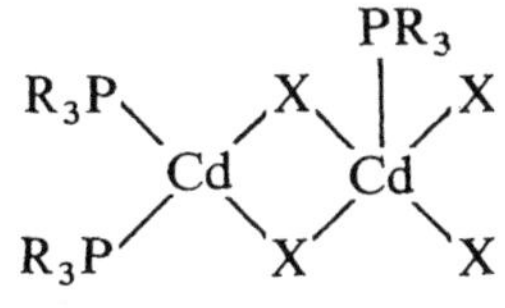

is the most likely solid-state structure.

Very recently subtle differences in δ_{iso} ^{113}Cd, in ^{113}Cd shielding anisotropies and asymmetry parameters have been reported for three structural isomers of the $[S_4Cd_{10}(SPh)_{16}]^{4-}$ cluster anion [100]. On the basis of X-ray crystal structure data these ^{113}Cd CP/MAS results have been rationalised as due to basically three different types of geometrical distortions of the CdS_4 coordination environment. Finally, for solid CdTe substantial anisotropy of J(^{113}Cd^{125}Te) ($J_{\perp} = 765 \pm 80$ Hz, $J_{\parallel} = 435 \pm 120$ Hz) has been determined from single crystal spectra [101].

re 3

The application of solution- and solid-state ^{113}Cd NMR, including biochemical aspects, was reviewed in 1988 [75]. Biochemistry is not the subject of this chapter, therefore, a few general considerations will have to suffice. It is quite clear that a routine application of ^{113}Cd NMR to probe metal binding sites of bio-molecules critically depends on a thorough understanding of ^{113}Cd shielding patterns and their dependence on various bonding situations. It is equally clear that this also requires knowledge of relative orientations of shielding tensor components—and this can only be obtained from a multitude of single-crystal ^{113}Cd NMR studies of suitable model compounds. Several such studies have been mentioned in the previous paragraph [89–93]. These studies alone do not provide a sufficiently tight network of well-characterised ^{113}Cd shielding "circumstances", and several more purpose-built Cd-coordination model compounds with O-, N-, S-coordination sites have been investigated [102–108]. Armed with the available bulk of information on ^{113}Cd-shielding, several Cd-substituted metalloproteins have been investigated by solid-state ^{113}Cd NMR methods [109–111].

3.4 ^{77}Se, ^{125}Te

With the noticeable exception of inorganic non-oxide chalcogene phases, both crystalline and amorphous, which constitute an autonomous field of activities, the chemical role of selenium and tellurium (at least in molecular chemistry) can be described as mainly representing homologues of related oxygen or sulfur derivatives. Often such Se- or Te-homologues are considerably less stable than the respective O- or S-compounds. This situation is reflected by the currently available ^{77}Se and ^{125}Te MAS and CP/MAS data: while ^{77}Se and ^{125}Te have been firmly established for purposes of probing short-range order in inorganic phases, not many applications of ^{77}Se, ^{125}Te MAS and CP/MAS experiments in the field of molecular chemistry have been reported. This lack of NMR activity seems rather wasterful since the properties of these two nuclei are such that for the majority of cases a particularly informative regime of isotropic chemical shifts, shielding anisotropies and scalar and dipolar interactions is obtained and experimentally straightforward to explore.

With a view to demonstrating the potential usefulness of ^{77}Se CP/MAS as a probe for molecular inorganic and organic selenium compounds, a range of Se compounds, like $[Se_4][HS_2O_7]$, $(Me_4N)(SeO_2Cl)$, H_2SeO_3, $(NH_4)_2SeO_4$, $(NH_4)_2SeCl_6$, Me_3SeJ, $(NH_2)_2CSe$ and its camphor-inclusion derivative have been studied [112]. The largest ^{77}Se shielding anisotropy (approximately 1000 ppm) was observed for the SeO_2Cl^- anion, even for the SeO_4^{2-} anion with Se in almost perfectly tetrahedral coordination an asymmetric, though much smaller, ^{77}Se shielding tensor was found. The ^{77}Se resonances of Me_3SeJ and $(Me_4N)(SeO_2Cl)$ display broadening and/or fine structure which was ascribed to dipolar and scalar interactions between ^{77}Se and the neighbouring $^{35/37}Cl$ and ^{127}J nuclei, respectively. The X-ray crystal structure of selenourea, $(NH_2)_2CSe$, shows nine inequivalent molecules per asymmetric unit and its ^{77}Se CP/MAS spectrum displays five ^{77}Se resonances in a 1:1:3:2:2 intensity ratio, while the selenourea/camphor inclusion compound gives rise to only one ^{77}Se resonance. This selection of selenium compounds indeed demonstrates that the ^{77}Se chemical shift range and shielding anisotropies will easily pick up even small geometric distortions, but at the same time the magnitude of ^{77}Se shielding parameters is such that convenient and easily accessible MAS and CP/MAS conditions are granted. The X-ray crystal structure of the Se-containing macrocycle 1,3,7,9,13,15-hexaselenacyclooctadecane shows a highly irregular and distorted geometry with six inequivalent Se sites per molecule. Accordingly, the ^{77}Se CP/MAS spectrum displays six resonances, some of which are accompanied by satellites due to $^2J(^{77}Se^{77}Se)$ [113]. Tentatively, the two lowest frequency resonances (which also happen to show the largest scalar coupling $^2J(^{77}Se^{77}Se) = 128$ Hz) have been assigned to the two Se sites in the macrocycle with the largest Se-C-Se bond angle. Several similar Se macrocycles have been characterised by ^{77}Se CP/MAS and X-ray crystallography [114] and have been considered as potentially useful complexing agents for soft metal ions. Solid adducts of $Cu(SO_3CF_3)$ and $Hg(CN)_2$ with 1,5,9,13-tetraselenacyclohexadecane have been synthesized [115]. Their respective ^{77}Se CP/MAS spectra show broad resonances for the Cu-compound, the ^{77}Se CP/MAS spectrum of the Hg-compound displays scalar coupling $^1J(^{199}Hg^{77}Se)$. However, X-ray crystallography showed that Cu and Hg are not encrypted inside the macrocycle but form adducts where one metal center is surrounded by four macrocycle molecules.

The single-crystal X-ray structure of Ag_7PSe_6 is unknown, and in this case ^{77}Se MAS spectroscopy (with and without ^{31}P high power decoupling) in conjunction with ^{31}P MAS results can provide useful nearest-neighbour-coordination constraints to allow structure refinement from powder diffraction methods [116]. Three chemically distinct Se-species have been identified in Ag_7PSe_6, one of which displays two distinct $^1J(^{77}Se^{31}P)$ scalar coupling patterns. This latter aspect was later confirmed by using ^{77}Se-^{31}P double quantum MAS experiments [82]. The binary system P-Se forms glasses from a wide range of compositions (0–52 atom % P). Such PSe-glasses have been characterised by ^{31}P-^{77}Se MAS, and ^{31}P-^{77}Se SEDOR in order to unravel the statistics of local Se-Se versus Se-P bonding [117, 118]. On the basis of these results the preferential formation of P-Se-P clusters has been ruled out. Fairly early, the favourable properties of ^{77}Se and ^{125}Te were exploited to consider shielding anisotropies and the anisotropy of heteronuclear scalar coupling in the crystalline binary phases ZnSe, CdSe, HgSe, PbSe, ZnTe [119]; CdTe, HgTe and PbTe [120]. These studies took advantage of a situation where the magnitude of the various scalar interactions greatly exceeds the magnitude of heteronuclear dipole interactions. Substantial anisotropy in J has been reported, for example: CdTe $^1J(^{113}Cd^{125}Te) = 655$ Hz and $J_\perp = 765$ Hz, $J_\parallel = 435$ Hz; HgTe $^1J(^{199}Hg^{125}Te) = 5080$ Hz and $J_\perp = 6720$ Hz, $J_\parallel = 1800$ Hz; CdTe and HgTe are isostructural. It seems somewhat surprising that these early results on single crystals and static powders have not been re-addressed by MAS techniques. ^{125}Te MAS has been used to characterise a wider range of compositions for the semiconductor system $Hg_{1-x}Cd_xTe$ ($x = 0.05$–0.95) [121]. As a function of composition, the ^{125}Te resonances cover a range of approximately 700 ppm. Based on relative intensities of these resonances a short range order model has been derived which seems to suggest that Hg and Cd prefer very different sites in this ternary system, leading to a largely undisturbed CdTe-lattice and to Hg-rich clusters.

^{125}Te CP/MAS results have so far only been reported for a small number of organotellurium compounds: $Me_3TeCl \cdot H_2O$, Me_3TeNO_3 and Me_3TeJ [122] as well as $Te(OH)_6$. Compared to analogous selenium compounds, the ^{125}Te shielding anisotropies are larger than for ^{77}Se, as expected. For $Me_3TeCl \cdot H_2O$ a symmetric splitting pattern of the ^{125}Te resonance has been found which can only be explained by scalar coupling of ^{125}Te with two equivalent Cl neighbours ($^1J(^{125}Te^{35}Cl) = 106 \pm 5$ Hz). For Me_3TeJ the ^{125}Te MAS resonance is very broad and dominated by unresolved dipolar and scalar interaction(s) between ^{125}Te and ^{127}J. Several further compounds containing Me_3Te^+ or Ph_3Te^+ cations have been investigated [123]. As expected, the phenyl compounds suffer from long proton T_1s and in all cases broadening of the ^{125}Te resonances due to a combination of scalar and dipolar interactions with the anion X (X = Cl, Br, J) has been observed. These experimental findings have been taken as proof that the cation/anion interactions in compounds of the type R_3TeX must be of considerably covalent character.

3.5 ^{119}Sn

Amongst all the so-far-mentioned, less-commonly used spin-1/2 nuclei, ^{119}Sn is certainly the one which almost qualifies as a "commonly used spin-1/2 nucleus". This fact has to be ascribed to the favourable combination of spectroscopic properties *and* chemical reasons: the element tin is rather prominently present in all branches of inorganic and organometallic chemistry and other than, for example, in the case of ^{195}Pt or ^{89}Y the NMR properties of the nucleus ^{119}Sn account for the relative ease of acquiring ^{119}Sn MAS and CP/MAS spectra. This statement is certainly applicable to the vast majority of Sn(IV) compounds (which also happens to be the most common oxidation state), while Sn(II) compounds can occasionally fall into the very-large-shielding-anisotropy category. Only tin in highly symmetrical surrounding will yield ^{119}Sn MAS or CP/MAS spectra in the fast spinning regime, in all other cases we have to operate in the slow-spinning regime. For ^{119}Sn in these cases the ratio of spinning frequencies to ^{119}Sn chemical-shift anisotropies is usually such that high-resolution information can easily be obtained while anisotropic interactions are only partially averaged by MAS. In addition, the ^{119}Sn chemical-shift range is sufficiently large to allow monitoring of rather subtle effects and the presence of a second spin-1/2 isotope, ^{117}Sn, helps exploiting scalar coupling interactions. The first ^{119}Sn CP/MAS spectra were reported in 1978 [124] and the following years have seen an ever increasing number of reported applications of ^{119}Sn MAS and CP/MAS spectroscopy. In the case of ^{113}Cd, well-defined, i.e. purpose-defined areas for the application of high-resolution solid-state NMR methodology have been established, while in the case of ^{119}Sn a much wider range of chemistry has been covered, with no (not yet) firmly defined prefered chemistry-derived application profile. This situation is necessarily reflected in the following overview which can only attempt to bundle up small packages of related applications.

One prevalent focus of attention in the organotin chemistry of compounds of the type

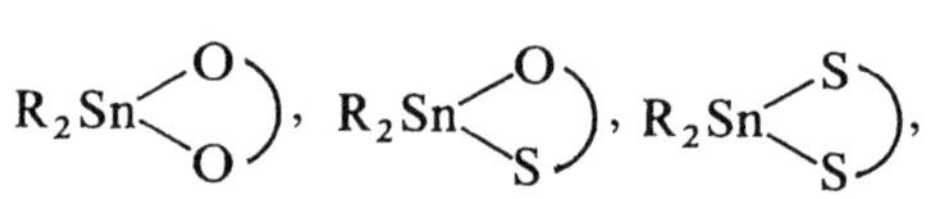

most of which are also of technical importance, is the question of intermolecular association in the solid state, ranging from monomolecular to 5- and 6-fold coordinated tin in dimeric, oligomeric and polymeric structures. Often solution-state ^{119}Sn NMR is of no help since these solid-state structures are not retained in solution. Several papers have been devoted to this problem [125–129], combining ^{119}Sn CP/MAS results with X-ray diffraction data and/or solution-state NMR results. In the case of polymeric compounds $[R_2SnO]_n$ [125] it seems justified to correlate δ_{iso} ^{119}Sn and large, highly asymmetric ^{119}Sn shielding tensors with a distorted *cis*-R_2SnO_3 fivefold coordination. For the above mentioned type of compounds the situation is far more complicated [126–129]. It is straightforward to distinguish tetrahedral tin from higher-coordinated tin species, both by δ_{iso} ^{119}Sn and by narrow ^{119}Sn shielding patterns for tetrahedral tin. Distinction of penta- and hexacoordinated species is severely hampered by overlapping ^{119}Sn shift ranges and by similar shielding anisotropy patterns for both kinds of distorted coordination. For example, ^{119}Sn CP/MAS data for tin atoms in nearly regular octahedral coordination like $[Me_2Sn(DMSO)_4]\cdot[N(SO_2Me)_2]_2$ [130] are in fact very similar to data for some penta-coordinated R_2SnO_3-species. Nevertheless, some consistent trends for ^{119}Sn shielding patterns for this type of compound, supported by X-ray diffraction data, do emerge: there is a steady increase of the width of the ^{119}Sn shielding anisotropy patterns on going from sulfur-only to oxygen-only coordination (ranging from approximatly 200 ppm to approximately 1000 ppm), but both isotropic ^{119}Sn chemical shifts and asymmetry parameters are by far less straightforward to rationalise in terms of coordination patterns. If ^{119}Sn shielding data were to serve as reliable tools for such assignments of coordination patterns, then it seems necessary to first establish a better knowledge of relative shielding tensor orientations via single crystal studies, similar to what has been attempted in the case of ^{113}Cd.

Organotin compounds of the type R_3SnX, where X is a mono- or bidentate oxygen-bearing ligand, also tend to polymerise in the solid state, leading to mostly polymeric zig-zag chains with tin in trigonal-bipyramidal *trans*-R_3SnO_2 geometry. Again, in solution, this polymeric structure is often destroyed, and chemical shift differences $\Delta\delta_{iso}$ ^{119}Sn of 150–200 ppm between the solid and the solution state are found. Such polymeric chains with tin in *trans*-R_3SnO_2 coordination are known for a variety of ligands X, like, e.g., X = OH, carboxylate, carbonate and the corresponding ^{119}Sn CP/MAS spectra invariably show large shielding anisotropies with axial or close to axial symmetry, whereby the most deshielded unique tensor component will usually coincide, or nearly coincide with the O-Sn-O molecular axis [131–135]. Very similar shielding patterns are also found for ^{119}Sn in trigonal-bipyramidal *trans*-R_3SnN_2 coordination, both for monomolecular species like $[Me_3Sn(NH_3)_2][N(SO_2Me)_2]$ [130] and for polymeric compounds where CN^- anions are the chain-building moiety [136, 137].

Depending on the steric requirements of the organic ligand R, triorganyl tin fluorides R_3SnF are either monomeric (R = mesityl) or polymeric (R = Me, Ph, ^iBu) in the solid state, the polymeric compounds forming chains with approximately linear F-Sn-F bonds. ^{119}Sn CP/MAS spectra of both types of compounds R_3SnF have been reported [138–140]. Due to the interaction of ^{119}Sn with either one (R = mesityl) or with two equivalent ^{19}F neighbours (R = Me, Ph, ^iBu) the ^{119}Sn CP/MAS spectra give rise to splitting patterns of an AX or AX_2 spin system, respectively. Under the condition that σ and J are axially symmetric, with their unique tensor components along the Sn-F bond and thus coaxial with the dipole coupling tensor, both spin systems AX and AX_2 have been analysed [139], yielding an estimate for the anisotropy in $J(^{119}Sn^{19}F)$ of $\Delta J \approx 3$ kHz or less; the accuracy of this analysis depends on the availability of sufficiently accurately determined bond distances r_{Sn-F} from X-ray diffraction.

Whether or not solid organotin compounds R_3SnX, R_2SnX_2 with X = Cl, Br, J are strongly intermolecularly associated has long been a matter of controversy, based on diverging interpretations of X-ray crystal structure determinations or ^{119}Sn Mössbauer data. ^{119}Sn CP/MAS spectra of such compounds strongly support the view that (other than for the corresponding fluorine derivatives) di- and triorganyl tin derivatives of the higher, less electronegative halogenes Cl, Br, J do not form strong intermolecular Sn-X bonds in the solid state [141–143]: firstly, only moderate to very small ^{119}Sn chemical shift differences between the solid state and solutions in non-coordinating solvents are observed and secondly, and more importantly, for several such chlorine derivatives R_3SnCl and R_2SnCl_2, asymmetric splittings of the ^{119}Sn resonances have been found. These splitting patterns can only be explained by a combination of residual dipolar and scalar coupling ^{119}Sn-$^{35/37}Cl$ with one (R_3SnCl) or two (R_2SnCl_2) chlorine atoms, respectively. With some simplifying assumptions, like neglecting ^{37}Cl, perturbation theory allows an estimation of the isotropic scalar coupling $^1J(^{119}Sn^{35}Cl)$ to be approximately 220 Hz for Ph_3SnCl. This analysis has been extended to the case of ^{119}Sn interacting with two (assumed equivalent) ^{35}Cl nuclei [143] and yields an isotropic coupling constant $^1J(^{119}Sn^{35}Cl) = 276$ Hz and an estimated anisotropy ΔJ in the range 400 to 800 Hz for (acac)$SnCl_2$ (acac = acetylacetonato). Qualitative considerations have been used to estimate the isotropic scalar coupling constants $^1J(^{119}Sn^{14}N)$ and $^1J(^{119}Sn^{35}Cl)$ for some compounds R_2ClSnX, with X = oxinate, thiooxinate where intramolecular N-Sn interaction leads to a strongly distorted trigonal-bipyramidal tin coordination $R_2ClSnNE$ (E = O or S) [144].

Scalar coupling of ^{119}Sn with ^{11}B has also been observed [145–147] in the ^{119}Sn CP/MAS spectra of several boron-containing organotin compounds, both coupling over one bond [145] and long-range coupling in oxoniastannaboratoles [146]. In both cases the observed splitting is, within experimental error, consistent with the high-field approximation and $J(^{119}Sn^{11}B)$ can be read directly from the ^{119}Sn CP/MAS spectra, thus representing the somewhat unusual case where extraction of scalar coupling information from solid-state NMR spectra is more straightforward than from solution-state NMR.

Both Sn(II) and Sn(IV) in less common bonding situations have been investigated by ^{119}Sn CP/MAS. Sn-Cl-groups built into macrocyclic molecules provide Lewis-acidic bonding sites in these host molecules which then are able to encrypt, e.g., Cl^- anions. The ^{119}Sn CP/MAS spectrum of such a host-guest complex with two Sn-Cl sites in the macrocycle shows that the encrypted Cl^- is strongly bound to only one of the two Sn-Cl sites [148] while in solution a chloride jump between the two tin sites occurs. The ^{119}Sn CP/MAS spectrum of tin (II) bisfluorosulfate, $Sn(OSO_2F)_2$ shows a substantially different isotropic ^{119}Sn chemical shift from what is observed in solution and thus corroborates that $Sn(OSO_2F)_2$ is partially ionised in solution [149]. Crystallographically inequivalent tin sites in bis-η^5-cyclopentadienyl-tin(II) derivatives are easily identified by ^{119}Sn CP/MAS [150, 151] and there seems to be a correlation between the asymmetry of the ^{119}Sn shielding tensor and the angle between the two sandwich-forming cyclopentadienyl-ligands in such stannocenes, ranging from parallel arrangements to large deviations from paralle. However, the role of the Sn(II) electron lone pair in these stannocenes is not fully understood.

X-ray crystallography shows that triorganyltin carbonates $(R_3Sn)_2CO_3$ have polymeric zig-zag chain structures with both trigonal-bipyramidal R_3SnO_2- and tetrahedral R_3SnO-sites, ^{119}Sn CP/MAS confirms this picture. In the case of $(Me_3Sn)_2CO_3$, variable temperature $^{1}H/^{119}Sn$ CP spin dynamics and ^{13}C CP/MAS revealed the presence of a motional process and a $2\pi/3$ propeller jump of the Me_3Sn groups around the Sn-O axes has been proposed [135]. Such a $2\pi/3$ propeller jump mechanism has been confirmed for several other solid trimethyltin compounds by 1D and 2D variable temperature ^{13}C CP/MAS, and ^{119}Sn rotor-synchronised 2D exchange spectroscopy was used to support the proposed motional mechanism [152]. Selective ^{119}Sn excitation (SELDOM), combined with TOSS and followed by a mixing time served as a 1D analogue. Selective ^{119}Sn excitation in itself can be useful to obtain suitable ^{119}Sn CP/MAS data for shielding tensor analysis in cases where the presence of more than one tin site with large differences in their respective shielding anisotropies would normally cause problems. This point is illustrated in Fig. 9 where the normal ^{119}Sn CP/MAS spectrum of $(Me_3Sn)_2CO_3$ is compared to selective excitation of the tetrahedral tin-site ^{119}Sn resonance at lower spinning frequency.

The ^{119}Sn MAS spectra of a series of rare-earth stannates $Ln_2Sn_2O_7$ (Ln = La, Pr, Nd, Sm, Eu, Tm, Yb, Ln, Y) and of solid solutions $Y_{2-y}Ln_ySn_2O_7$ have been reported [153, 154]. Most of these rare-earth stannates are paramagnetic compounds, and the accordingly large (paramagnetic) next-nearest-neighbour-induced ^{119}Sn chemical shifts as a function of composition, plus relative intensities, of the different ^{119}Sn resonances allowed the determination of the composition limits of some solid solutions, like, e.g., $Y_{2-y}Sm_ySn_2O_7$. For the paramagnetic stannates $Ln_2Sn_2O_7$ (Ln = Yb, Nd), the ^{119}Sn MAS spinning sideband manifold is overwhelmed by susceptibility broadening and thus precludes the determination of ^{119}Sn chemical shielding tensor parameters, Susceptibility matching by using erbium nitrate solutions has been shown to

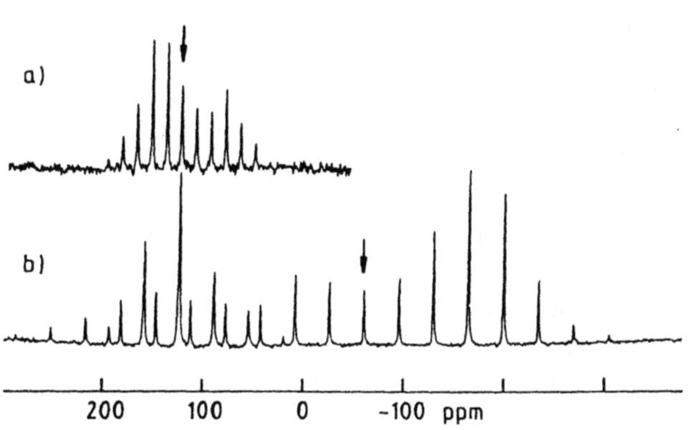

Fig. 9a, b. 111.9 MHz ^{119}Sn CP/MAS of $(Me_3Sn)_2CO_3$ [152], showing two ^{119}Sn resonances for tin in tetrahedral (higher frequency) and in trigonal-bipyramidal (lower frequency) coordination. **a** SELDOM selective excitation for the higher frequency resonance, $\nu_{rot} = 1.6$ kHz, 820 transients; **b** CP/MAS, 2940 transients, $\nu_{rot} = 3.9$ kHz

achieve a substantial reduction in broadening and can help to recover shielding information [155].

Diamagnetic ternary tin oxides $MSnO_3$ and M_2SnO_4 (M = Li, Na, K, Ca, Mg, Sr, Ba, Zn, Cd) have been investigated by ^{119}Sn MAS [156]. In all these phases tin resides in more or less distorted octahedral SnO_6-environments of different structure-types, both with edge- and with vertex-sharing SnO_6-octahedra. Small geometrical distortions lead to substantial ^{119}Sn shielding anisotropies, in some cases scalar coupling $^2J(^{119}Sn^{117}Sn)$ has been observed, and isotropic ^{119}Sn chemical shifts follow trends like an increase in shielding with increasing size of the cation M in M_2SnO_4 and $MSnO_3$. The ^{119}Sn MAS spectra of SnO_2 and SnO have been reported [157]. As expected, the ^{119}Sn shielding anisotropy for SnO is large and scalar coupling $J(^{119}Sn^{117}Sn)$ of approximately 8300 Hz has been reported. ^{119}Sn MAS spectra of $CaSnO_3$ and SnO_2 are shown in Fig. 10.

Where observable, homonuclear scalar coupling ^{119}Sn-^{119}Sn (and heteronuclear scalar coupling ^{119}Sn-^{117}Sn) is a useful source of information in the field of inorganic tin chemistry. In the field of organotin chemistry this parameter becomes even more important, both with respect to structure assignment purposes and with respect to considerations of bonding properties. The ^{119}Sn CP/MAS spectrum of $R_2Sn = SnR_2$ ($R = (Me_3Si)_2CH$) [158] reveals a comparatively small one-bond scalar coupling $^1J(^{119}Sn^{117}Sn) = 1340$ Hz; on the basis of this information it was concluded that this formal Sn-Sn double bond should be considered as a very weak dative interaction between the two "monomers" R_2Sn. Satellites due to $^{119/117}Sn$ scalar interactions have been reported for $C(SnMe_3)_4$, Me_3Sn-C_6H_4-$SnMe_3$ and for a tetrastannacyclohexane [159]. Two-bond scalar coupling interactions $^2J(^{119}Sn^{119/117}Sn)$ have attracted considerable attention. Data have been obtained from the ^{119}Sn CP/MAS spectra of several compounds $(R_3Sn)_2E$ and $(R_2SnE)_3$ with E = O, S, Se [160]. Advantages of low external magnetic field strengths for the recovery of scalar coupling information from ^{119}Sn CP/MAS spectra have been discussed for the dimethyltin chalcogenides $(Me_2SnE)_3$, E = S, Se, Te [161]. $^2J(^{119}Sn^{119}Sn)$ data have been reported for a large number of organotin chalcogenides, a correlation of Sn-E-Sn bond angles with the magnitude of J has been claimed [162].

In solution-state ^{119}Sn NMR it is a straightforward and widely practised task to extract isotropic one- and two-bond coupling constants $^{1,2}J(^{119}Sn^{119/117}Sn)$. This task becomes rather less straightforward for the solid state, given the combined effects of homonuclear dipole interactions, isotropic scalar coupling and anisotropy in J under MAS conditions—regardless of whether we consider interactions between isochronous ^{119}Sn nuclei or between chemically inequivalent ^{119}Sn nuclei (plus heteronuclear interactions ^{119}Sn-^{117}Sn of similar magnitudes). Strategies to overcome some of these problems have been outlined. A rotation-synchronised CPMG pulse sequence can help to measure homonuclear J-coupling [160]; the use of double-quantum filtering has been proposed for the same purpose and a preliminary report of the ^{119}Sn INADEQUATE experiment on $[Me_2SnS]_3$ has been given [163].

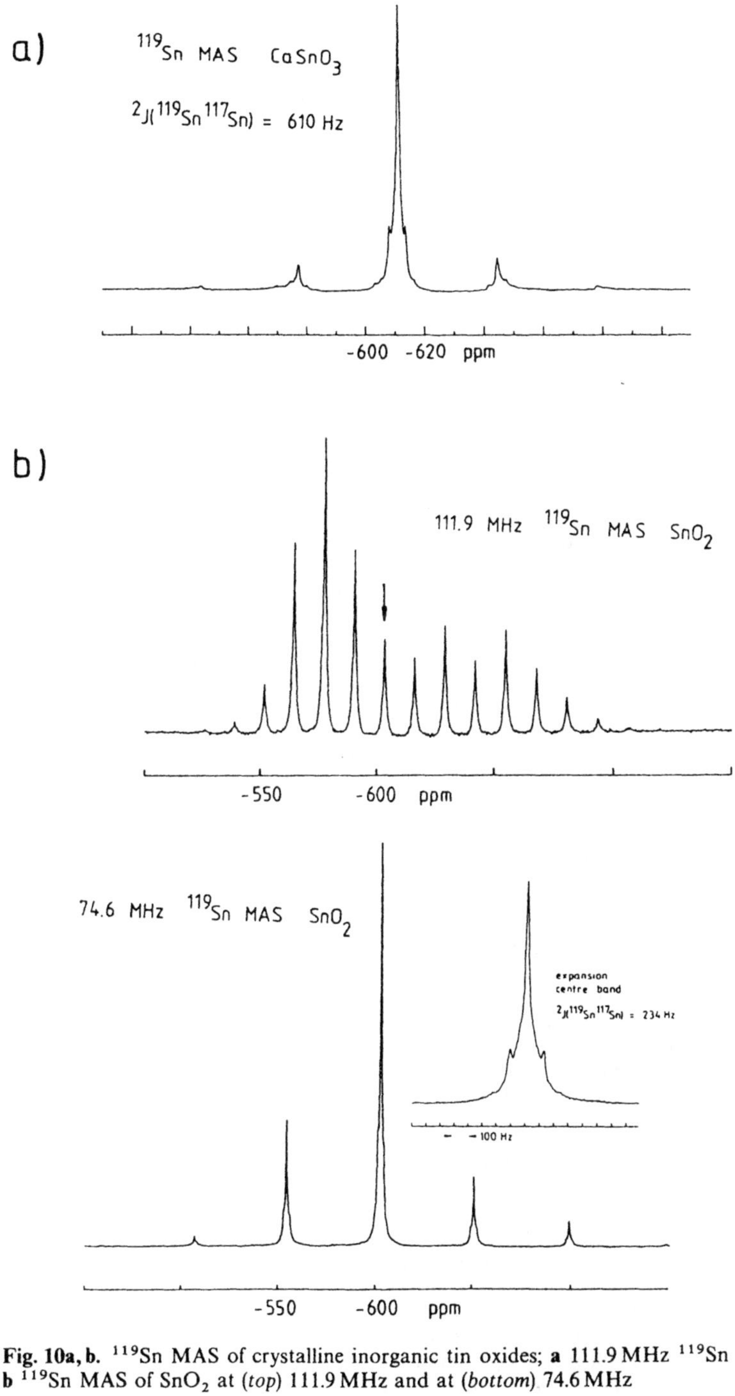

Fig. 10a, b. ^{119}Sn MAS of crystalline inorganic tin oxides; **a** 111.9 MHz ^{119}Sn MAS of $CaSnO_3$; **b** ^{119}Sn MAS of SnO_2 at (*top*) 111.9 MHz and at (*bottom*) 74.6 MHz

Acknowledgments. Our work has been generously supported by the Deutsche Forschungsgemeinschaft, the Fonds der Chemischen Industrie and the Alexander von Humboldt Foundation. It is my pleasure to express my gratitude to those who work(ed) with me: I thank F. Ambrosius, R. Challoner, J. Kümmerlen, L. Merwin, T. Schaller and E. Sendermann for their collaboration. I am most grateful to F. Seifert, Bayerisches Geoinstitut, for his continued support and encouragement.

4 References

1. Engelhardt G, Michel D (1987) In: High-resolution solid-state NMR of silicates and zeolites, John Wiley, New York
2. Collins MJ, Ratcliffe CJ, Ripmeester JA (1986) J. Magn. Reson. 68: 172
3. Merwin LH, Sebald A (1990) J. Magn. Reson. 88: 167
4. Merwin LH, Sebald A (1992) J. Magn. Reson. 97: 628
5. Charles NG, Griffith EAH, Rodesiler PF, Amma EL (1983) Inorg. Chem. 22: 2717
6. Harris RK, Sebald A (1987) Magn. Reson. Chem. 25: 1058
7. Collins MJ, Ripmeester JA, Sawyer JF (1987) J. Am. Chem. Soc. 109: 4113
8. Lee F, Gabe E, Tse JS, Ripmeester JA (1988) J. Am. Chem. Soc. 110: 6014
9. Merwin LH, Sebald A (1992) Solid State NMR 1: 45
10. Sebald A, Merwin LH, Schaller T, Knöller W (1992) J. Magn. Reson. 96: 159
11. Challoner R, Schaller T, Sebald A (1993) J. Magn. Reson. 101: 106
12. Kléin Douwel CH, Maas WEJR, Veeman WS, Werumeus Buning GH, Vankan JMJ (1990) Macromolecules 23: 406
13. Mehring M (1983) In: Principles of high resolution NMR in solids, Springer, Berlin Heidelberg New York
14. Bodenhausen G, Freeman R, Morris GA (1976) J. Magn. Reson. 23: 171
15. Tekely P, Brondeau J, Elbayed K, Retournard A, Canet D (1988) J. Magn. Reson. 80: 509
16. Franke D, Hudalla C, Maxwell R, Eckert H (1992) J. Phys. Chem. 96: 7506
17. Fyfe CA, Grondey H, Mueller KT, Wong-Moon KC, Markus T (1992) J. Am. Chem. Soc. 114: 5876
18. Vega AJ (1992) J. Magn. Reson 96: 50
19. Vega AJ (1992) Solid State NMR 1: 17
20. Takegoshi K, McDowell CA (1986) J. Magn. Reson. 67: 356
21. Bax A, Howkins L, Maciel GE (1984) J. Magn. Reson. 59: 530
22. Sebald A, Seifert F; unpublished results
23. Thompson AR, Oldfield E (1987) J. Chem. Soc. Chem. Commun. 27
24. Battle PD, Montez B, Oldfield E (1988) J. Chem. Soc. Chem. Commun. 584
25. Dupree R, Smith ME (1988) Chem. Phys. Lett. 148: 41
26. Kremer RK, Mattausch HJ, Simon A, Steuernagel S, Smith ME (1992) J. Solid State Chem. 96: 237
27. Dupree R, Gencten, Paul DMK (1992) Physica C 193: 81
28. Grey CP, Smith ME, Cheetham AK, Dobson CM, Dupree R (1990) J. Am. Chem. Soc. 112: 4670
29. Wu J, Boyle TJ, Shreeve JL, Ziller JW, Evans WJ (1993) Inorg. Chem., 32: 1130
30. Granger P (1991) In: Pregosin PS (ed) Transition Metal Nuclear Magnetic Resonance. Elsevier, Amsterdam, p 273 (Studies in inorganic chemistry, Vol 13)
31. Brun E, Oeser J, Staub HH, Telschow CG (1954) Phys. Rev. 93: 172
32. Sogo PB, Jefries CD (1954) Phys. Rev. 93: 174
33. Looser H, Brinkmann D (1985) J. Magn. Reson. 64: 76
34. Becker KD, von Goldammer E (1990) Chem. Phys. 48: 193
35. Jansen M (1987) Angrew. Chem. 99: 1136

36. Villa M, Chiodelli G, Magistris A, Licheri G (1986) J. Chem. Phys. 85: 2392
37. Plischke JK, Benesi AJ, Vannice MA (1992) J. Phys. Chem. 96: 3799
38. Brevard C, Thouvenot R (1991) In: Pregosin PS (ed) Transition Metal Nuclear Magnetic Resonance, Elsevier, Amsterdam, p 82 (Studies in inorganic chemistry, Vol 13)
39. Rehder D (1987) In: Mason J (ed) Multinuclear NMR, Plenum, New York, p 479
40. Knight CTG, Turner GL, Kirkpatrick RJ, Oldfield E (1986) J. Am. Chem. Soc. 108: 7426
41. Dechter JJ, Kowalewski J (1984) J. Magn. Reson. 59: 146
42. Schramm W, Schöllhorn R, Eckert H, Müller-Warmuth W (1983) Mater. Res. Bull. 18: 1283
43. Eckert H, Müller-Warmuth W, Schramm W, Schöllhorn R (1984) Solid State Ionics 13: 1
44. Eckert H, Müller-Warmuth W, Hamann W, Krebs B (1984) J. Noncryst. Solids 65: 53
45. Eckert H, Müller-Warmuth W (1985) J. Noncryst. Solids 70: 199
46. Hinton JF, Metz KR, Millett FS (1981) J. Magn. Reson. 44: 217
47. Hinton JF, Metz KR, Turner GL, Bennett DL, Millett FS (1985) J. Magn. Reson. 64: 120
48. Metz KR, Hinton JF (1982) J. Magn. Reson. 50: 212
49. Metz KR, Hinton JF (1982) J. Am. Chem. Soc. 104: 6206
50. Alla M, Freude D, Lippmaa E, Prager D (1984) Z. Phys. Chem. 265: 825
51. Goodfellow RJ (1987) In: Mason J (ed) Multinuclear NMR, Plenum Press, New York, pp 521–562
52. Pregosin PS (1991) In: Pregosin PS (ed) Transition Metal Nuclear Magnetic Resonance, Vol. 13 of: Studies in Inorganic Chemistry, Elsevier, Amsterdam, p 216
53. Sparks SW, Ellis PD (1986) J. Am. Chem. Soc. 108: 3215
54. Hayashi S, Hayamizu K (1992) Magn. Reson. Chem. 30: 658
55. Harris RK, Reams P, Packer KJ (1986) J. Mol. Struct. 141: 13
56. Harris RK, Reams P, Packer KJ (1986) J. Chem. Soc. Dalton Trans. 1015
57. Doddrell DM, Barron PF, Clegg DE, Bowie C (1982) J. Chem. Soc. Chem. Commun. 575
58. Harris RK, McNaught IJ, Reams P, Packer KJ (1991) Magn. Reson. Chem. 29: S60
59. Duer MJ, Khan MS, Kakkar AK (1992) Solid State NMR 1: 13
60. Davies JA, Dutremez S (1992) Coord. Chem. Rev. 114: 61 and references given therein
61. Lindner E, Fawzi R, Mayer HA, Eichele K, Hiller W (1992) Organometallics 11: 1033
62. Power WP, Wasylishen RE (1992) Inorg. Chem. 31: 2176
63. Kubo A, McDowell CA (1990) J. Chem. Phys. 92: 7156
64. Santos RA, Gruff ES, Koch SA, Harbison GS (1991) J. Am. Chem. Soc. 113: 469
65. Natan MJ, Millikan CF, Wright JG, O'Halloran TV (1990) J. Am. Chem. Soc. 112: 3255
66. Harris RK (1986): In 'Nuclear magnetic resonance spectroscopy—A physicochemical view'. Longman, Harlow, p 155
67. Burrows HD, Geraldes CFGC, Pinheiro TJT, Harris RK, Sebald A (1988) Liqu. Cryst. 3: 853
68. Wrackmeyer B, Horchler K, Sebald A, Merwin LH (1990) Magn. Reson. Chem. 28: 465
69. Wrackmeyer B, Sebald A, Merwin LH (1991) Magn. Reson. Chem. 29: 260
70. Harris RK, Sebald A (1989) Magn. Reson. Chem. 27: 81
71. Challoner R, Sebald A, Müller D, Steuernagel S (1993), to be published
72. Sebald A, Harris RK (1990) Organometallics 9: 2096
73. Ascenso JR, Harris RK, Granger P (1986) J. Organomet. Chem. 301: C 23
74. Goodfellow RJ (1987) In: Mason J (ed) Multinuclear NMR, Plenum, New York, p 563
75. Summers MF (1988) Coord. Chem. Rev. 86: 43
76. Ellis PD (1983) Science 221: 1141
77. Herron N, Wang Y, Eckert H (1990) J. Am. Chem. Soc. 112: 1322
78. Franke DR, Eckert H (1991) J. Phys. Chem. 95: 331
79. Franke D, Maxwell R, Lathrop D, Eckert H (1991) J. Am. Chem. Soc. 113: 4822
80. Franke D, Banks K, Maxwell R, Eckert H (1992) J. Phys. Chem. 96: 1906
81. Franke D, Banks K, Eckert H (1992) J. Phys. Chem. 96: 11048
82. Franke D, Hudalla C, Eckert H (1992) Solid State NMR 1: 33
83. Cheung TTP, Worthington LE, DuBois Murphy P, Gerstein BC (1980) J. Magn. Reson. 41: 158
84. Sardashti M, Maciel GE (1987) J. Magn. Reson. 72: 467
85. DuBois Murphy P, Stevens WC, Cheung TTP, Lacelle S, Gerstein BC, Kurtz DM (1981) J. Am. Chem. Soc. 103: 4400
86. DuBois Murphy P, Gerstein BC (1981) J. Am. Chem. Soc. 103: 3282
87. Mennitt PG, Shatlock MP, Bartuska VJ, Maciel GE (1981) J. Phys. Chem. 85: 2087
88. Ackerman JJH, Orr TV, Bartuska VJ, Maciel GE (1979) J. Am. Chem. Soc. 101: 341
89. Honkonen RS, Doty FD, Ellis PD (1983) J. Am. Chem. Soc. 105: 4163

90. Honkonen RS, Ellis PD (1984) J. Am. Chem. Soc. 106: 5488
91. Marchetti PS, Honkonen RS, Ellis PD (1987) J. Magn. Reson. 71: 294
92. Kennedy MA, Ellis PD, Jakobsen HJ (1990) Inorg. Chem. 29: 550.
93. Marchetti PS, Bank S, Bell TW, Kennedy MA, Ellis PD (1989) J. Am. Chem. Soc. 111: 2063
94. Bank S, Bank JF, Ellis PD (1989) J. Phys. Chem. 93: 4847
95. Eichele K, Wasyishen RE (1992) Angew. Chem. 104: 1263
96. Nishikiori S, Ratcliffe CI, Ripmeester JA (1990) Can. J. Chem. 68: 2270
97. Nishikiori S, Ratcliffe CI, Ripmeester JA (1991) J. Chem. Soc. Chem. Commun. 735
98. Nishikiori S, Ratcliffe CI, Ripmeester JA (1992) J. Am. Chem. Soc. 114: 8590
99. Kessler JM, Reeder JH, Vac R, Yeung C, Nelson JH, Frye JS, Alcock NW (1991) Magn. Reson. Chem. 29: S94
100. Lee GSH, Fisher KJ, Vassallo AM, Hanna JV, Dance IG (1993) Inorg. Chem. 32: 66
101. Nolle A (1979) Z. Physik B Condens. Matter 34: 175
102. Griffith EAH, Li HY, Amma EL (1988) Inorg. Chim. Acta 148: 203
103. Griffith EAH, Charles NG, Lewinski K, Amma EL, Rodesiler PF (1987) Inorg. Chem. 26: 3983
104. Kennedy MA, Sessler JL, Murai T, Ellis PD (1990) Inorg. Chem. 29: 1050
105. Kennedy MA, Ellis PD (1990) Inorg. Chem. 29: 541
106. Rivera E, Kennedy MA, Adams RD, Ellis PD (1990) J. Am. Chem. Soc. 112: 1400
107. Rivera E, Ellis PD (1992) Inorg. Chem. 31: 2096
108. Honkonen RS, Marchetti PS, Ellis PD (1986) J. Am. Chem. Soc. 108: 912
109. Marchetti PS, Bhattacharyya L, Ellis PD, Brewer CF (1988) J. Magn. Reson. 80: 417
110. Kennedy MA, Ellis PD (1989) J. Am. Chem. Soc. 111: 3195
111. Rivera E, Kennedy MA, Ellis PD (1989) In: Waugh JS (ed) Advances in Magnetic Resonance, Vol 13, Academic Press, New York p 257
112. Collins MJ, Ratcliffe CI, Ripmeester JA (1986) J. Magn. Reson. 68: 172
113. Pinto BM, Batchelor RJ, Johnston BD, Einstein FWB, Gay ID (1988) J. Am. Chem. Soc. 110: 2990
114. Batchelor RJ, Einstein FWB, Gay ID, Gu JH, Johnston BD, Pinto BM (1989) J. Am. Chem. Soc. 111: 6582
115. Batchelor RJ, Einstein FWB, Gay ID, Gu JH, Pinto BM (1991) J. Organomet. Chem. 411: 147
116. Maxwell R, Lathrop D, Franke D, Eckert H (1990) Angew. Chem. Int. Ed. Engl. 29: 882
117. Lathrop D, Eckert H (1989) J. Phys. Chem. 93: 7895
118. Lathrop D, Eckert H (1990) J. Am. Chem. Soc. 112: 9017
119. Koch W, Lutz O, Nolle A (1978) Z. Physik A 289: 17
120. Balz R, Haller M, Hertler WE, Lutz O, Nolle A, Schafitel R (1980) J. Magn. Reson. 40: 9
121. Zax DB, Vega S, Yellin N, Zamir D (1987) Chem. Phys. Lett. 138: 105
122. Collins MJ, Ripmeester JA (1987) J. Am. Chem. Soc. 109: 4113
123. Collins MJ, Ripmeester JA, Sawyer JF (1988) J. Am. Chem. Soc. 110: 8583
124. Lippmaa ET, Alla MA, Pekh TJ, Engelhardt G (1978) J. Am. Chem. Soc. 100: 1929
125. Harris RK, Sebald A (1987) J. Organomet. Chem. 331: C9
126. Davies AG, Slater SD, Povey DC, Smith GW (1988) J. Organomet. Chem. 352: 283
127. Bates PA, Hursthouse MB, Davies AG, Slater SG (1989) J. Organomet. Chem. 363: 45
128. Grindley TB, Curtis RD, Thangarasa R, Wasylishen RE (1990) Can. J. Chem. 68: 2102
129. Grindley TB, Wasylishen RE, Thangarasa R, Power WP, Curtis RD (1991) Can. J. Chem. 70: 205
130. Blaschette A, Hippel I, Krahl J, Wieland E, Jones PG, Sebald A (1992) J. Organomet. Chem. 437: 279
131. Harris RK, Packer KJ, Reams P (1985) J. Magn. Reson. 61: 564
132. Harris RK, Packer KJ, Reams P, Sebald A (1987) J. Magn. Reson. 72: 385
133. Komoroski RA, Parker RG, Mazany AM, Early TA (1987) J. Magn. Reson. 73: 389
134. Molloy KC (1988) Inorg. Chim. Acta 141: 151
135. Kümmerlen J, Sebald A, Reuter H (1992) J. Organomet. Chem. 427: 309
136. Apperley DC, Davies NA, Harris RK, Brimah AK, Eller S, Fischer RD (1990) Organometallics 9: 2672
137. Behrens U, Brimah AK, Soliman TM, Fischer RD, Apperley DC, Davies NA, Harris RK (1992) Organometallics 11: 1718
138. Bai H, Harris RK, Reuter H (1991) J. Organomet. Chem. 408: 167
139. Bai H, Harris RK (1992) J. Magn. Reson. 96: 24
140. Harris RK, Reams P, Packer KJ (1986) J. Mol. Struct. 141: 13
141. Harris RK, Sebald A, Furlani P, Tagliavini G (1988) Organometallics 7: 388

142. Harris RK (1988) J. Magn. Reson. 78: 389
143. Apperley DC, Bai H, Harris RK (1989) Mol. Phys. 68: 1277
144. Lyčka A, Holeček J, Sebald A, Tkáč I (1991) J. Organomet. Chem. 409: 331
145. Wrackmeyer B, Kupce E, Kümmerlen J (1992) Magn. Reson. Chem. 30: 403
146. Wrackmeyer B, Wagner K, Sebald A, Merwin LH, Boese R (1991) Magn. Reson. Chem. 29: S3
147. Wrackmeyer B, Kehr G, Sebald A, Kümmerlen J (1992) Chem. Ber. 125: 1597
148. Newcomb M, Horner JH, Blanda MT, Squattrito PJ (1989) J. Am. Chem. Soc. 111: 6294
149. Adams DC, Birchall T, Faggiani R, Gillespie RJ, Vekris JE (1991) Can. J. Chem. 69: 2122
150. Janiak C, Schumann H, Stader C, Wrackmeyer B, Zuckermann JJ (1988) Chem. Ber. 121: 1745
151. Wrackmeyer B, Kupce E, Kehr G, Sebald A (1992) Magn. Reson. Chem. 30: 964
152. Kümmerlen J, Sebald A (1993) J. Am. Chem. Soc. 115: 1134
153. Cheetham AK, Dobson CM, Grey CP, Jakeman RJB (1987) Nature 328: 706
154. Grey CP, Dobson CM, Cheetham AK, Jakeman RJB (1989) J. Am. Chem. Soc. 111: 505
155. Grey CP, Dobson CM, Cheetham AK (1992) J. Magn. Reson. 98: 414
156. Clayden NJ, Dobson CM, Fern A (1989) J. Chem. Soc. Dalton Trans. 843
157. Cossement C, Darville J, Gilles JM, Nagy JB, Fernandez C, Amoureux JP (1992) Magn. Reson. Chem. 30: 263
158. Zilm KW, Lawless GA, Merrill RM, Millar JM, Webb GG (1987) J. Am. Chem. Soc. 109: 7236
159. Harris RK, Mitchell TN, Nesbitt GJ (1985) Magn. Reson. Chem. 23: 1080
160. Harris RK, Sebald A (1989) Magn. Reson. Chem. 27: 81
161. Gay ID, Jones CHW, Sharma RD (1989) J. Magn. Reson. 84: 501
162. Lockhart TP, Puff H, Schuh W, Reuter H, Mitchell TN (1989) J. Organomet. Chem. 366: 61
163. Gay ID, Jones CHW, Sharma RD (1991) J. Magn. Reson. 91: 186

Satellite Transition Spectroscopy of Quadrupolar Nuclei

Christian Jäger

Max-Planck-Institut für Polymerforschung, Ackermannweg 10, D-55128 Mainz, FRG

Table of Contents

NMR Basic Principles and Progress, Vol. 31

During the last five years the availability of fast MAS probes lead to a major progress in high resolution NMR of quadrupolar nuclei. It is now possible to measure the quadrupole interaction of polycrystalline solids precisely using the satellite transitions. But furthermore, a considerable gain in spectral resolution especially for $I = 5/2$ and $I = 9/2$ nuclei is now available just by employing the ordinary MAS technique with single pulse excitation when observing the MAS lineshape of certain satellite transitions rather than the central transition MAS pattern. This gain in spectral resolution enables a tremendous progress for improved quantitative studies of disordered solids such as glasses and ceramics exhibiting distributions of the interaction parameters.

1 Introduction and Scope

Magic-angle sample spinning (MAS) has had a major impact on several aspects of the development of NMR and its application to solid state physics and chemistry. MAS was first used to average out dipolar broadening in ^{23}Na NMR of monocrystalline sodium chloride, as published in 1958 by Andrew, Bradbury and Eades [1]. After that many more examples were published in the early days including applications to other nuclei such as ^{1}H [2] and ^{31}P [3]. However, successful applications of MAS required either a substantial motional narrowing in the case of strong homonuclear dipole broadening, or a dominant inhomogeneous line broadening interaction such as the chemical shift anisotropy.

Later on, the combination of MAS with multiple-pulse sequences (CRAMPS) [4] allowed the study of compounds with large dipolar couplings. Using more sophisticated multiple-pulse sequences such as BR-24 [5] it was possible to resolve even the indirect spin-spin coupling of ammonium groups in solids as early as 1981 [6].

However, since most of the nuclei of the periodic table possess an electric quadrupole moment the NMR line shape is often governed by first- and second-order quadrupole effects [7–9]. Again, the use of MAS allowed a major progress by the end of the 1970s in NMR of quadrupolar nuclei. Almost simultaneously Behrens [10] and Samoson [11] were the first to calculate second-order quadrupole MAS spectra of the central transition (CT). In addition, Behrens showed the effect on the CT line shapes of a sample spinning at angles other than the magic angle, a result which was confirmed by other groups [e.g. 12]. It was proved soon afterwards, that MAS allows the detection of octahedrally and tetrahedrally coordinated Al species in the NMR spectra separated by their different chemical shifts in CT MAS spectra of polycrystalline solids and even in glasses using superconducting solenoids [13–15].

At that time the spinning speeds of the MAS probes were typically of the order of several kHz, causing severe problems for quantitative measurements (e.g. the determination of the relative amounts of differently coordinated species) because of the overlap of centre and spinning sidebands. However, the recent availability of commercial high-speed spinning probes is the prerequisite for a considerable progress in high resolution NMR of quadrupolar nuclei in disordered solids with explicit use of the satellite transitions (ST).

It is the aim of this paper to review this recent development and to show what can be achieved in materials science and inorganic chemistry in this way. For this reason Sect. 2 gives a brief review of some of the most important applications of NMR of quadrupolar nuclei for structural investigations. In addition, the main problems of NMR of quadrupolar nuclei are illustrated as well. This is followed by the description of both the theoretical and experimental details of the alternative method Satellite Transition Spectroscopy (SATRAS) in Sect. 3. Section 4 describes some recent applications of the method and is divided according to the main strategies for the applications of SATRAS. It will be shown that tremendous progress for structural investigations of disordered solids is possible based on the simple MAS experiment and single-pulse irradiation only, which cannot be achieved so easily with other methods such as Double Rotation (DOR) [16, 17] and Dynamic-Angle Spinning (DAS) [18]. An outlook for new applications is given in Sect. 5.

In this review only half-integer spins are considered with the special attention to ^{27}Al, ^{11}B, ^{7}Li, ^{23}Na and ^{51}V NMR. The most important nucleus with integer spin number (^{2}D) plays an extraordinary role especially in organic and polymer chemistry. It was considered in Volume 30 of this series and is not included here.

2 NMR of Quadrupolar Nuclei

When looking up the periodic table one immediately finds that the vast majority of NMR accessible nuclei posses a spin $I > 1/2$. Specifically, among 188 NMR active isotopes, only 29 have a spin $I = 1/2$. Hence, all other nuclei posses an electric quadrupole moment which causes their spectra to be broadened by quadrupole effects in the case of non-cubic symmetry. In the following, pure quadrupole resonance NQR will not be considered. Table 1 summarizes some of the quadrupole nuclei and their spin numbers [19]:

Consequently, any real application of NMR in inorganic chemistry, solid-state physics or material science in general has to deal with these quadrupole effects. But is was the development of superconducting solenoids in conjunction with MAS and DOR/DAS which made the application of NMR to these nuclei not only possible but also very important. As a result, the number of papers reporting on NMR of quadrupolar nuclei has rapidly increased over the last few years.

Table 1. Selection of some quadrupole nuclei listed according to their spin number [19]. The second column provides the number of nuclei with spin I whereas the third column gives some examples

Spin	Number of nuclei	Nucleus
1	3	^{2}D, ^{6}Li, ^{14}N
3/2	32	^{7}Li, ^{9}Be, ^{11}B, ^{21}Ne, ^{23}Na, ^{63}Cu, ^{69,71}Ga, ^{87}Rb, ...
2	—	—
5/2	22	^{17}O, ^{25}Mg, ^{27}Al, ^{47}Ti, ^{67}Zn, ^{91}Zr, ...
3	1	^{10}B
7/2	17	^{43}Ca, ^{51}V, ^{123}Sb, ^{133}Cs, ...
4	—	
9/2	7	^{73}Ge, ^{83}Kr, ^{87}Sr, ^{93}Nb, ^{115}In, ^{209}Bi, ...
5	1	(^{138}La)
6	1	(^{50}V)
7	1	(^{176}Lu)

It is far beyond the scope of this paper to give a complete review of all the applications of NMR. Hence, the aim is merely to provide the necessary background such that the reader understands on one hand what can be done by NMR, and on the other hand, the problems which arise and sometimes severely prevent one from getting quantitative results. This forms the basis of a discussion of SATRAS and the promise of this technique. The main attention is directed towards disordered solids in the following but not exclusively since SATRAS can be applied advantageously to polycrystalline solids as well.

2.1 Applications

During the last few years several reviews have been published reporting on applications of NMR of quadrupolar nuclei [20–24]. Each is a comprehensive paper on a specific topic, e.g., on zeolites [20, 21] or on noncrystalline solids [22]. Others discuss the physics and the experiments in more detail [25]. And besides this there is a vast variety of papers reporting on applications in inorganic chemistry, mineralogy, solid state physics....

Here, progress in materials science during the last decade is illustrated by considering the various nuclei separately, focussing in particular on the most commonly used nuclei listed in Table 1.

$I = 3/2$ nuclei. ^{11}B and ^{23}Na are certainly the most widely investigated nuclei as far as spectroscopy is concerned.

The most important applications of ^{11}B NMR are structural investigations of glasses, since tetrahedral BO_4 and nearly planar BO_3 are easily distinguishable because of their different quadrupole coupling constants. For BO_3 units it is about 2.6....2.7 MHz, with the asymmetry depending on whether terminal oxygens are present or not. Hence typical second-order quadrupole patterns result in contrast to the BO_4 unit. Static experiments are advantageous in order to quantify the relative amount of BO_3 and BO_4 units. For more details and applications see the comprehensive review [22] and the references therein. But also crystalline borates [e.g., 26] and extrinsic semiconductors Si[B] [27] have been investigated.

^{23}Na is also easy to investigate. The line shapes exhibit second-order quadrupole couplings in most cases and almost very little differences between the chemical shifts. In glasses the small shift of the peak positions of the very broad, structureless and slightly asymmetric lines can be only tentatively attributed to certain structural units (e.g., [22, 28]). However, in crystalline solids with known structure, the second-order powder pattern can be fitted in order to obtain the quadrupole parameters (e.g., [28–30]). Since ordinary CT MAS is not very successul because of those reasons, other sophisticated methods have become well established. DOR has been applied first to ^{23}Na [16] and is an extremely valuable tool for studying the structure, e.g., of sodalites [31] and zeolites [20, 21, 32] since second-order quadrupolar broadening is completely removed. However, the quadrupolar parameters can be recovered by carrying out DOR at different B_0 fields. In addition, ^{23}Na Nutation NMR is also a very suitable method for investigations of zeolites [33–35] and of sodium ion exchange in NASICONs [36], to mention only few examples.

^{7}Li is another prominent $I = 3/2$ nucleus. Due to the very small quadrupole coupling constant ($C_Q \approx 75$ kHz [37]) only first-order quadrupole effects are observed. Since chemical shift effects can be neglected in most cases the CT is mainly broadened by the dipole interaction and the ST can be observed easily. For that reason, ^{7}Li is often used for studying dynamic properties (diffusion) by measuring the temperature dependence of either the dipolar line width of the CT or of the quadrupole coupling (e.g., in $LiNbO_3$ [38, 39]) or of the relaxation times T_1 and $T_{1\rho}$ [40, 41] which also holds true for glasses [21, 42, 43]. Structural results are of minor importance and can only be obtained with difficulty by exploiting the dipolar line shape (Li pair formation because of Pake splitting [44]) or if species with different quadrupole couplings are present (e.g., after crystallisation [45]).

Chemical shifts and quadrupole parameters of inorganic rubidium salts have been published recently using ^{87}Rb DAS and VAS NMR [46]. Finally, ^{69}Ga and ^{71}Ga NMR has been used to study group III-V semiconductors and the main line broadening (quadrupole and exchange interaction) has shown to be related to the crystal structure [68].

$I = 5/2$ nuclei. Among those nuclei ^{17}O and ^{27}Al are widely used for structural studies. ^{17}O NMR is particularly useful since the geometry of the bonding is directly reflected in the quadrupole interaction. 2D DAS spectroscopy revealed intermediate-range order in the distribution of the inter-tetrahedral (Si-O-Si) bond angles and a high degree of order in the disposition of oxygen atoms around the network-modifying cations in silicate glasses [47]. And there are more papers on applications of DAS and DOR to various polycrystalline materials and minerals [48–50] where inequivalent lattice sites are easily resolved. However, the standard NMR techniques with or without MAS have also been succesfully applied to glasses and can distinguish between bridging and non-bridging oxygens due to different quadrupole couplings [22, 28]. Furthermore, zeolites [51] and many other materials e.g. earth alkaline earth metasilicates [52] or aluminium oxides [53] have been investigated (for superconductors see, e.g., [54]). The necessary isotopic enrichment is the only drawback of ^{17}O NMR.

^{27}Al NMR is certainly the most important technique, for studying the microstructure of disordered materials. Only glasses and ceramics will be considered here. A summary of data for various polycrystalline solids including the references has been published recently [25] and applications to zeolites have been reviewed in [22]. At the time of writing this review a special issue of Applied Magnetic Resonance was published with applications of ^{27}Al NMR for characterisations of amorphous and polycrystalline solids including catalysts, minerals, alumina hydrates and thermal transformations of kaolinite [55]. As pointed out in the introduction, the major success of ^{27}Al MAS NMR at high fields is based on the opportunity of resolving the signals of differently coordinated Al species in disordered materials. This has been applied to many differently composed glasses (see review [21]). There are also studies of the microstructure of ceramics such as AlN [56] and silicon aluminium oxynitride ceramics (sialon) [57]. In the latter, ^{27}Al allowed unequivocal identification of specific mixed aluminium local coordinations AlO_yN_{4-y} ($y = 1$, 2 or 3) but no quantification. Technologically important amorphous anodic alumina has been studied by NMR [58] suggesting an inhomogeneous film material with aluminium present both within microcrystalline and within more disordered film regions. Variations of the Al coordination appear to be related to the presence of OH groups. Finally as pointed out earlier for ^{23}Na and ^{87}Rb NMR, ^{27}Al DOR and DAS experiments play an important role for investigations of molecular sieves [59, 60] and, e.g., of sodalites [31].

$I = 7/2$ *nuclei.* For the most prominent among these nuclei, ^{51}V, chemical-shift anisotropy has to be taken into account in addition to the second-order quadrupole interaction. Several papers report on the determination of both tensors and their relative orientation using either single crystals of V_2O_5 [61] or powders [62]. Amorphous V_2O_5 has been investigated finding V^{5+} ions in both fourfold VO_4 and fivefold VO_5 coordination [63]. Other papers report on e.g., vanadium-contaminated aluminas and aluminosilicate gels [64], bismuth vanadates [65] and other vanadium-oxygen compounds [66, 67].

$I = 9/2$ *nuclei.* So far only a few papers have been published reporting on NMR of these nuclei. Han et al. [68] reported on ^{115}In NMR (abundance: 95.72%) in various III-V semiconductors with either a cubic blend structure (InP, InSb, InAs) and expected zero electric field gradient (EFG) or a hexagonal wurtzite structure (InN) with moderate EFGs. The chemical shifts have been reported but no severe second-order quadrupole broadening was found.

^{93}Nb is another interesting candidate because of 100% natural abundance and a large gyromagnetic ratio [19]. But again there are only few papers since the large quadrupole interaction severely broadens the CT pattern. Douglass et al. [69] investigated $LiNbO_3$, Li_3NbO_4, $LiNb_3O_8$ and some samples of the solid solution Li_2O-Nb_2O_5. The second-order broadening of the CT can be quite large reaching several hundred kHz. Only for Li_3NbO_4 a rather narrow CT was obtained. Using a stepped-frequency echo method they showed that the whole pattern reaches up to 3 MHz and more. Recently a single crystal study of $KNbO_3$ has been published [70]. The quadrupole parameters of the $^{93}NbO_6$-site were found to be $C_Q = 23.5$ MHz with an asymmetry $\eta = 0.8$. In another ^{93}Nb NMR study [71] of the $LiNbO_3$-$LiTaO_3$ solid solution system, the quadrupole coupling constant was determined using the cw technique. For $LiNbO_3$ a $C_Q = 22.03$ MHz was found [37]. 2D Nutation spectra of $LiNbO_3$ and NbO_2F have been presented for methodological reasons [72].

2.2 Problems in NMR of Quadrupolar Nuclei

Despite the distinct progress that has been made since the early 1980s by MAS and the application of high magnetic fields produced by superconducting solenoids there are still a number of problems. Especially if glasses and other disordered systems are investigated, quantitative results are very difficult to obtain. The central problem is the quadrupole interaction itself. The main reason is that MAS cannot average out second-order effects since they contain both second- and fourth-order polynomials. The consequences and difficulties can be summarized as follows and will be discussed for ^{27}Al NMR:

1. As pointed out first by Behrens [10] and Samoson [11], MAS narrows second-order quadrupolar broadening of the CT only by about a factor of three compared with the static line shape. Furthermore, the centre of gravity of the second-order quadrupole patterns are shifted from the correct isotropic chemical shift position. Depending on the strength of the quadrupole coupling and the magnetic field, the second-order MAS powder pattern width can reach from several hundred Hz up to some kHz which clearly can result in overlapping spinning sidebands.

2. If differently coordinated species are present in polycrystalline materials and slow speed MAS (3–5 kHz) is used, as was often the case, centre bands and spinning sidebands of the resonances overlap. For example, in ^{27}Al NMR the isotropic chemical shift range between 4- and 6-coordinated Al species is about 70 ppm which corresponds to 5–6 kHz at $B_0 \approx 7$ T. Hence, the described superpositions occur as long as the spinning frequency is less than about 10 kHz and significant second-order quadrupole broadening is present. Quantitative results can only be obtained by fitting the line shape including the effects of spinning at finite speed (see Sects 3.2 and 4).

3. The problems are even worse in disordered materials. Here, distributions of the interaction parameters, in particular of the quadrupole interaction, broaden the lines additionally. But more importantly, distinct features of second-order powder patterns such as poles and shoulders are washed out. What is observed is a broad asymmetric CT MAS line with a sharp rise at one side and broad wings at the other side of the spectrum (see, e.g., Figs. 1, 9 and 10). If now several resonances are present a correct deconvolution of the lines (determination of the relative amount of the different species) is only possible once the quadrupole interaction *and* its distribution are known for each of the lines (see Fig. 10, spectra at the top). This is the main reason why only in few favourite cases correct quantitative results were obtained despite the fact that many papers on ^{27}Al NMR of disordered materials have been published.

4. The observation of different AlO_X polyhedra with the same coordination number is another problem that can hardly be solved by CT MAS NMR especially in disordered solids. Mullites [73] represent one important example with two different AlO_4 groups. Concerning NMR, the spectral resolution because of the second-order quadrupolar broadening is simply not good enough to

resolve the small difference in chemical shifts [e.g., 74, 75] in order to quantify the relative amount of those two units.

^{27}Al MAS NMR spectra ($B_0 = 7.1$ T) of a mullite $3Al_2O_3$, $2SiO_2$ are shown in Fig. 1 as function of the spinning speed in order to illustrate these problems.

All the features described above are found. Even at the highest spinning speed of 11 kHz the AlO_4 signal shows only a kind of splitting, but any serious attempt to quantify the relative amounts will fail since it requires knowledge of the quadrupole parameters and their distributions for the two AlO_4 lines. It is remarkable that for the slow spinning speed of 3.5 kHz an 'oscillating' baseline arises. This is a superposition of the ST MAS sidebands (of both AlO_4 and AlO_6 signals) in conjunction with folding back of the same ST MAS lines

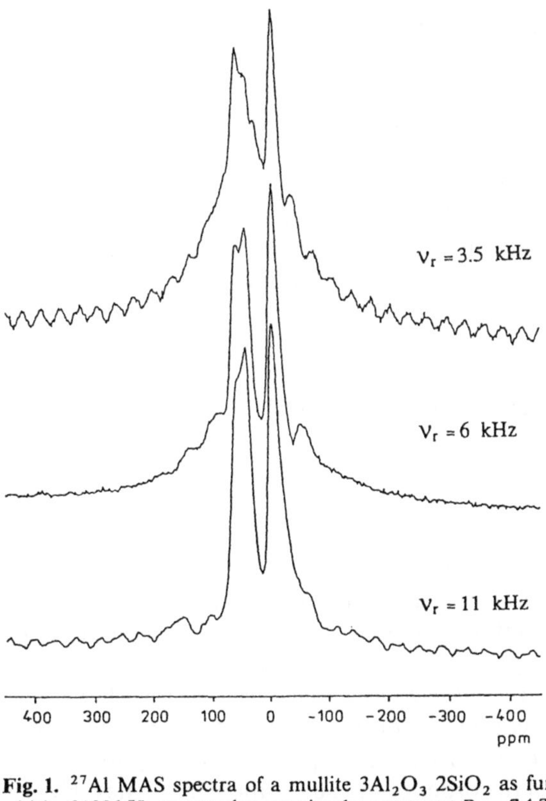

Fig. 1. ^{27}Al MAS spectra of a mullite $3Al_2O_3$ $2SiO_2$ as function of the spinning speed. A sweep width of 100 kHz was used to acquire the spectra at $B_0 = 7.1$ T. For slow spinning at 3.5 kHz a strong overlap of spinning sidebands and centre bands of the AlO_6 (*right*) and the two AlO_4 units (*left*) is observed for the CT. Furthermore, a coincidental superposition of ST spinning sidebands because of the small sweep width is also observed. But spinning at medium speed (middle spectrum) is the normal situation. However, for fast MAS (*bottom spectrum*) those inner ST MAS pattern are present again. This is now the typical behaviour because different orders of the ST spinning sidebands are separated by 11 kHz which is equivalent to ~ 140 ppm at 7.1 T. At the same time the separation of the AlO_6 and the two AlO_4CT signals is only ~ 70 ppm.

because of the small acquisition window which was 100 kHz. It should be noted that this is a coincidence since by changing the speed to ~6 kHz their superposition is removed; the normal situation. On the other hand, there are distinct ST MAS lines at the MAS speed of 11 kHz. At these spinning speeds, different orders of the sidebands for AlO_4 and AlO_6 are almost completely separated and hence these ST MAS lines are observed.

One way out of these severe limitations is certainly the application of higher and higher magnetic field strengths since the second order broadening is inversely proportional to the applied magnetic field. At present magnets are available producing fields of up to 14.1 T. But most laboratories are equipped with weaker magnets (7.1 T). This situation is the starting point for the consideration of Satellite Transition Spectroscopy SATRAS.

Certainly for polycrystalline solids DOR and DAS can be used because the second-order broadening is completely suppressed and distributions of chemical shifts are not present [17, 18]. But if disordered materials are to be investigated, the limiting factor for DOR seems to be the isotropic chemical shift distribution which may reach 1.5...2 kHz for ^{27}Al at $B_0 = 7$ T. With respect to DAS, the often very short ^{27}Al T_1 relaxation times may limit the application of that technique. On the other hand, fast MAS probes as the prerequisite for SATRAS are almost standard equipment today whereas this is not (yet) the case for DOR and DAS. Additionally, SATRAS is an easy experiment since it only employs single-pulse irradiation and MAS and is therefore routine.

3 Satellite Transition Spectroscopy (SATRAS)

3.1 Brief History

As pointed out in Sect. 2.1. ST of quadrupolar nuclei (e.g., ^7Li) have been observed and used for studies of structural and dynamic properties in solids in several cases. However, this is confined to cases where the quadrupole coupling was not too large and hence the spectra could be acquired easily even for static samples. This may be typical for ^7Li, but in general, one has to deal with medium or large quadrupole interactions causing second order broadening of all transitions. Under these circumstances it was often impossible to acquire the ST because of their spectral spread of the order of the quadrupole frequency. Hence, all attention was paid to the CT if NMR of quadrupolar nuclei was used for investigations of solids, even under MAS conditions.

In retrospect it was Samson [76] who first calculated second order quadrupolar ST MAS patterns, their overall "width" and quadrupole shift in 1985. The main features of SATRAS are indeed published in that paper. In those days it was the opportunity to obtain the correct ^{27}Al chemical shifts in polycrystalline solids with just one experiment which made the observation of

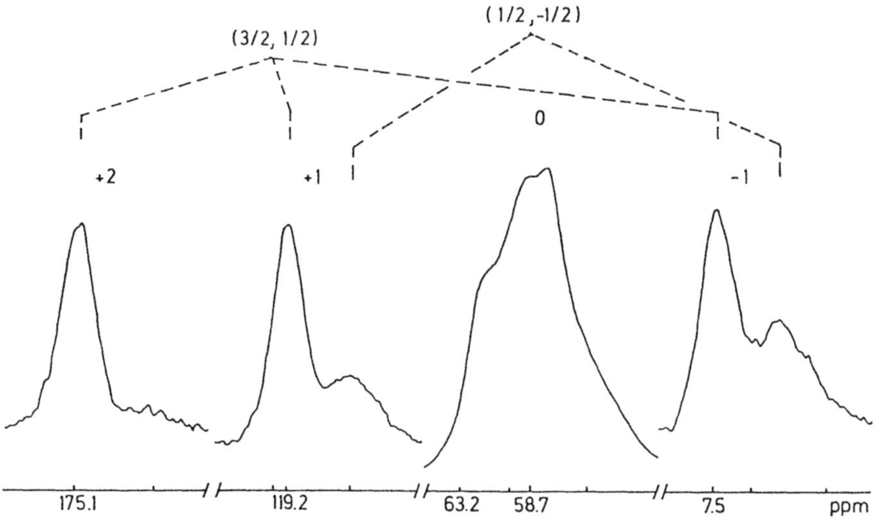

Fig. 2. The ^{27}Al NMR spectrum of the mineral albite. The rotation sidebands $N = +2, +1, -1$ are shown $15 \times$ magnified. The position of the isotropic chemical shift (63.2 ppm) is shifted by $+4.5$ ppm from the central transition line centre of gravity, but almost coincides with the average of the rotation sideband position of the (3/2, 1/2) transition. (Reproduced from Ref. [76] (Fig. 1) with permission of Elsevier Science Publishers BV)

the inner ST so interesting. Samoson determined the ^{27}Al chemical shift in the mineral albite by this method. This first example is shown in Fig. 2.

Otherwise, measurements at two distinct different magnetic fields are required in order to determine this quantity. But there have been only a few papers since that time reporting on this technique [76–78], as a result of slow MAS spinning at several kHz and the associated problems mentioned in Sect. 2.2.

More attention to the ST has been paid since the end of the 1980s especially by the group at Aarhus [79–84] who took advantage of high speed MAS probes [85]. The quadrupole interaction has been determined in various polycrystalline solids and applications to ^{7}Li, ^{17}O, ^{23}Na ^{27}Al and ^{51}V have been published. There are also papers by other authors who took into account the ST for accurate quantification of the CT spectra. This is necessary in order to obtain the relative amount of AlO_4 and AlO_6 species in solids [86–89].

Apart from Ref. [76], the gain in resolution was also mentioned explicitly in Ref. [78]. The conclusion section (^{27}Al NMR study of vesuvianite) there ends with: "The results also illustrate the increase in resolution for ^{27}Al (and other $I = 5/2$ nuclides) obtained by observing the spinning sidebands of the $\pm(3/2, 1/2)$ satellite transitions. This technique appears to be useful for natural samples with low Fe contents and will be most useful for those phases with multiple Al sites that yield a signal that overlaps in the peaks for the central transition." But no further comments on the quadrupole interaction were given.

Since then it has taken several years for this gain in spectral resolution to be exploited both in polycrystalline solids but more importantly in disordered systems [90–94] especially because the high spinning speed of 10–15 kHz allows the separation of neighbouring orders of the ^{27}Al ST MAS spinning sidebands even if signals of differently coordinated Al species are present. This simplifies the data processing considerably.

Finally, a recent version uses the phenomenon of the quadrupole induced shift of the centres of gravity. The ^{11}B ST MAS lines of BO_3 and BO_4 units can be separated in this way [95]. Hence, both the quadrupole interaction and the correct chemical shifts can be determined with high accuracy (see Sect. 4.3).

3.2 Theory

3.2.1 Spectra of Polycrystalline Solids

The way of calculating the MAS line shapes for any allowed transition of the resonant nucleus at finite MAS speed has been described by several authors [82, 88] and will be explained here only briefly. Many more details can be found in the comprehensive paper by Skibstedt et al. [82] which also contains some hints for simplifications of the numeric procedures.

Under MAS conditions and considering the quadrupole interaction only, the time dependent Hamiltonian can be expressed as follows in the high field limit $H_Z \gg H_Q$ [82]:

$$H(t) = H_Q^{(1)}(t) + H_Q^{(2)}(t) \tag{1}$$

In the secular approximation the first- and second-order quadrupole Hamiltonian take the form:

$$H_Q^{(1)}(t) = \frac{\omega_Q^{(1)}(t)}{6}[3I_z^2 - I(I+1)] \tag{2}$$

and

$$H_Q^{(2)}(t) = \frac{\omega_Q^{(21)}(t)}{36}[-8I_z^2 + 4I(I+1) - 1]I_z + \frac{\omega_Q^{(22)}(t)}{36}[-2I_z^2 + 2I(I+1) - 1]I_z \tag{3}$$

It should be noted that factors 1/6 and 1/36 have been used in Eqs. (2) and (3) in contrast to Ref. [82]. This ensures that the ordinary relation between the quadrupole frequency ν_Q and the quadrupole coupling constant C_Q holds [7] (see also [25]):

$$\nu_Q = \frac{\omega_Q}{2\pi} = \frac{3C_Q}{2I(2I-1)} \tag{4}$$

In this case the quadrupole frequency is a measure for the spectral range of the ST. For an asymmetry $\eta = 0$ it corresponds to the frequency difference between the poles of the inner ST ($\pm 3/2$, $\pm 1/2$). The spectral spread for any ST can be estimated easily.

With the necessary changes the time-dependent factors are given then by:

$$\omega_Q^{(1)}(t) = \frac{12\pi C_Q}{4I(2I-1)} \sum_{n=1}^{2} [P_n^{(0)} \cos(n\gamma + \omega_r t) + R_n^{(0)} \sin(n\gamma + \omega_r t)]$$

$$\omega_Q^{(2q)}(t) = \frac{18}{\omega_L} \left(\frac{2\pi C_Q}{4I(2I-1)} \right)^2 \left\{ C^{(q)} + \sum_{n=1}^{4} [P_n^{(q)} \cos(n\gamma + \omega_r t) + R_n^{(q)} \sin(n\gamma + \omega_r t)] \right\} \tag{5}$$

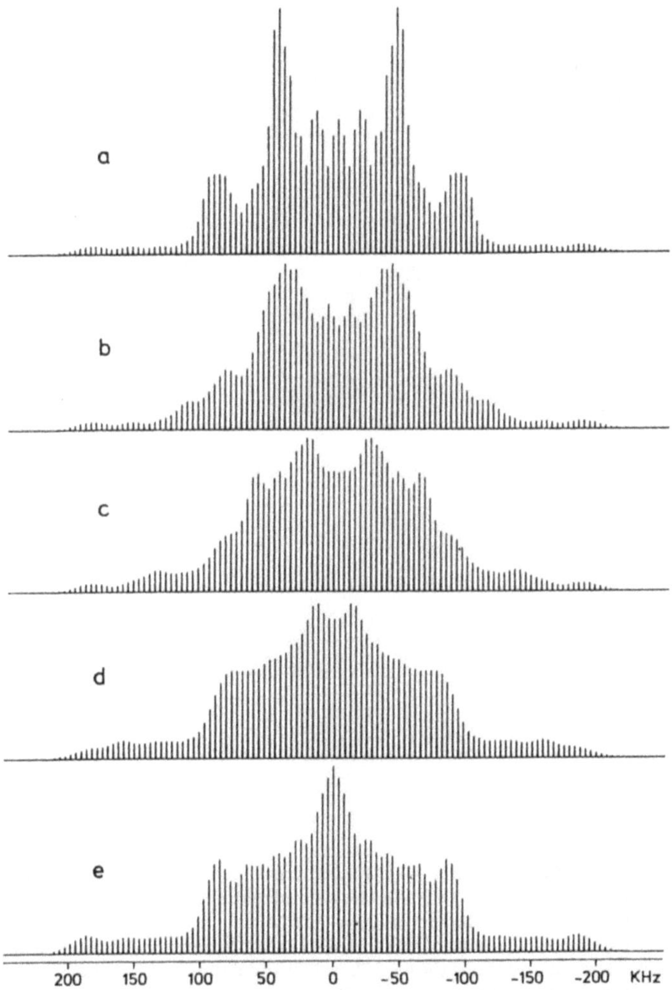

Fig. 3a–e. Simulated MAS NMR spectra of ^{27}Al ($\pm 5/2$, $\pm 3/2$) and ($\pm 3/2$, $\pm 1/2$) satellite transitions corresponding to $C_Q = 0.688$ MHz (C_Q value for $Ca_3Al_2O_66H_2O$) and **a** $\eta = 0.00$, **b** $\eta = 0.25$, **c** $\eta = 0.50$, **d** $\eta = 0.75$, and **e** $\eta = 1.00$. The spinning speed and line widths are the same as those used in Fig. 3b of Ref. [79]. (Reproduced from ref. [79] (Fig. 4) by permission of Academic Press)

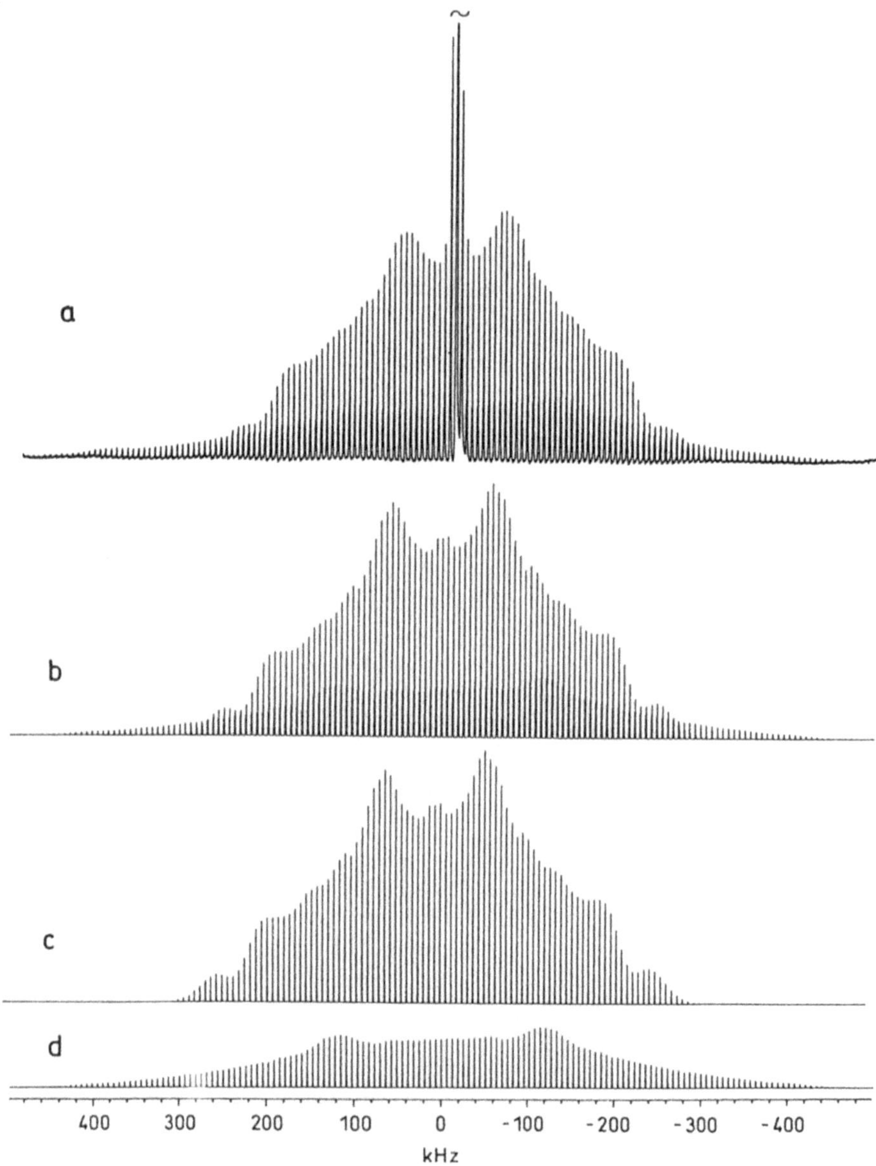

Fig. 4a–d. ^{27}Al (104.2 MHz) MAS NMR spectra of β-$LiAlO_2$. **a** Experimental spectrum employing $\tau_p = 2\,\mu s$, $\omega_1/2\pi = 30$ kHz, and $\nu_r = 6380$ Hz. **b** Least-squares simulated spectrum for all satellite transitions in **a** using the experimental parameters for **a**, $C_Q = 1.86$ MHz, $\eta = 0.56$, $Q = 180$ with an offset of -45 kHz, and Gaussian line broadenings of 400 and 500 Hz for the inner and outer satellite transitions, respectively. Separate simulations of **c** the ($\pm 3/2$, $\pm 1/2$) and **d** the ($\pm 5/2$, $\pm 3/2$) satellite transitions using the parameters for **b**. (Reproduced from Ref. [82] (Fig. 11) with permission of Academic Press

As usually the common definitions hold for the quadrupole coupling constant ($C_Q = e^2qQ/h$), the Larmor ($\nu_L = \omega_L/2\pi$) and spinning frequencies ($\nu_r = \omega_r/2\pi$). The relevant coefficients $C^{(q)}$, $P_n^{(q)}$ and $R_n^{(q)}$ are listed in Table 1 of Ref. [82].

Thus the spectrum for the ($\pm m$, $\pm(m-1)$) transition is obtained by:

$$\frac{\omega_{m,m-1}(t)}{2\pi} = \langle m|H(t)|m\rangle - \langle m-1|H(t)|m-1\rangle \tag{6}$$

Furthermore, Skibstedt et al. [82] also included effects due to finite pulse widths (nonideal pulse) and the Q-factor of the probe in their simulation/fit software. The latter is particularly important since the LC circuit of the probe has a finite bandwidth. For that reason the signal is dampd by about two for frequency offsets of ca. $\pm 400 \ldots 500$ kHz as a typical measure of high speed MAS probes. Especially for large spectral widths of that order this effect has to be considered. Another opportunity for correcting these effects is explained in Sect. 3.3.

Figure 3 shows the results of the simulations of the ^{27}Al ST MAS sidebands as a function of the asymmetry η.

It is clear from this that if the ST MAS can be observed then the relevant quadrupole parameters can be derived with high accuracy. The ^{27}Al MAS NMR spectrum of β-$LiAlO_2$ is shown in Fig. 4 as an example and is analysed in terms of the quadrupole interaction parameters and the contributions of the inner and outer ST to the total spectrum.

So far only the quadrupole interaction has been taken into account. Anisotropic chemical shifts can be neglected for light nuclei such as ^{7}Li, ^{23}Na and ^{27}Al. On the other hand ^{51}V exhibits an anisotropic chemical shift which can no longer be neglected compared with the quadrupole coupling. CT line shape simulations for static powders have been published for ^{85}Rb, ^{87}Rb [96] and ^{133}Cs [97]. However, the influence of the two tensors (chemical shift and quadrupole interaction) with different sizes, symmetries and relative orientations of their principal axes systems on the ST MAS spectra has been worked out in detail in Ref. [83]. Examples are considered in Sect. 4.1.2.

3.2.2 *Influence of Distributions of Interaction Parameters*

Because of the lack of long range order in disordered solids, all relevant interactions exhibit distributions of their parameters. For distributions of isotropic chemical shifts, it is quite reasonable to assume gaussian-like broadenings for both centre bands and spinning sidebands for a fit of e.g. ^{31}P and ^{29}Si MAS NMR spectra of disordered solid (e.g., [98, 99]).

However, the situation is quite different for second-order effects. It is well known that the ^{27}Al CT MAS line shape is glasses in strongly asymmetric (see e.g., Figs. 1, 9 and 10). The influence of distributions of second-order quadrupole couplings on both CT and ST MAS line shapes has been described recently

[94] and uses the following simple model [100] for describing distributions of the quadrupole interaction:

(1) Two of the principal axes values of the EFG-tensor (V_{zz}, V_{xx}) are distributed in a gaussian-like fashion with equal distribution widths.
(2) Since $V_{xx} + V_{yy} + V_{zz} = 0$ is to be fulfilled, V_{yy} has to be calculated accordingly.
(3) In order to fulfil $|V_{xx}| \leqq |V_{yy}| \leqq |V_{zz}|$, a reordering of the principal axes values is sometimes necessary.

Using this model the MAS line shapes of any of the transitions can be calculated as pointed out in Sect. 3.2.1. Figure 5 shows both the ST MAS and CT MAS line shapes as function of the relative distribution width for a mean quadrupole frequency of 1 MHz (reordered V_{zz} in frequency units) using a starting value of the asymmetry $\eta = 0.5$. The Larmor frequency is 104.26 MHz and a spinning speed of 11.5 kHz is assumed. For increasing distribution widths the CT MAS line shape changes step by step and results in the typical strongly asymmetric pattern as being observed always in glasses and disordered materials (mullites, spinels, ceramics).

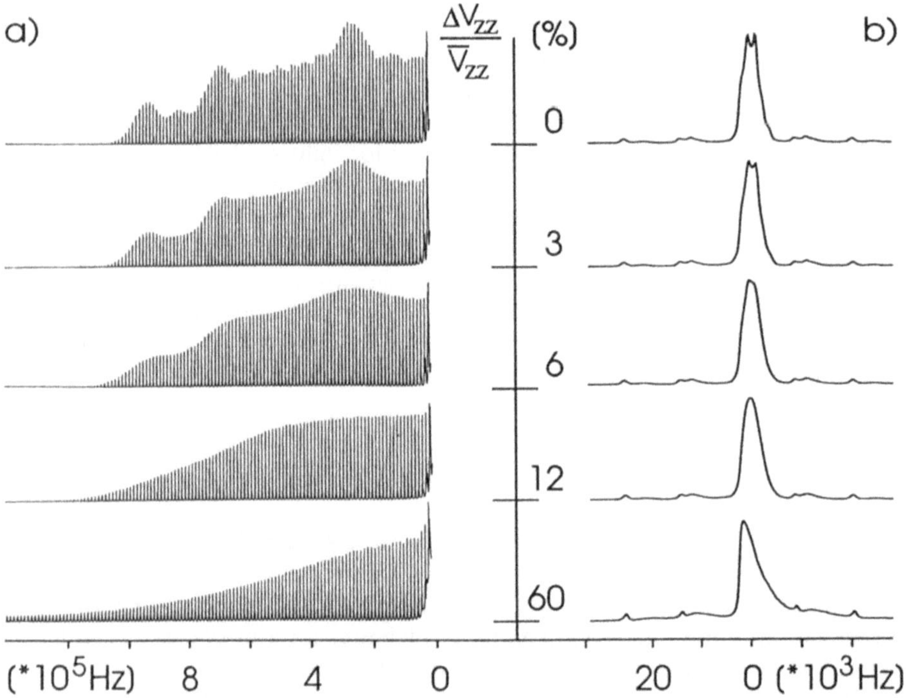

Fig. 5. Simulations of the resulting CT MAS and ST MAS line shapes (inner ST) for a mean quadrupole frequency of 1 MHz as a function of the relative distribution widths of 0%, 3%, 6%, 12% and 60%. **a** ST MAS pattern (only one half of the ST is shown). **b** CT MAS spectra. A resonance frequency of 104.26 MHz and a MAS speed of 11.5 kHz has been assumed. (Reproduced from Ref. [94] (Fig. 2) with permission of Elsevier Science Publishers BV)

The small ST MAS sidebands appear in Fig. 5 to the left and right of the CT. Their amplitude envelopes are shown on the left-hand side of Fig. 5 as a function of the relative distribution width. The influence of the distribution width on the envelope of the ST MAS spinning sidebands is apparent and much more distinct compared with the changes of the shape of the CT MAS centre band. Provided that these spinning sidebands can be acquired accurately even for large quadrupole couplings, they represent a very sensitive source of information about the quadrupole interaction and its distribution width.

The simple model requires a starting value for the asymmetry η. Of course the starting value of η determines the shape of the envelope for small distribution widths as has been shown in Ref. [94]. Certainly this allows the extraction of this parameter by the experiment, if necessary. However, as far as our investigations on glasses and other disordered solids are concerned, the large relative distribution width of about 50% leads to only minor effects for differently assumed η which can be neglected in most cases.

The relative distribution width can readily be estimated from the spectra using these graphs. The more uniform the decrease of the amplitudes of the ST MAS lines, the larger the distribution is. On the other hand, if any distinct features (poles or shoulders of the static ST pattern) can be recognized, then starting values for $V_{zz}(\nu_Q)$, its distribution and η can be obtained.

3.3 Experimental Details and Requirements

The only prerequisite for doing SATRAS is a high-speed MAS probe. The necessary spinning frequency is determined by the quadrupole interaction strength, the applied B_0 field and the difference in the chemical shifts if various resonances (species) are present. As pointed out in Sect. 2 the lower limit for the spinning is about 10 kHz for ^{27}Al at 7.1 T. If stronger fields are used, still higher MAS frequencies have to be used since the chemical shift varies linearly with the increase of the external B_0 field. However, in disordered solids this is compensated to some extend by the second order quadrupolar broadening which narrows the ($\pm 3/2, \pm 1/2$) ST MAS pattern. Hence spinning at 10...15 kHz at $B_0 \approx 7...11$ T is quite sufficient for many ^{27}Al applications. For other nuclei with smaller quadrupole frequencies (e.g. ^{7}Li) slower spinning speeds can be applied and are desirable in order to have enough spinning sidebands for quantitative calculations.

In contrast to ordinary CT MAS spectroscopy, the precise setting of the magic angle and an extremely stable spinning frequency are of particular importance for SATRAS. If the angle is not set accurately the spinning sidebands are broadened mainly by first order effects but also by dipolar couplings and anisotropic chemical shifts. This would result in a severe loss of the spectral resolution, which is one of the most important advantages of the method (Sect. 4). (In fact an intentional missetting of the magic angle has been used in [101] to get rid of the ST MAS spinning sidebands). Furthermore, if the spinning is not stable, then the sidebandes are additionally broadened; an effect which is linear with the order of the spinning sideband. Hence, an instability of only few Hz would result in some hundred Hz broadening for high orders of spinning sidebands.

Another crucial question is concerned with the correct data acquisition. There are basically two approaches for getting accurate results which will be explained now. Taking the quadrupole frequency as defined by Eq. (4) and neglecting the asymmetry η a spectral width of

$$\nu_{obs} = (2m - 1)\nu_Q \tag{7}$$

has to be acquired with m being the number of the desired pair of $(\pm m, \pm(m-1))$ ST. Since modern high speed MAS probes have typical bandwidths of ca. ± 500 kHz (signal damping of about 2) the data acquisition and processing depends on the size of the necessary ν_{obs}. The first approach [82] takes both the bandwidth of the probe (Q-factor) and the finite rf excitation explicitly into account when simulating or fitting the experimental data. In other words, the actual spectrum is compared with the theoretically modified simulations (additional frequency dependent amplitude damping). This works particularly well for $\nu_{obs} \leqq 400 \ldots 500$ kHz as an estimate. Concerning ^{27}Al NMR, quadrupolar couplings up to about 3 MHz can be measured easily by SATRAS. However, for larger quadrupolar coupling constants another approach is useful [93, 94] in which two spectra are measured. One experiment is carried out by setting the rf irradiation exactly at the CT, whereas the second spectrum is acquired with an frequency offset of about 600 kHz...1 MHz after carefully retuning the probe to the new rf frequency. Both spectra are corrected for the actual frequency characteristics of the probe (LC circuit) including the finite pulse length which is obtained by measuring the signal height of the CT as function of the offset [93] before both spectra are combined to give the complete, now undistorted ST MAS pattern. This procedure is shown in Fig. 6 (^{27}Al NMR of polycrystalline aluminoborate $9Al_2O_3 \cdot 2B_2O_3$).

The spectra are obtained by single-pulse irradiation. In most cases pulse length of 0.6.... 1 μs are used. This ensures correct quantification of the data in terms of relative numbers of species in different structural groups (e.g., [25, 89]). The strong baseline distortions can be corrected using, e.g., a cubic spline method provided with the spectrometer software package. It can easily be applied as long as the baseline between neighbouring orders of the spinning sidebands is well resolved.

Another important point is the exact tuning of the probe. If only a small quadrupole interaction is present (no second-order quadrupolar ST MAS line broadening), a slight mistuning is not crucial, since the amplitudes of the ST MAS lines are symmetric around the CT. Hence, a mistuning can be corrected if necessary. But for larger quadrupole couplings those ST MAS amplitudes (and shapes) are no longer symmetric around the CT as pointed out first in Ref. [112] for static powder patterns for integer spins. The same effect occurs in the off-resonance experiment and if in addition to the second-order quadrupole

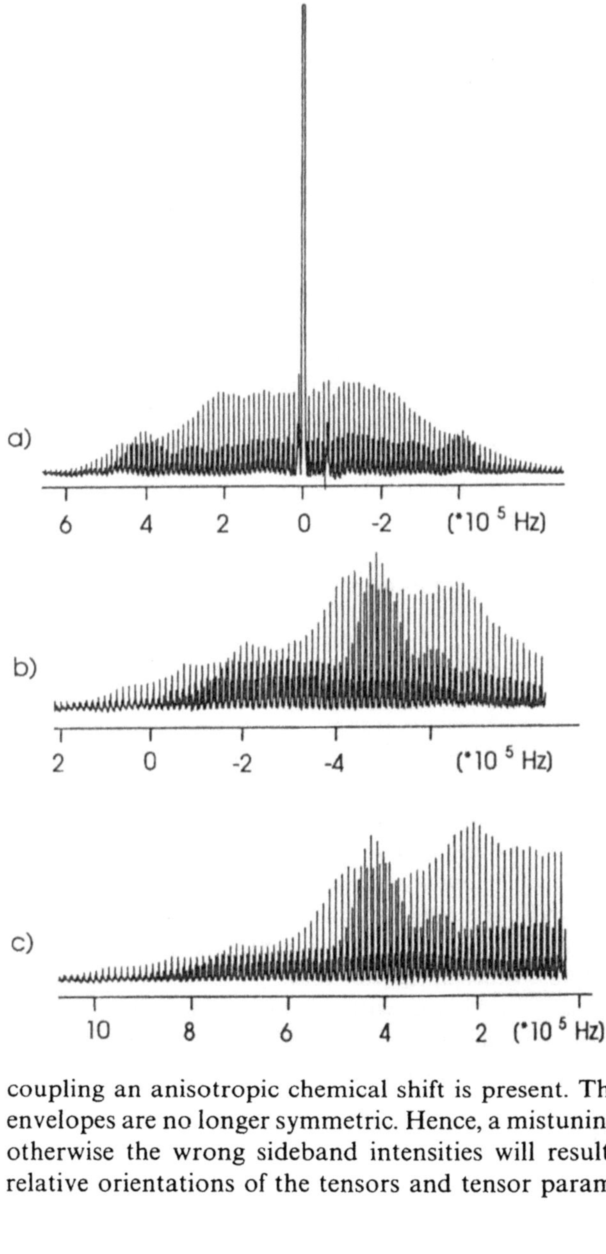

Fig. 6a–c. Experimental procedure for obtaining one half of the MAS satellite transition. **a** On-resonance experiment. The rf frequency is set to the central transition. The almost symmetric decay mainly due to the MAS probe characteristics is obvious. **b** Off-resonance experiment (900 kHz) with carefully retuned probe. The zero point of the scale indicates the position of the rf frequency. Compared to **a** the changed line shape caused by the finite rf bandwidth of the probe is obvious. **c** Reconstructed ST MAS pattern as explained in the experimental section. (Reproduced from Ref. [93]) (Fig. 1) with permission of Elsevier Science Publishers BV)

coupling an anisotropic chemical shift is present. Then the ST MAS sideband envelopes are no longer symmetric. Hence, a mistuning must be avoided because otherwise the wrong sideband intensities will result in incorrect results (e.g., relative orientations of the tensors and tensor parameters).

3.4. *Major Strategies of Applications*

The *determination of the quadrupole interaction* via analysis of the amplitudes of the ST MAS spinning sidebands is the most straightforward important application of the technique. Using MAS a considerable increase in spectral resolution is obtained since dipolar broadenings are removed. This is particularly interesting, e.g., for ^{7}Li, but also for other nuclei. Besides this it is also possible to determine the relative orientation of chemical shift and quadrupole tensors in polycrystalline solids (Sect. 4.1.2).

But SATRAS has other important advantages. Samoson [76] has calculated the quadrupole induced shift and overall width of the MAS spinning sidebands (infinite spinning speed) of any allowed transition $(\pm m, \pm(m-1))$ of a nucleus with spin I:

$$\nu^{(2)}(m) = \nu_L \sigma_{QS}(m) + \Delta(m) g(\alpha, \beta, \eta) \qquad (8)$$

where $g(\alpha, \beta, \eta)$ determines the particular MAS line shape, and α and β describe the orientation of the spinner with respect to the principal axis system of the EFG tensor, characterized by the strongest component C_Q/eQ and the asymmetry η. The relative line width of the ST MAS pattern compared with the CT is determined by the ratio of the coefficients of the second term in Eq. (8) and is:

$$\frac{\Delta(m)}{\Delta(1/2)} = \frac{6I(I+1) - 34m(m-1) - 13}{6I(I+1) - \frac{9}{2}} \qquad (9)$$

In the same fashion the ratio of the $\sigma_{QS}(m)$ gives the relative quadrupole shift compared with the CT:

$$\frac{\sigma_{QS}(m)}{\sigma_{QS}(1/2)} = \frac{I(I+1) - 9m(m-1) - 3}{I(I+1) - \frac{3}{4}} \qquad (10)$$

Samoson's results for the quadrupole shift of the centres of gravity and the relative overall widths are listed in Table 2.

Although, those numbers have been obtained in the limit of infinite spinning speed, they nevertheless explain two more advantages of SATRAS. For certain spin numbers there is a *considerable gain in spectral resolution* just by observing the MAS lines of a particular ST rather than the CT. $I = 5/2$ nuclei (^{27}Al, ^{17}O) are of particular interest for many reasons. The second order quadrupole line width of the inner ST is about 3.4 times narrower. The $(\pm 5/2, \pm 3/2)$ transitions of $I = 9/2$ nuclei (e.g. ^{93}Nb) are even better with a theoretical gain of a factor 18. In fact, it is this gain in spectral resolution which enables a tremendous improvement of structural studies in disordered materials since the limits of

Table 2. Relative line width (first number) and relative quadrupole shift (second number) of the $(\pm m, \pm(m-1))$ ST for half integer spins I (adopted from [76]) Negative numbers mean reversed line shape or a shift in opposite direction compared with the CT

	$m = 3/2$	$m = 5/2$	$m = 7/2$	$m = 9/2$
$I = 3/2$	−0.8889/−2.000			
$I = 5/2$	0.2917/−0.1250	−1.8333/−3.5000		
$I = 7/2$	0.6222/0.4000	−0.5111/1.4000	−2.4000/−4.4000	
$I = 9/2$	0.7639/0.6250	0.0556/−0.5000	−1.1250/−2.3750	−2.7778/−5.000

spectral resolution can be reached which are governed by isotropic chemical shift distributions (Sect. 4.2).

On the other hand *line separation caused by the quadrupole shift* is possible. The prerequisites are either distinct differences in the quadrupole coupling constants for the different species or a large value for the relative quadrupole shift compared with the CT. ^{11}B NMR in glasses is an interesting application for the first case and will be considered in Sect. 4.3.

Summarising, the determination of the quadrupole parameters and of the chemical shift are the main goals in all cases.

4 Applications of SATRAS

4.1 Determination of the Quadrupole Parameters in Polycrystalline Solids

It has been shown especially by the group at Aarhus that the quadrupole interaction parameters can advantageously be detemind using the ST. This enables the opportunity of structural characterisations of various polycrystalline solids by the quadrupole interaction in addition to the isotropic chemical shift, which has been nearly exclusively used so far, particularly for ^{27}Al NMR. The advantage of using the ST lies in the fact that the quadrupole effects are much more sensitive to changes of bonding distances and angles than the isotropic chemical shift. This is immediately evident when comparing the quadrupole couplings which can reach several MHz with the frequency separation of $\sim$6 kHz of the isotropic chemical shifts of differently coordinated Al species (e.g., [9] and also Table 3). In the following and according to the subject of this review the only papers that are included are those where explicit use of the ST has been made for determining the quadrupole and chemical shift parameters. This may be compared with a recent review [25] where ^{27}Al data are summarised mainly based on CT MAS investigations.

4.1.1 Dominating Quadrupole Interaction: ^{7}Li, ^{17}O, ^{23}Na, ^{27}Al

Except for cubic symmetry the quadrupole interaction is the dominating interaction for those nuclei. Only recently, a first report appeared on ^{27}Al chemical shift anisotropy of one of the pentacoordinated sites in $AlPO_4$-21 from the analysis of the DOR spinning sidebands [102]. Structural studies of various compounds such as calcium aluminates, aluminate hydrates, several aluminosilicate sodalites and other compounds have been published [79–84, 93, 103] explicitly using the ST MAS sidebands. The chemistry of alumina cements is an interesting and important application of this technique since it has been shown recently that major progress can be made exploiting the ST [84]. Based on this approach, it will be exciting to follow the progress in the future when real samples are investigated rather than model compounds.

On the other hand, for other nuclei without distinctly different chemical shifts for inequivalent sites it is only the quadrupole interaction that can be exploited for structural studies. This especially holds for nuclei such as ^{7}Li, ^{11}B and ^{23}Na. Using MAS which averages out dipolar interactions and thus enhances the resolution, the ST can provide precise information on the local environment around the nuclei in inorganic solids (e.g., of the exchangeable cage Li^{+} and Na^{+} cations and the aluminium framework in above mentioned sodalites [80]) but also in minerals.

The ^{27}Al MAS spectrum of β-$LiAlSO_2$ is shown in Fig. 4 as an example of the method. The data for the quadrupole interactions and chemical shifts of various solids obtained by SATRAS are summarised in Table 3.

Table 3. Collection of ^{17}O,^{27}Al,^{23}Na and ^{51}V quadrupole interaction and chemical shift data published so far using the ST explicitly. Isotropic shifts are referenced to 1m aqueuous LiCl, NaCl and $AlCl_3 \cdot 6H_2O$ for ^{7}Li, ^{23}Na and ^{27}Al [80]. ^{17}O shifts are referenced to H_2O [82] and ^{51}V to liquid $VOCl_3$

Compound	C_Q(MHz)	η	δ_{iso}(ppm)		ref.
$Li_{7.6}Na_{0.4}(AlSiO_4)_6Cl_{1.9}$	0.041	0.10	-0.7 ± 0.2	^{7}Li	[80]
$0.7H_2O$	0.98	0.59	71.9 ± 0.2	^{27}Al	
$Li_{7.3}Na_{0.3}(AlSiO_4)_6Br_{1.8}$	0.050	0.32	-1.2 ± 0.2	^{7}Li	[80]
$1.1H_2O$	0.71	0.61	70.9 ± 0.2	^{27}Al	
$Na_{8.0}(AlSiO_4)_6Cl_{1.8}$	—	—	-8.8	^{23}Na	[80]
$0.4H_2O$	0.94 ± 0.05	0.32 ± 0.05	62.9 ± 0.2	^{27}Al	
$Na_{7.7}(AlSiO_4)_6Br_{1.8}$	0.72 ± 0.05	0.12 ± 0.05	-9.9 ± 0.2	^{23}Na	[80]
$0.4H_2O$	0.81 ± 0.05	0.29 ± 0.05	61.8 ± 0.2	^{27}Al	
$Na_{7.8}(AlSiO_4)_6I_{1.7}$	1.73 ± 0.05	0.06 ± 0.05	-20.6 ± 0.2	^{23}Na	[80]
$0.6H_2O$	0.57 ± 0.05	0.34 ± 0.05	60.4 ± 0.2	^{27}Al	
AlOOH (OAl_4 site)	1.15 ± 0.03	0.13 ± 0.03	70.0 ± 0.5	^{17}O	[82]
$NaNO_3$	0.337 ± 0.002	0.00 ± 0.03	-8.0 ± 0.1	^{23}Na	[79, 82]
$NaNO_2$	1.09 ± 0.03	0.11 ± 0.03	-8.1 ± 0.2	^{23}Na	[82]
NaN_3(25°C)	0.297 ± 0.003	0.12 ± 0.02	-3.8 ± 0.1	^{23}Na	[82]
$CH_3COOHNa \cdot 3H_2O$	0.779 ± 0.007	0.38 ± 0.02	2.1 ± 0.1	^{23}Na	[82]
$Na_2C_2O_4$	2.5 ± 0.03	0.74 ± 0.02	17.9 ± 0.5	^{23}Na	[82]
$KAl(SO_4)_2 \cdot 12H_2O$	0.400 ± 0.006	0.00 ± 0.03	-4.1 ± 0.1	^{27}Al	[82]
$NH_4Al(SO_4)_2 \cdot 12H_2O$	0.456 ± 0.012	0.00 ± 0.03	-0.4 ± 0.1	^{27}Al	[82]
α-Al_2O_3	2.38 ± 0.01	0.00 ± 0.02	16.0 ± 0.2	^{27}Al	[79,82]
$Ca_3Al_2O_6 \cdot 6H_2O$	0.688 ± 0.006	0.00 ± 0.03		^{27}Al	[79]
β-$LiAlO_2$	1.86 ± 0.03	0.56 ± 0.03	83.0 ± 0.1	^{27}Al	[82]
CA_2	6.25 ± 0.05	0.88 ± 0.02	75.5 ± 0.5	$^{27}Al(1)$	[84]
	9.55 ± 0.05	0.82 ± 0.02	69.5 ± 0.5	$^{27}Al(2)$	
CA	2.50	0.20	81.9	$^{27}Al(1)$	[84]
	2.60	0.75	83.8	$^{27}Al(2)$	
	2.60	0.95	86.2	$^{27}Al(3)$	
	3.32	0.53	82.7	$^{27}Al(4)$	

Table 3. (*Contd.*)

Compound	C_Q(MHz)	η	δ_{iso}(ppm)		ref.
	3.37	0.39	81.6	^{27}Al(5)	
	4.30	0.47	81.2	^{27}Al(6)	
$C_{12}A_7$	9.7 ± 0.2	0.40 ± 0.10	85.9 ± 1.0	^{27}Al(1)	[84]
	3.8 ± 0.2	0.70 ± 0.10	80.2 ± 0.3	^{27}Al(2)	
C_3A	8.69 ± 0.05	0.32 ± 0.02	79.5 ± 0.5	^{27}Al(1)	[84]
	9.30 ± 0.05	0.54 ± 0.02	78.3 ± 0.5	^{27}Al(2)	
CAH_{10}	$2.4\pm0.2^{2)}$		10.2 ± 0.3	^{27}Al	[84]
C_3AH_6	0.705 ± 0.010	0.09 ± 0.02	12.36 ± 0.06	^{27}Al	[84]
C_4AH_{13}	$1.8\pm0.2^{2)}$		10.2 ± 0.2	^{27}Al	[84]
γ-AH_3	1.97 ± 0.07	0.73 ± 0.04	10.4 ± 0.3	^{27}Al(1)	[84]
	4.45 ± 0.05	0.44 ± 0.03	11.5 ± 0.3	^{27}Al(2)	
$C_6AS'_3H_{32}$	0.360 ± 0.010	0.19 ± 0.03	13.1 ± 0.1	^{27}Al	[84]
$C_4AS'H_{12}$	$1.7\pm0.2^{2)}$		11.8 ± 0.2	^{27}Al	[84]
$NH_4VO_3^{3)}$	2.91 ± 0.07	0.30 ± 0.03	δ_{iso}: -563.7 ± 0.5 δ_σ: 237 ± 5 η_σ: 0.71 ± 0.03	^{51}V	[83]
$V_2O_5^{3)}$	0.797 ± 0.020	0.00 ± 0.05	δ_{iso}: -609.0 ± 0.5 δ_σ: 645 ± 5 η_σ: 0.11 ± 0.05	^{51}V	[83]
$9Al_2O_3\cdot2B_2O_3$	1.36	0.43	70	AlO_4	[93]
	1.08	0.0	52	AlO_5(1)	
	1.42	0.49	43	AlO_5(2)	
	0.91	0.44	8	AlO_6	
$(Mg,Fe)Al_3SiBO_9$	3.50 ± 0.10	0.50 ± 0.05	9.0 ± 1.0	Al1	[103]

1) shorthand notation according to [84]: C=CaO, A=Al_2O_3, S=SiO_2; F=Fe_2O; S' = SO_3; H=H_2O; D=D_2O

2) second order quadrupolar effect parameter SOQE = $C_Q\sqrt{1+\eta^2/3}$

3) For relative orientations of the EFG and chemical shift tensors see [83]

4.1.2 Quadrupole Interaction and Chemical Shift: ^{51}V NMR

As pointed out in Sect. 3.1 for heavier nuclei such as ^{51}V the anisotropy of the chemical shift must be taken into account in addition to the quadrupole interaction. Skibstedt et al. [83] have solved this problem for MAS including all the ST. Both the sizes and symmetries of the chemical shift and EFG tensors and their relative orientations determine the envelopes of the ST MAS spinning sidebands which are now by no means any longer symmetric with respect to the CT (see Fig. 7). The experimental consequence is that the probe has to be tuned extremely carefully and as exactly as possible to the frequency of the CT (Sect. 3.3). This technique has been applied to NH_4VO_3 and V_2O_5 where all relevant quadrupole and chemical shift parameters have been determined using a seven parameter fit, including the relative orientations of the two tensors (see Table 3).

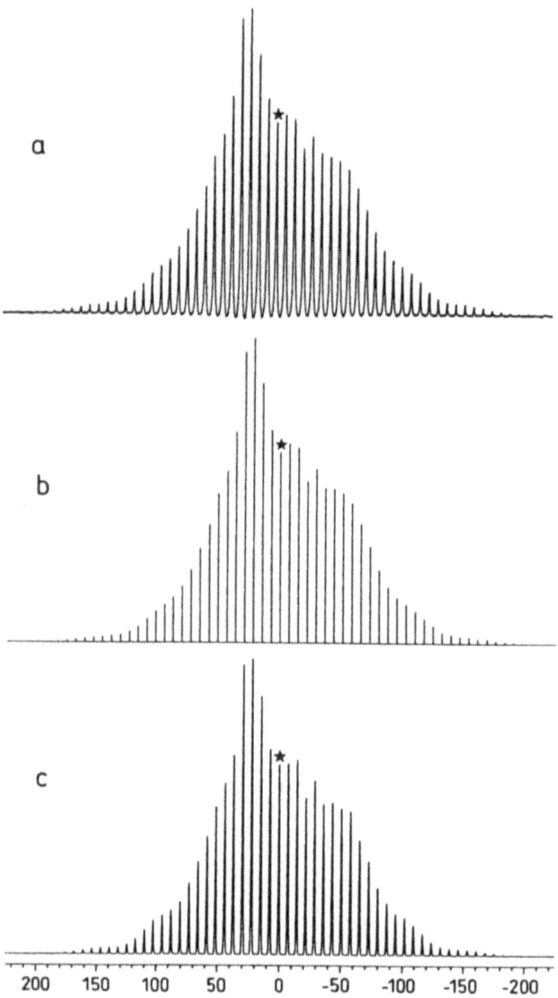

Fig. 7a–c. ^{51}V MAS NMR spectrum (105.15 MHz) of the central and satellite transitions for V_2O_5. **a** Experimental spectrum (ν_r = 7300 Hz). **b** Stick plot of 54 integrated SSB intensities for the spectrum in **a** used for the seven parameter fit. **c** Simulated MAS spectrum corresponding to the optimized parameters from the fitting procedure (Table 2 of Ref. [83]). A Gaussian line width of 1000 Hz has been employed for the SSBs in **c**. The isotropic peak is marked with an *asterisk*. (Reproduced from Ref. [83] (Fig. 3) with permission by Elsevier Science Publishers BV)

4.2 Increased Spectral Resolution

The exploitation of the gain in spectral resolution is the most interesting application of SATRAS especially for amorphous and disordered solids. The observation of certain ST MAS patterns rather than the CT MAS line shape yields an additional line narrowing for various spin numbers I just by applying the same MAS experiment. This effect can be used, e.g., for separating the ST MAS lines of differently coordinated ^{27}Al species which overlap in the CT. As a consequence a considerable simplification of the calculations results as will be explained in the following.

Hence, if SATRAS is understood to be the experiment where both CT and all ST are acquired at the same time, it contains informations of independent experiments although it is only a single experiment. This is easily explained, e.g., for ^{27}Al. The spectral width of the CT MAS centre band pattern is governed by the applied B_0 field for a given quadrupole coupling. In contrast to this, the spectral width of a single inner ST MAS pattern is only ~ 30% and corresponds thus to a measurement carried out at a virtual field of 3.4 B_0 apart from the changed ST line shape. Or in other words, a magnet which produces 3.4 B_0 must be available in order to achieve the same spectral resolution in the CT.

4.2.1 $I = 5/2$ Nuclei: ^{27}Al and ^{17}O

It has recently been shown that ^{27}Al quadrupole couplings $C_Q \approx 10$ MHz can be determined using strong B_0 fields (11...14 T) and high spinning speeds of more than 16 kHz in polycrystalline $3CaO \cdot Al_2O_3$ [81] and minerals [104]. Exploiting the above-mentioned gain in resolution of the ST MAS lines and provided only one ^{27}Al site is present, it is straightforward to predict that C_Q of up to 30 MHz could be measured in this way. This is indeed a fascinating opportunity.

However, the main advantage of SATRAS in terms of the increased resolution is that the signals of 4-, 5-, and 6-coordinated Al species can be resolved almost completely from each other not only in polycrystalline solids, but more importantly in glasses. This enables, for the first time, a correct determination of all relevant parameters including distribution widths and relative amounts of the differently coordinated species in disordered solids even if severe superpositions of the lines are present in the CT. In addition, crystallographically different species with the same coordination number can be resolved using SATRAS also permitting a correct quantification (see Sect. 5).

4.2.1.1 Polycrystalline Solids

The determination of the quadrupole and chemical shift parameters using the ST was the subject of Sect. 4.1. In applications to complicated materials, however, the situation can be difficult. There can be, e.g., mixtures of different materials or samples containing several crystalline phases. In both cases the increased spectral resolution of the ST can be exploited to obtain all the relevant information of the quadrupole interaction and the chemical shift.

Even in single phase materials the presence of differently coordinated species is the normal case. Polycrystalline alumoborate $9Al_2O_3 \cdot 2B_2O_3$ is a textbook example with four crystallographically inequivalent sites; one AlO_4, two AlO_5 and another AlO_6 unit [105]. It has been shown recently [93], that the ST resonances of the AlO_4, the AlO_6 and the two AlO_5 polyhedra are completely

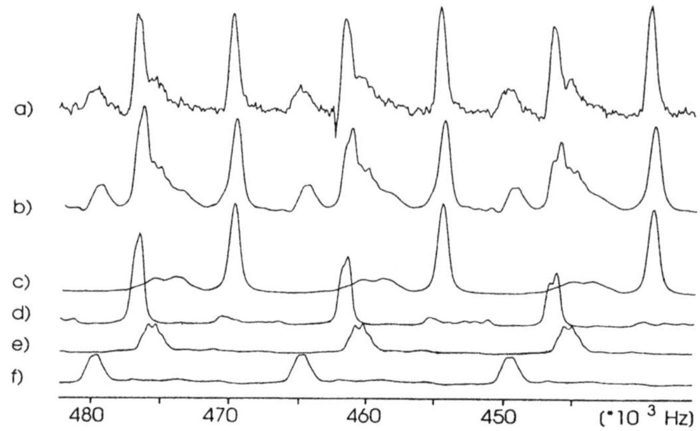

Fig. 8a–f. Comparison of the experimental 156 MHz spectrum and calculated line shape with deconvolution of the latter into the four different ST MAS lines (all transitions included) according to the values for the quadrupole interaction and chemical shift of Table 3: **a** Experimental spectrum; **b** total theoretical spectrum; **c** AlO_6 signal, besides the *strong* ($\pm 3/2$, $\pm 1/2$) *line* also the outer ST MAS line ($\pm 5/2$, $\pm 3/2$) is observed (*small broad line*); **d** $AlO_5(1)$ signal; **e** $AlO_5(2)$ resonance; **f** AlO_4 signal. (Reproduced from Ref. [93] (Fig. 5) with permission by Elsevier Science Publishers BV)

separated and the signals of the two pentacoordinated Al species are nearly resolved in the same fashion (see Fig. 8). Both the isotropic chemical shifts and the quadrupole parameters obtained from the ST MAS sidebands envelopes and the CT are listed in Table 3. As mentioned above the advantage of the resolved ST MAS lines was used to determine these parameters separately, at least for AlO_4 and AlO_6.

4.2.1.2 Glasses and ceramics

In glasses, the situation is very different from polycrystalline solids for severals reasons. First of all, there are distributions of the quadrupole interaction parameters. It has been shown in Ref. [94] and here in Sect. 3.2.2. that the CT MAS line exhibits a strong asymmetric and structureless pattern in that case. Second, there is also a considerable distribution of isotropic chemical shifts which additionally broaden the structureless CT line. It is in fact this superposition of two different distributions in the CT which makes correct quantifications in glasses and ceramics nearly impossible, especially if several differently coordinated species are present.

The key to understand the advantage of SATRAS is that both interactions and especially their distributions are influenced differently. Whereas the chemical shift distribution remains unchanged in the ST MAS lines, the influence of distributions of the second order quadrupole interaction is suppressed in the same way as for a constant quadrupole interaction. Hence, a considerable line

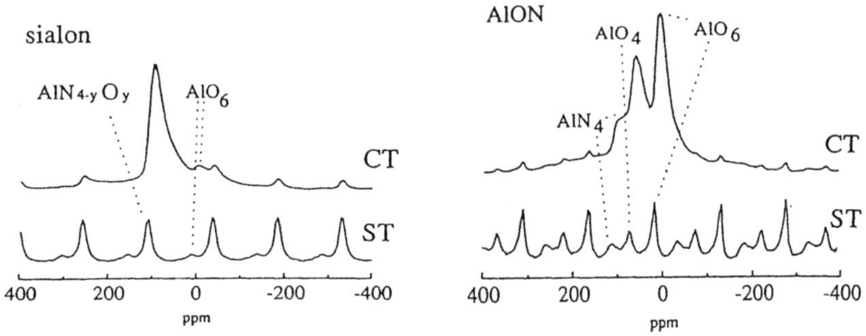

Fig. 9. ^{27}Al MAS NMR spectra of two different ceramics ($B_0 = 8.5$T, $\nu_r = 13.5$ kHz). The distinctly better resolution of the various Al signals in the inner ST MAS spectra is obvious. The signals of the various Al species of the aluminium oxynitride (AlON) are almost completely resolved and symmetric in contrast to the well known problems with the CT MAS lines. In the case of the sialon the assignment $AlN_{4-y}O_y$ indicates that mixed first coordination around the tetrahedrally coordinated Al is likely to occur ([157], see also Sects. 2.1 and 5). The measurements were carried out during a visit to the University of Warwick (Prof. R. Dupree).

narrowing follows even in solids exhibiting distributions of those parameters. This is demonstrated in Fig. 9 showing the ^{27}Al MAS spectra of two ceramics.

The gain of resolution between the ST MAS and CT MAS lines is obvious, leading to almost completely resolved ST lines for AlN_4, AlO_4 and AlO_6 signals for the AlON sample and for $AlN_{4-y}O_y$ and AlO_6 for the sialon. Furthermore, these ST MAS lines are symmetric in contrast to the strongly asymmetric CT MAS centre bands. The straightforward conclusion is that obviously the chemical shift distribution now determines the shape and width of these lines, whereas the distribution of the quadrupole parameters is no longer dominating. In other words, SATRAS approaches the limits of resolution in disordered solids governed by the chemical shift distribution. For this reason, SATRAS is an extraordinary experiment, since it combines two requirements which normally exclude each other. On the one hand, one is interested in suppressing second-order quadrupolar effects on the NMR line shape almost completely in order to achieve high spectral resolution. However, since quadrupolar effects are very sensitive to structural changes of the coordination polyhedra (bonding angles and distances) this interaction should be conserved. The way to combine these contradictory requirements is to observe the MAS line shapes of the inner ST since the envelope of the ST MAS sidebands is still determined by the (unchanged) quadrupole interaction itself whereas the chemical shift distribution determines the individual MAS line width.

It has been shown that SATRAS can successfully be applied to many different systems such as fluoroaluminate glasses with low contents of phosphates and K_2O-SiO_2-Al_2O_3, PbO-B_2O_3-Al_2O_3 and Bi_2O-B_2O_3-Al_2O_3 glasses [92].

As shown in Fig. 10 the ST MAS signals of 4-, 5- and 6-coordinated Al species are well resolved although a field of 'only' 7.1 T has been applied. This

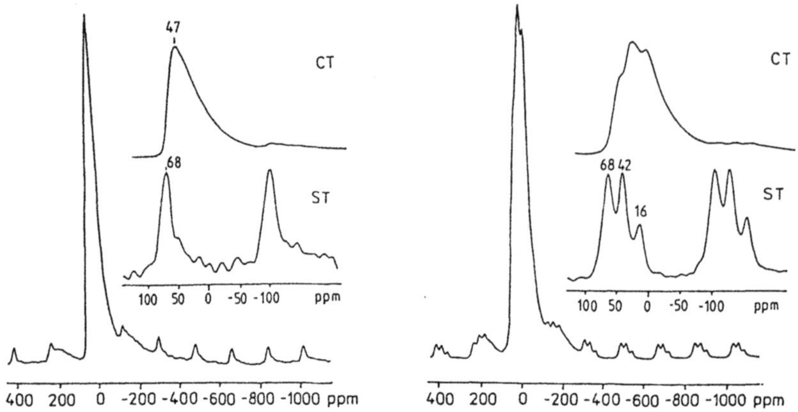

Fig. 10. NMR spectra of ternary glasses with low content of Al_2O_3 ($\nu_0 = 78{,}3$ MHz). The CT spectra are broad and asymmetric because of distributions of second-order quadrupole effects. However, the ST MAS lines are much better resolved, thus simplifying quantitative calculations. For further details see text. (*Left*) 4 mol % Al_2O_3; (*right*) 17 mol% Al_2O_3. (Reproduced from Ref. [91] (Fig. 5) with permission of Academic Press.)

fact proves the potential of the method. It has turned out that samples with only 4 mol% Al_2O_3 can be measured without difficulty. Furthermore, the mean C_Q, the correct mean chemical shifts (obtained with just one experiment at a single B_0 field) and the relative Al content of the various structural units can be determined using the ST rather than the CT, where the strongly asymmetric line shape does not allow a correct quantification without knowledge of the quadrupole interaction.

In the same way, amorphous SiO_2 was investigated by ^{17}O SATRAS [106]. The increased spectral resolution has been proven and the chemical shift and quadrupole coupling obtained by this single experiment agree well with earlier multifield studies.

However, using SATRAS, all relevant parameters including the distributions widths of the interactions and the relative amount of Al in 4-, 5-, and 6-fold co-ordinations in glasses can be determined with high accuracy as has been shown recently [94]. (Al_2O_3-P_2O_5-B_2O_3 glass). This can only be achieved by acquiring the whole inner ST using the experimental method as explained in Sect. 3.3. Because of the increased spectral resolution the ST MAS lines of the differently coordinated Al species are almost completely resolved from each other. Hence, the envelope (intensity distribution) of the individual resonance lines can be determined easily and no severe overlap has to be taken into account as for the CT. Consequently, the interaction parameters can be determined separately for the individual signals. They are listed in Table 4.

For the first time a reliable fit of the CT MAS pattern is possible based on these results via the ST MAS envelopes. Figures 11 and 12 show the results of these calculations. In Fig. 11 an expanded plot of the CT region is shown to-

Table 4. Values for the mean quadrupole interaction (reordered V_{zz}, given in frequency units as explained in Sect. 4.1), the distribution widths, the mean chemical shift, its distribution width in ppm ($\langle \Delta\delta_{iso} \rangle$) and the relative intensities for the AlO_4, AlO_5 and AlO_6 units of a Al_2O_3-P_2O_5-B_2O_3 glass. Numbers in brackets are the results of a fit of only the CT MAS pattern with the starting values determined by the ST MAS spectra. (Reproduced from Ref. [94])

	AlO_4	AlO_5	AlO_6
V_{zz} [kHz]	610 (607)	680 (668)	510 (555)
ΔV_{zz} [kHz]	300 (300)	350 (350)	300 (310)
error of quadrupole parameters	≦5%	≦5%	≈10%
δ_{iso} [ppm]	35.5	5.5	−19
$\langle \Delta\delta_{iso} \rangle$ [ppm]	14	11	10
intensity	0.40 (0.40)	0.42 (0.44)	0.18 (0.16)

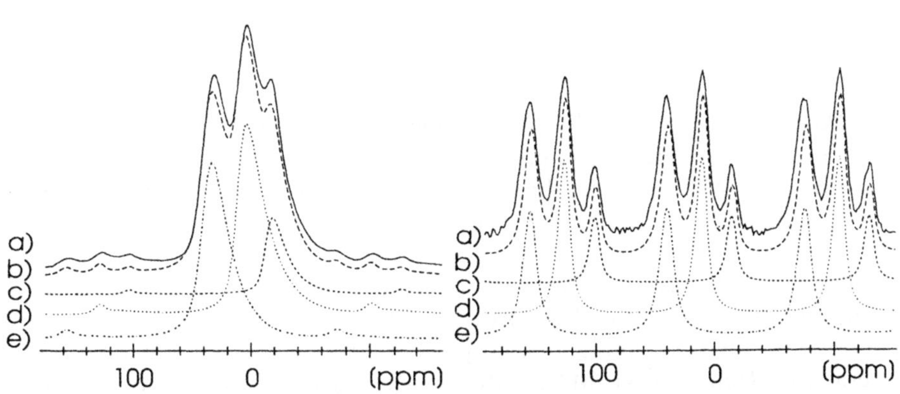

Fig. 11a–e. Comparison of the experimental (**a**) and simulated (**b**) CT region (*top spectrum*) and of several ST MAS spinning sidebands (*bottom spectrum*) of a Al_2O_3-P_2O_5-B_2O_3 glass. 4- (**e**), 5- (**d**) and 6-coordinated Al (**c**) always occurs from the left to the right within each ST MAS sideband order as is the case for the CT MAS pattern. For the simulations the results of Table 4 have been used. Whereas a severe overlap of the CT MAS lines is present, the ST MAS lines are almost completely resolved thus showing the increased spectral resolution. [Reproduced from Ref. [94] (Fig. 5) with permission of Elsevier Science Publishers BV)

gether with the line shape simulation/fit including the three different Al pattern. Figure 12 in turn compares the complete experimental ST pattern with the calculation.

SATRAS is not restricted to the examples shown here. It is simply a new experiment and so far only the first applications have been published.

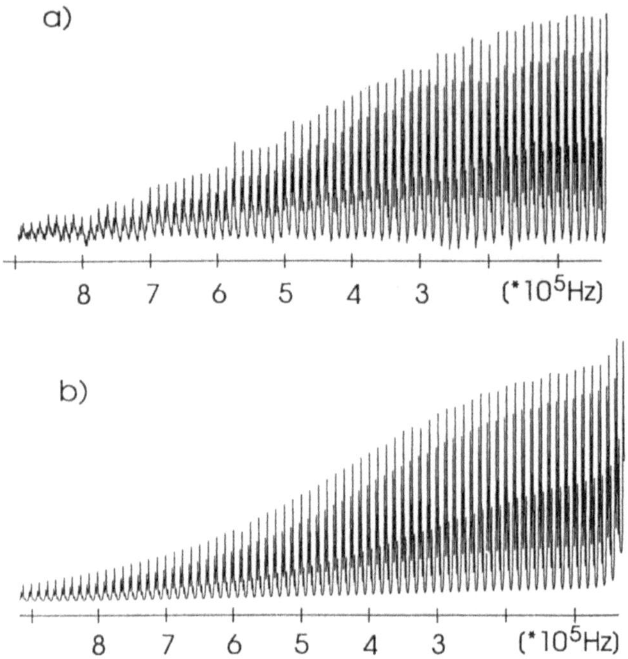

Fig. 12a,b. Comparison of the experimental ST MAS line shape (**a**) and the best simulation using the results listed in Table 1 of Ref [94] (**b**). The coincidence between the two spectra is obvious. See also Fig. 10 of Ref. [9] (Reproduced from Ref. [94] (Fig. 7) with permission of Elsevier Science Publishers BV)

NMR does not require three-dimensional translational symmetries of a unit cell in order to provide reasonably narrow MAS spectra. If the increased spectral resolution of SATRAS is combined with that, many other interesting applications are possible e.g. the investigation of crystallisation phenomena in glasses [107].

The general behaviour of ST envelopes in glasses shows a continuous decrease of the amplitudes because of the distributions of the interaction parameters. In contrast to this crystalline phases possess fixed values of the quadrupole interaction. Hence, a ST MAS pattern of a crystal phase clearly shows the symmetry of the EFG tensor by exhibiting the typical pattern as shown e.g. in Fig. 3. The example shown in Fig. 13 is certainly an easy one, but our efforts have proven that crystallisation phenomena in glasses can be detected to less than one per thousand (Al content in crystal phase) by this technique [108].

4.2.3 $I = 9/2$ Nuclei: ^{93}Nb

$I = 9/2$ nuclei are very attractive candidates for exploiting the increases spectral resolution provided the ($\pm 5/2$, $\pm 3/2$) ST are observed. According to Table 2 the MAS patterns of those ST are about 18 times narrower than the CT MAS line. This is in fact observed for ^{93}Nb MAS NMR as shown for $LiNbO_3$ in Fig. 14 [109].

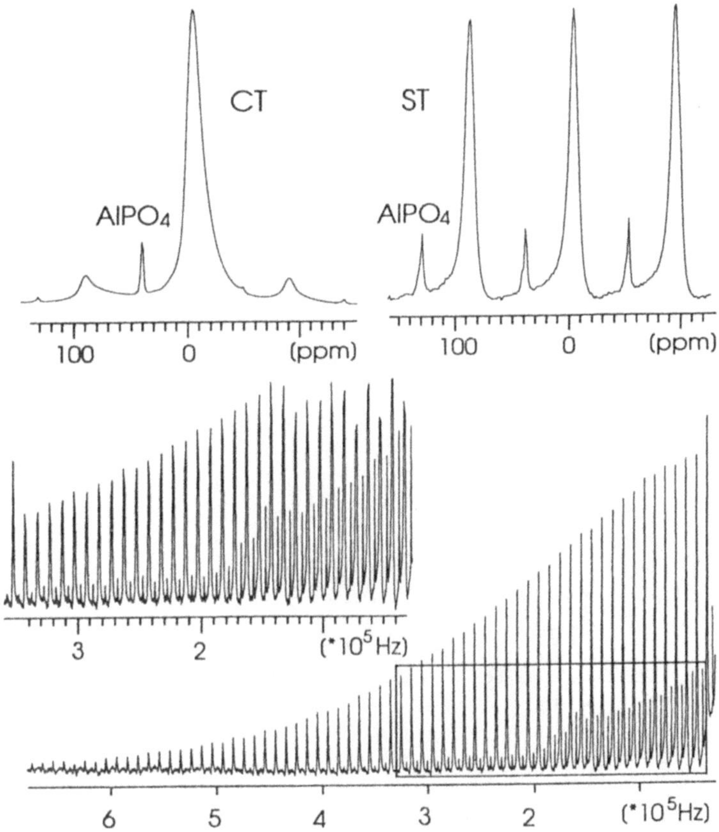

Fig. 13: ^{27}Al MAS NMR study of the crystallisation of a Na_2O-P_2O_5-Al_2O_3 glass ($B_0 = 11.7$T). In contrast to the Al signals of the glass which shows a continuous decrease of the ST MAS envelope, different crystalline $AlPO_4$ modifications are found. They can be distinguished by the pronounced changes of the ST MAS envelopes

The CT MAS line shape can be fitted reasonably well to a single NbO_6 site with $\nu_Q = 941$ kHz and an asymmetry $\eta = 0$. This quadrupole frequency corresponds to $C_Q = 22.58$ MHz using Eq. (4) which agrees well with that of a cw experiment [71] and a single crystal study [37] ($C_Q = 22.02$ MHz). However, the fit can be further improved by taking a very small distribution of only 2% of the quadrupole coupling into account, which results in a mean value of $\nu_Q = 935$ kHz ($C_Q = 22.44$ MHz) and is also reflected by the ST MAS lines. This is shown in Fig. 14 where the experimental spectrum (solid line) is compared with that simulation (broken line). The presence of the small distribution of the quadrupole coupling may be explained by a minor non-stoichiometry since stoichiometric $LiNbO_3$ is very difficult to grow. The extremely narrow ($\pm 5/2$, $\pm 3/2$) ST MAS sidebands are found to dominate the ST pattern far away from the CT at about ± 7000 ppm (top spectrum in Fig. 14). This is because of the

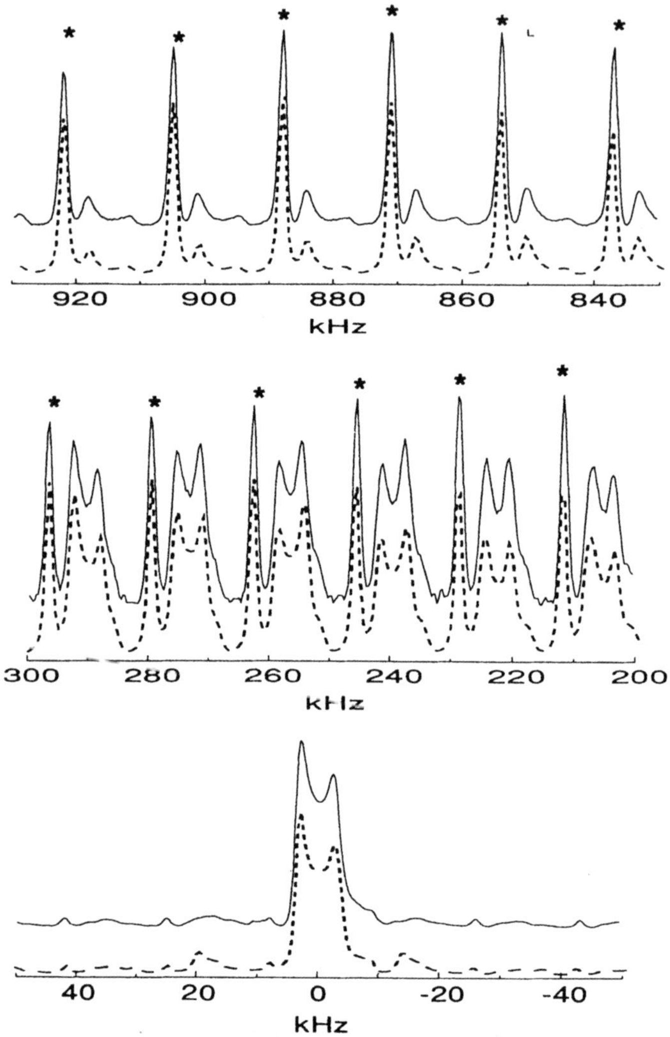

Fig. 14. Demonstration of the significantly improved resolution of the ($\pm$ 5/2, $\pm$ 3/2) ST MAS lines of $LiNbO_3$ (the sample was prepared by Dr. G. Corradyi, Research Laboratory for Crystal Physics, Hungarian Academy of Sciences, Budapest). The experimental spectrum (*solid line*) is compared with the simulation (*broken line*) including all ST using the quadrupole parameters obtained from a fit of the CT MAS line (mean $\nu_Q = 935$ kHz or $C_Q = 22.44$ MHz with a distribution width of 2%, $\eta = 0$). For away from the CT transition region the ($\pm$ 5/2, $\pm$ 3/2) ST MAS lines (marked by *stars*) dominate in the spectrum (*top spectrum*) and can be exploited for detailed structural studies (see text)

above-mentioned quadrupole frequency and symmetry. This extreme line narrowing can be exploited, e.g., for more detailed structural investigations, e.g., of non-stoichiometry or of defects which are present as described in various papers on $LiNbO_3$ [69, 110, 111] or for a more precise determination of distributions of the EFG tensor in solid solution systems [71].

4.3 Quadrupole Shift

In the previous sections the advantages of the increased spectral resolution have been shown. The basic idea was to exploit the differences of the chemical shifts of the various lines while suppressing the second-order quadrupolar broadening effects.

However, in ^{11}B NMR for example, the situation is quite different. There are only minor differences in the chemical shifts of the nearly planar BO_3 and tetrahedral BO_4 units. But on the other hand, the quadrupole coupling constants of the two units are extremely different. Typical ν_Q values of about 50 kHz and 2.6....2.7 MHz are observed, respectively (e.g., [22] and references therein). Consequently, the CT pattern of the BO_4 group is mainly governed by the dipolar broadening, whereas a BO_3 unit exhibits a typical second order powder pattern. But the two resonances overlap and need to be separated for quantifying the spectra. This is an established method for static cw spectroscopy, but has two disadvantages. First, any possible differences in the chemical shifts cannot be resolved and, secondly, the exact quadrupole parameters of the BO_4 units can only be estimated because of the dipolar line broadening. Of course MAS can be applied to the CT pattern. But it does not improve the situation much because both lines are narrowed and still overlap (see top of Fig. 15).

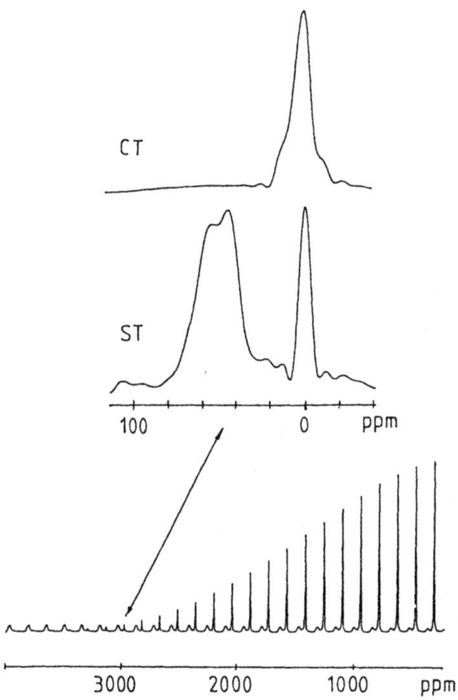

Fig. 15. Application of ^{11}B SATRAS for investigations of a Na_2O-B_2O_3-SiO_2 glass (spectra provided by L. van Wüllen [95]). As proved by the bottom spectrum the ST MAS lines of the BO_4 and BO_3 units are separated because of their different quadrupole shift. Both CT MAS and a selected ST MAS pair are compared by plotting those spectra on an identical frequency scale

In fact, a complete separation of both lines is only possible for $B_0 \geqq 14\,T$ [26]. Since such strong magnets are not available in most laboratories one has to look for another opportunity. The goal of resolving the BO_3 and BO_4 resonances even at the 'low' field of 7.1 T is achieved by acquiring the ST using high-speed MAS. Since the quadrupole shift of the ST MAS lines is twice as large and also in the opposite direction compared with the CT pattern, the BO_3 and BO_4 lines are separated because of their different quadrupole coupling (Fig. 15, bottom spectrum [95]). Both the CT MAS centre band and a selected single ST MAS sideband are also shown for comparison plotted in the same frequency scale. Clearly, there is no gain in resolution but the lines are separated and can now be analysed independently.

The advantages are straightforward as has been shown recently [95]. Since the lines are completely separated their centres of gravity can be determined with high accuracy. This enables studies of the small changes of the isotropic chemical shifts as function of the glass composition which also applies to the quadrupole interaction. Hence both parameters can be used for further discussions of structural details in glasses.

Of course, the whole procedure is not restricted to $I = 3/2$ nuclei. Referring to Table 2 many other suitable combinations of transitions are possible to exploit provided the prerequisites mentioned here are fulfilled.

5 Summary and Outlook

SATRAS is a powerful method for improved investigations of the atomic scale structure of both polycrystalline and disordered solids. The determination of the quadrupole interaction and the chemical shift is the main objective using MAS and single-pulse irradiation. Hence, SATRAS is an easy experiment provided a fast MAS probe is available. Under these circumstances the quadrupole interaction can be exploited as an additional source of structural information supplementary to the chemical shift in polycrystalline solids.

However, the main advantage of SATRAS for $I = 5/2$ and $I = 9/2$ nuclei is the 'natural' gift of a considerable additional suppression of second-order quadrupolar line broadening of certain inner ST compared with the CT MAS line. This additional line narrowing (under the same MAS condition as for the CT and nothing else) leads to nearly completely resolved ST MAS lines of differently coordinated Al species for example. In addition, it simplifies the simulation and fit of the ST MAS considerably since it allows an independent treatment of each of the Al resonances even in disordered solids. It is important to note that all interaction parameters such as the mean chemical shifts and the quadrupole interactions including their distribution widths can be obtained with just one experiment at one magnetic field strength. Most importantly, the old problem of getting the correct relative numbers of differently coordinated Al species, e.g., in glasses and ceramics is now solved using the ST as explained

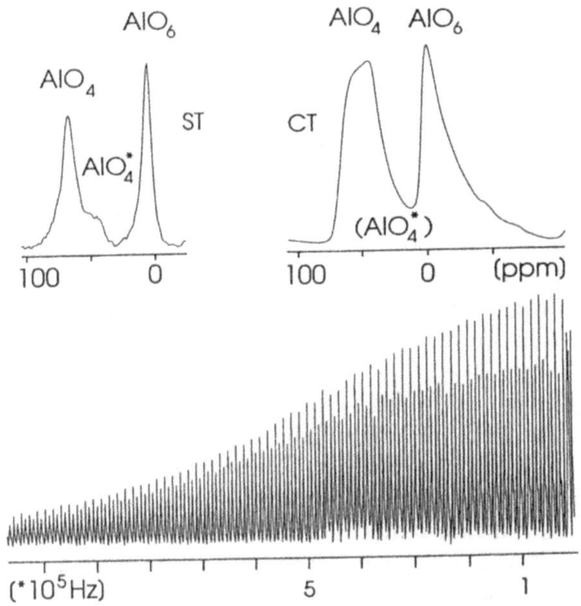

Fig. 16. ^{27}Al MAS NMR spectra ($B_0 = 11.7\,T$) of the same mullite sample as in Fig. 1. showing the possible progress using SATRAS. Although a much stronger B_0 field has been applied, the CT MAS patterns of the three Al coordinations are not better resolved than with the lower field of 7.1 T. However, the ST MAS lines of the AlO_6 and the two AlO_4 signals are completely resolved. More importantly, there is now a splitting between the two crystallographically different AlO_4 tetrahedra with AlO_4^* denoting those tetrahedra which connect the chains running along the crystallographic c-axis [73]. Those two AlO_4 signals superimpose in the CT MAS pattern, but they are much better resolved in the ST MAS pattern. This will allow the determination of the relative amount of Al in the different units to be made

in Sect. 4 and, indeed, reliable numbers are obtained by exploiting both ST and CT spectra simultaneously. Using the proposed model the asymmetric CT MAS line shape can be fitted without assumption on the basis of the ST MAS results.

The increased spectral resolution enables more opportunities. It is now possible to resolve the ST MAS lines of crystallographically different polyhedra with the same coordination number as it has been shown recently for the two 5-coordinated Al species of polycrystalline alumoborate $9Al_2O_3 \cdot 2B_2O_3$ [92]. Mullites are another example with two different tetrahedrally coordinated Al polyhedra [73]. In fact, the problems of NMR of quadrupolar nuclei have been visualised by presenting the 'ordinary' CT MAS pattern of a mullite sample in Fig. 1 at the beginning of this review. The achievable progress using SATRAS is documented in Fig. 16 showing that it is possible to distinguish between the two tetrahedra spectroscopically. Whereas it is impossible to quantify the relative amount of Al in the two AlO_4 units using the CT, the far better resolution of the inner ST MAS lines will give these numbers; work on which is in progress now. Obviously, SATRAS is the only method for obtaining this number in powder samples.

SATRAS can always be applied if distributions of the quadrupole parameters are present. Hence, for these cases a considerable improvement of structural investigations can be predicted, especially if one is interested in correct quantitative deconvolutions of the spectra. Referring to Sect. 2.1 and to $I = 5/2$ nuclei, SATRAS can be applied preferably, e.g., for quantitative investigations of thermal transformations of kaolinite, of the sol-gel route for preparing various materials (mullites, ceramics), of the formation of amorphous alumina films and even of specific mixed aluminium local coordinations AlO_yN_{4-y} ($y = 1, 2$ or 3) in ceramics. In all these cases it is the increased spectral resolution in conjunction with the high MAS spinning speed which enables the distinct progress using this new approach to high resolution NMR of quadrupolar nuclei.

Acknowledgements. It is a pleasure to thank my friends and colleagues G. Kunath, Dr. P. Losso and Dr. D. Ehrt (Friedrich-Schiller-Universität Jena) for their support and providing some of the samples. The fruitful collaboration with Dr. S. Steuernagel (Bruker Analytische Meßtechnik, Rheinstetten) and Prof. W. Müller-Warmuth, C. Mundus and L. van Wüllen (Westfälische Wilhelms-Universität, Münster) is acknowledged. The financial support of the Deutsche Forschungsgemeinschaft within the Heisenberg programme is gratefully acknowledged which made among many other benefits- a visit to the University of Warwick (Prof. R. Dupree) possible where the ^{17}O NMR experiments and the investigations of the AlON and sialon ceramics were carried out. I would also like to thank Dr. N. C. Nielsen and Dr. J. Skibsted (University of Aarhus) and Dr. A. Samoson (Tallinn, Estonia) for providing their figures used in this review. Finally, critical comments by Dr. H. Titman (Max-Planck-Institut für Polymerforschung, Mainz) and Dr. M. E. Smith (University of Kent, Canterbury) are gratefully acknowledged.

6 References

1. Andrew ER, Bradbury A, Eades RG (1958) Nature 182: 1659; Andrew ER (1970) Nuclear magnetic resonance in rapidly rotating solids. In: Coogan CK, Ham NS, Stuart SN, Pilbrow JR, Wilson GHH (eds) Magnetic Resonance. Plenum, New York, and cited references therein
2. Schnabel B (1965) Exp. Techn. Phys. XIII: 289
3. Andrew ER, Bradbury A, Eades RG, Jenks GJ (1960) Nature 188: 1096
4. Gerstein BC, Dybowski CR (1985) Transient techniques in NMR of solids. Academic New York; Haeberlen U (1976) High resolution NMR in solids: Selective Averaging. Academic, New York
5. Burum D, Rhim WK (1979) J. Chem. Phys. 71: 944
6. Scheler G, Haubenreißer U, Rosenberger H (1981) J. Magn. Reson. 44: 134
7. Cohen MH, Reif F (1957) Solid State Physics 5: 321
8. Slichter CP (1989) Principles of magnetic resonance. Springer, Berlin Heidelberg New York
9. Engelhard G, Michel D (1987) High resolution solid state NMR of silicates and zeolites, Wiley, Chichester
10. Behrens HJ, B. Schnabel (1982) Physica B114: 185
11. Samoson A, Kundla E, Lippmaa E (1982) J. Magn. Reson. 49: 350
12. Ganapathy S, Schramm S, Oldfield E (1982) J. Chem. Phys. 77: 4360

13. Müller D, Gessner W, Behrens HJ, Scheler G (1981) Chem. Phys. Lett 79: 59
14. Müller D, Berger G, Grunze I, Ladwig G, Hallas E, Haubenreißer U (1983) Phys. Chem. Glasses 24: 37
15. Engelhard G, Nofz M, Forkel K, Wihsmann FG, Mägi M, Lippmaa E (1985) Phys. Chem. Glasses 26: 157
16. Samoson A, Lippmaa E, Pines A (1988) Mol. Phys. 65: 1013
17. Llor A, Virlet J (1988) Chem. Phys. Lett. 152: 248
18. Chmelka BF, Mueller KT, Pines A, Stebbins J, Wu Y, Zwanziger JW (1989) Nature 339: 42
19. Bruker Almanach (1990)
20. Klinowski J (1991) Chem. Rev. 91: 1459
21. Fyfe CA, Feng Y, Grondey H, Kokotailo GT, Gies H (1991) Chem. Rev. 91: 1525
22. Eckert H (1992) Progr. Nucl. Magn. Reson. Spectrosc. 24: 159
23. Akitt JW (1989) Progr. Nucl. Magn. Reson. Spectrosc. 21: 1
24. Dye JL, Ellaboudy AS, Kim J (1991) Solid state NMR of quadrupolar nuclei. In: Popov AI, Hallenga K (eds) Practical Spectroscopy Series, vol 11, Marcel Dekker, New York
25. Freude D, Haase J (1993) Quadrupole effects in solid state NMR, Heidelberg New York NMR—Basic Principles and Progress, vol 29. Springer, Berlin
26. Dec SF, Maciel GE (1990) J. Magn. Reson. 87: 153
27. Haase J, Oldfield E, Schmitt K (1992) Chem. Phys. Lett. 193: 274
28. Bunker BC, Tallant DR, Kirkpatrick RJ, Turner GL (1990) Phys. Chem. Glasses 31: 30
29. Ganapathy S, Schramm S, Oldfield E (1982) J. Chem. Phys. 77: 4360
30. Jäger C, Barth S, Feltz A (1989) Chem. Phys. Lett. 154: 45
31. Engelhard G, Koller H, Sieger P, Depmeier W, Samoson A (1992) Solid State Nucl. Magn. Reson. 1: 127
32. Jelinek R, Ozkar S, Ozin GA (1992) J. Phys. Chem. 96: 5949
33. Lin CF, Chao KJ (1990) J. Phys. Chem. 95: 9411
34. Veeman WS (1991) Z. Naturforsch. 47a: 353
35. Janssen R, Tijink GAH, Veeman WS, Maesen TLM, van Lent JF (1989) J. Phys. Chem. 93: 899
36. Ohki H, Nakamura N (1992) Z. Naturforsch. 47a: 319
37. Peterson GE, Bridenbaugh PM (1968) J. Chem. Phys. 48: 3402
38. Halstead TK (1970) J. Chem. Phys. 53: 3427
39. Baiqin L, Yening W, Ziran X (1988) J. Phys. C: Solid State Phys 21: L251
40. Strange JH, Rageb SM, Slade RCT (1991) Philos. Mag. A64: 1159
41. Strange JH, Rageb SM, Chadwick AV, Flack KW (1990) J. Chem. Soc. Faraday Trans 86: 1239
42. Grüne M, Meierkord H, Müller-Warmuth W, zum Hebel P, Krebs B, Wulf M (1989) Ber. Bunsenges. Phys. Chem 93: 1313
43. Wang B, Szu SP, Greenblatt M (1991) J. Non-Cryst. Solids 134: 249
44. Dupree R, Holland D, Mortuza MG (1990) J. Non-Cryst. Solids 116: 148
45. Eckert H, Zhang Z, Kennedy JH (1989) J. Non-Cryst. Solids 107: 271
46. Baltisberger JH, Gann SL, Wooten EW, Chang TH, Mueller KT, Pines A (1992) J. Am. Chem. Soc. 114: 7489
47. Farnan I, Grandinetti PJ, Baltisberger JH, Stebbins JF, Werner U, Eastman MA, Pines A (1992) Nature 358: 31
48. Mueller KT, Sun BQ, Chingas GC, Zwanziger JW, Terao T, Pines A (1990) J. Magn. Reson. 86: 470
49. Wu Y, Sun BQ, Pines A, Samoson A, Lippmaa E (1990) J. Magn. Reson. 89: 297
50. Mueller KT, Wu Y, Chmelka BF, Stebbins J, Pines A (1991) J. Am. Chem. Soc. 113: 33
51. Timken HKC, Turner GL, Gilson JP, Welsh LB, Oldfield E (1986) J. Am. Chem. Soc. 108: 7231
52. Timken HKC, Schramm SE, Kirkpatrick RJ, Oldfield E (1987) J. Phys. Chem. 91: 1054
53. Walter TH, Oldfield E (1989) J. Phys. Chem. 93: 6744
54. Brinkmann D (1993) High T_c superconductors. NMR—Basic Principles and Progress, vol 31, Springer, Berlin Heidelberg New York
55. Special Issue (1993) Appl. Magn. Reson. 4: 1–236
56. Haase J, Freude D, Fröhlich T, Himpel G, Kerbe F, Lippmaa E, Pfeifer H, Sarv P, Schäfer H, Seiffert B (1989) Chem. Phys. Lett. 156: 328
57. Smith ME (1992) J. Phys. Chem. 96: 1444
58. Farnan I, Dupree R, Forty AJ, Jeong YS, Thompson GE, Wood GC (1989) Philos. Mag. Lett. 59: 189
59. Wu Y, Chmelka BF, Pines A, Davies ME, Grobet PJ, Jacobs PA (1990) Nature 346: 550
60. Grobet PJ, Samoson A, Geerts H, Martens JA, Jacobs PA (1991) J. Phys. Chem. 95: 9620

61. Gornostansky SD, Stager CV (1967) J. Chem. Phys. 46: 4959
62. France PW (1991) J. Magn. Reson. 92: 30
63. Nabavi M, Sanchez C, Livage J (1991) Philos. Mag. B63: 941
64. Iyer PS, Eckert H, Occelli ML, Stencel JM (1991) ACS Symp. Ser. 452: 242
65. Hardcastle FD, Wachs IE, Eckert H, Jefferson DA (1991) J. Solid State Chem. 90: 194
66. Nabavi M, Taulelle F, Sanchez C, Verdaguer M (1991) J. Phys. Chem. Solids 51: 1375
67. Baugher JF, Taylor PC, Oja T, Bray JP (1969) J. Chem. Phys. 50: 4914
68. Han OH, Timken HKC, Oldfield E (1988) J. Chem. Phys. 89: 6046
69. Douglass DC, Petersson GE (1986) J. Am. Ceram. Soc. 69: 48
70. Pfändler S, Kind R, Dolinsek J (1992) 26th Congress Ampere on Magnetic Resonance, Athens, Greece, Conference Proceedings, p 245
71. Petersen GE, Carruthers JR, Carnevale A (1970) J. Chem. Phys. 53: 2436
72. Man P, Theveneau H, Papon P (1985) J. Magn. Reson. 64: 271
73. Angel RJ, Prewitt CT (1986) Am. Mineralogist 71: 1476; Angel RJ, McMullan RK, Prewitt CT (1991) Am. Mineralogist 76: 332
74. Merwin LH, Sebald A, Rager H, Schneider H (1991) Phys. Chem. Minerals 18: 47
75. Sanz J, Madami A, Serratosa JM, Moya JS, Aza S (1988) J. Am. Ceram. Soc. 10: C418
76. Samoson A (1985) Chem. Phys. Lett. 119: 29
77. Müller D, Gessner W, Samoson A, Lippmaa E (1986) Polyhedron 5: 779
78. Phillips BL, Allen FM, Kirkpatrick RJ (1987) Am. Mineralogist 72: 1190
79. Jakobsen HJ, Skibstedt J, Bildsøe H, Nielson NC (1989) J. Magn. Reson. 85: 173
80. Nielsen NC, Bildsøe H, Jakobsen HJ (1991) Zeolites 11: 622
81. Skibstedt J, Bildsøe H, Jakobsen HJ (1991) J. Magn. Reson. 92: 669
82. Skibstedt J, Nielsen NC, Bildsøe H, Jakobsen HJ (1991) J. Magn. Reson. 95: 88
83. Skibstedt J, Nielsen NC, Bildsøe H, Jakobsen HJ (1992) Chem. Phys. Lett. 188: 405
84. Skibstedt J, Henderson E, Jakobsen HJ (1993) Inorganic Chemistry (in press)
85. Jakobsen HJ, Daugaard P, Langer VJ (1988) J. Magn. Reson. 76: 162
86. Massiot D, Kahn-Harari A, Michel D, Müller D, Taulelle F (1990) Magn. Reson. Chem. 28: S82
87. Taulelle F, Bessada C, Massiot D (1992) J. Chim. Phys. 89: 379
88. Massiot D, Bessada C, Coutures JP, Taulelle F (1990) J. Magn. Reson. 90: 231
89. Alemany LB, Massiot D, Sherriff BL, Smith ME, Taulelle F (1991) Chem. Phys. Lett. 177: 301
90. Jäger C, Rocha J, Klinowski J (1992) Chem. Phys. Lett. 188: 208
91. Jäger C (1992) J. Magn. Reson. 99: 353
92. Jäger C, Müller-Warmuth W, Mundus C, van Wüllen L (1992) J. Non-Cryst. Solids 149: 209
93. Kunath G, Losso P, Steuernagel S, Schneider H, Jäger C (1992) Solid State Nucl. Magn. Reson. 1: 261
94. Jäger C, Kunath G, Losso P, Scheler G (1993) Solid State Nucl. Magn. Reson. 2: 73
95. van Wüllen L (1993) PhD Thesis, Westfälische Wilhelms-Universität Münster, FRG
96. Cheng JT, Edwards JC, Ellis PD (1990) J. Phys. Chem. 94: 553
97. Power WP, Wasylishen RE, Mooibroek S, Pettitt BA, Danchura W (1990) J. Phys. Chem. 94: 591
98. Dupree R, Holland D, Mortuza MG (1990) J. Non-Cryst. Solids 116: 148
99. Losso P, Schnabel B, Jäger C, Sternberg U, Stachel D, Smith DO (1992) J. Non-Cryst. Solids 143: 265
100. Sternberg U (1987), 3rd Otto-Schott Colloquium Jena, poster contribution
101. Alemany LB, Timken HKC, Johnson ID (1988) J. Magn. Reson. 80: 438
102. Samoson A (1992) 26th Congress Ampere on Magnetic Resonance, Athens, Greece Conference Proceedings, p 263
103. Smith ME, Steuernagel S (1992) Solid State Nucl. Magn. Reson. 1: 175
104. Fitzerald JJ, Dec SF, Hamza AI (1989) Am. Mineral. 74: 1405
105. Garsche M, Tillmanns E, Almen H, Schneider H, Kupcik V (1991) Eur. J. Mineral. 3: 793
106. Jäger C, Dupree R, Kohn SC, Mortuza MG (1993) J. Non-Cryst. Solids 155: 95
107. Jäger D, Ehrt D (1988) Exp. Techn. d Physik, 36, 349 (1988)
108. Jäger C, Ehrt D, unpublished data
109. Jäger C, Kunath G, Losso P (1992) 26th Congress Ampere on Magnetic Resonance Athens, Greece, Conference Proceedings, p 375
110. Schirmer OF, Thiemann O, Wöhlecke M (1991) J. Phys. Chem. Solids 52: 185
111. Donnerberg HJ, Tomlinson SM, Catlow (1991) J. Phys. Chem. Solids 52: 201
112. Jellison GE, Feller SA, Bray PJ (1977) J. Magn. Reson. 27: 121

NMR-NQR Studies of High-Temperature Superconductors

Detlef Brinkmann and Michael Mali

Physik-Institut, University of Zürich, Winterthurerstr. 190, CH-8057 Zürich, Switzerland

Table of Contents

This paper reviews recent nuclear magnetic resonance (NMR) and nuclear quadrupole resonance (NQR) investigations performed in high-temperature superconductors, in particular in members of the Y-Ba-Cu-O and La-Sr-Cu-O families. The review focuses on investigations of spin–lattice and spin–spin relaxation times, of Knight shifts and electric field gradients in these compounds. By discussing several representative experimental studies and some present-day theories it is demonstrated how NMR/NQR studies have contributed to a better understanding of the electronic properties of high-temperature superconductors.

1 Introduction

Nuclear Magnetic Resonance (NMR) played an important role in understanding "classical" superconductors such as aluminum [1] when the Bardeen-Cooper-Schrieffer (BCS) theory provided the first microscopic explanation of this extraordinary state of matter. Thus it was logical that soon after the discovery of the new high-temperature superconductors (HTSC) by G. Bednorz ad K. A. Müller [2], the NMR and nuclear quadrupole resonance (NQR) community devoted much activity to these new materials. The first results which were reported by mid-1987 [3–6] dealt with many of the important parameters which can be studied by NMR/NQR such as electric field gradient, Knight shift and relaxation. In the last few years the activity has steadily expanded.

Even seven years after their discovery there is no complete and accepted "theory" for the HTSC and intensive theoretical research is still going on. A major issue is the question of whether or not the HTSC are described by some sort of generalized BCS theory, one with pairing but possibly with an attractive interaction arising from some mechanism other than lattice vibrations. Another matter of debate is the role played by anti-ferromagnetic (AF) fluctuations and correlations present in these materials. Do the spin correlations mediate the formation of the Cooper pair? Even the normal state properties are not fully understood. For a comprehensive discussion of the various and partly controversial theoretical issues we refer to Ref. [7].

The power of NMR and NQR in the studies on the various aspects of superconductivity lies in the fact that these methods can obtain information on static and dynamic properties on an atomic scale. How this has been accomplished for "classical" superconductors has been reviewed in [8]. For the HTSC, an immense amount of detailed information on structural, electronic and superconducting aspects has been collected in recent years. Studies of the most intensively investigated structure, $YB_2Cu_3O_7$, were reviewed in [9, 10]; work on these and many other compounds is described in [11]. A summary of the present activities of many leading NMR/NQR laboratories in the field of HTSC research has recently been published [12].

Because of limited space, this review cannot deal with all facets of NMR/NQR studies in all compounds. Instead, we will restrict our discussion to the *present* research and in doing so we will deal mainly with members of the Y-Ba-Cu-O and the La-Sr-Cu-O families. Furthermore, we will focus on investigations of spin-lattice and spin–spin relaxation times and Knight shift together with a section on electric field gradients.

Rather than listing here the prominent issues of NMR/NQR research in HTSC, we will introduce them successively when discussing the NMR parameter that supplies information on the respective question and issue. Since each of these HTSC can be probed by various isotopes, in each chapter dealing with a certain NMR parameter, we will collect the most important studies for a particular compound utilizing the various isotopes. By doing so we try to

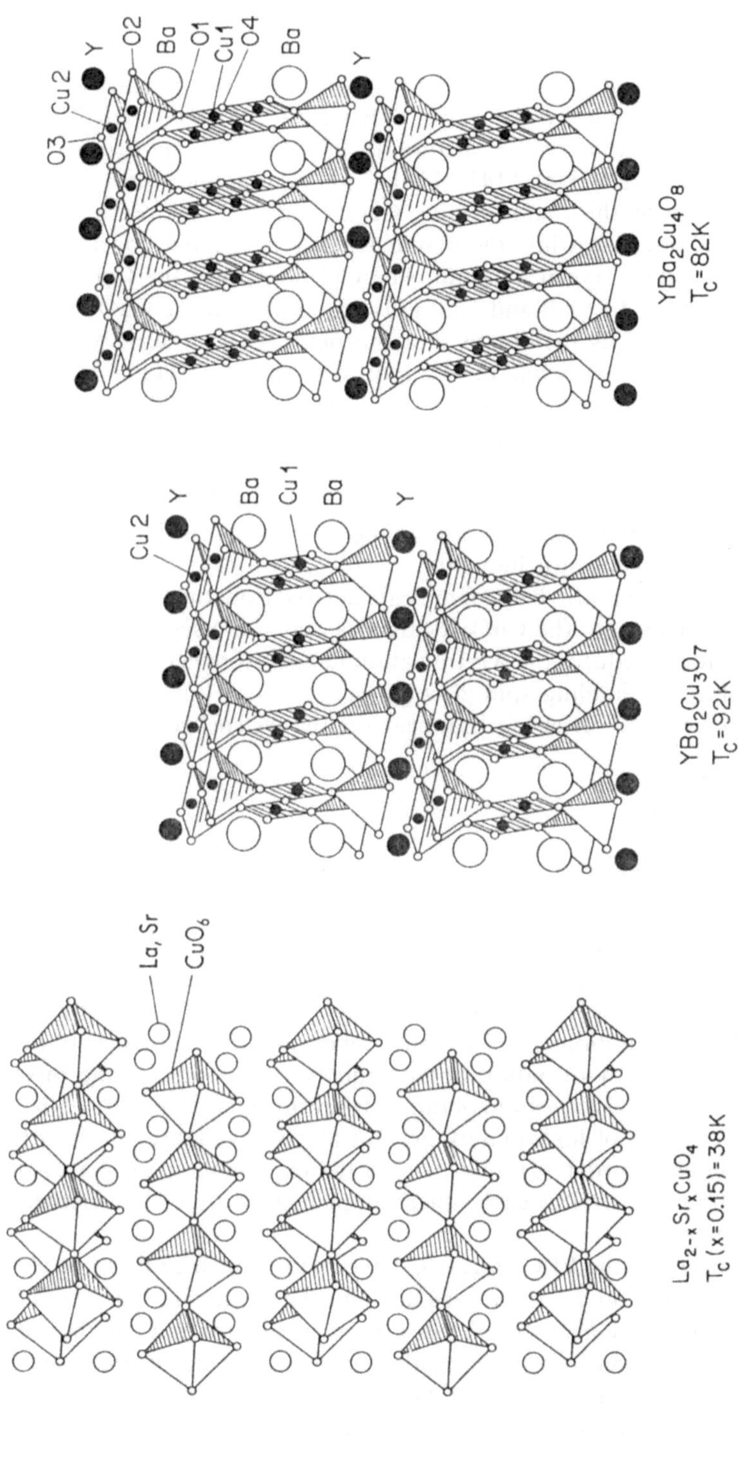

Fig. 1. Structure of $La_{2-x}Sr_xCuO_4$, $YBa_2Cu_3O_7$ and $YBa_2Cu_4O_8$

provide a report on the present state of the art of the HTSC NMR/NQR research.

Figure 1 sketches the structures of the most important compounds we are going to deal with. By doping with oxygen, $YB_2Cu_3O_{6+x}$ can be varied from an AF insulator with tetragonal structure ($x=0$) [13] to an orthogonal superconducting phase ($x \geqq 0.4$) [14]. There are two copper sites in the perovskite-like structure: the chain-forming Cu(1) ions sit at the center of an oxygen rhombus-like square while the planar Cu(2) ions are five-coordinated by an apically elongated oxygen pyramid. The O(4) ions form the Cu(1)-O(4) chains along the *b*-axis, the O(2) and O(3) ions are located in the CuO_2 plane and the apical O(1) bridges the chains and the planes. The Y ions separate CuO_2 planes along the *c*-axis while the Ba ions form the Ba-O(1) planes.

The orthorhombic unit cell of $YBa_2Cu_4O_8$ [15] can be considered as two $YBa_2Cu_3O_7$ unit cells joined chain-to-chain with the second cell displaced *b*/2 along the *b*-axis. Thus, instead of the single Cu-O chains of $YBa_2Cu_3O_7$, the $YBa_2Cu_4O_8$ structure contains double Cu-O chains which form an edge-sharing, square-planar network. The compound is distinguished by its high thermal stability and precise stoichiometry.

The compounds of the La-Sr-Cu-O family are derived from the AF La_2CuO_4. Essential building blocks are the CuO_6 octahedra which are separated by La and Sr. Doping with Sr reduces the AF transition temperature very rapidly and for $x \geqq 0.01$ a phase resembling spin-glass phase occurs at low temperature. For $x \geqq 0.08$ all traces of AF ordering disappear and the system becomes metallic and superconducting.

2 Electric Field Gradients

Before dealing with topics typical for HTSC, we will address a parameter, the electric field gradient (EFG) at a nuclear site, which is an important *static* property of any solid since it depends sensitively on the charge distribution around the nucleus in question. The EFG thus reflects structural aspects of the compound and provides information about ionicity and bonding. The most accurate values of the EFG tensor are obtained from single-crystal NMR. Since sufficiently large crystals of high quality are not always available one resorts to NMR of "oriented powder" samples (where the crystallographic *c*-axes of all the crystallites are parallel to each other) supplemented by a measurement of the NQR frequency.

Many calculations of the EFG in HTSC have been performed more or less based on a point-charge model where one distinguishes contributions from all the surrounding ions and from the valence electron distribution on the ion under consideration. Among the more reliable or nearly ab initio calculations we mention those by the groups of T. P. Das employing Hartree-Fock cluster investigations [16, 17] and K. H. Schwarz. The latter group has developed a method to compute EFGs from a full potential linearized augmented plane wave (LAPW) band structure calculation [18]. It should be emphasized that in these calculations there is no need for Sternheimer factors.

Results of an application to $YBa_2Cu_3O_7$, $YBa_2Cu_3O_6$ and $YBa_2Cu_3O_{6.5}$ [19] and to $YBa_2Cu_4O_8$ [20] have been reported. The good agreement with experimental values for $YBa_2Cu_3O_7$, $YBa_2Cu_3O_6$ and $YBa_2Cu_3O_{6.5}$ (except for the planar Cu(2) site) was discussed in [19]. In the meantime, the EFGs in $YBa_2Cu_4O_8$ for both Cu sites [21], all four O sites [22] and the Ba site [23] have been measured. Together with the recent experimental result for the Ba site in $YBa_2Cu_3O_7$ [24], it has now been established that the LAPW method within the local density approximation adequately describes the charge distributions for Cu(1), all O and Ba sites in $YBa_2Cu_3O_7$ and $YBa_2Cu_4O_8$. The theory reveals [20] that the Cu(1) EFG is dominated by the valence contribution, that the Ba EFG originates from the interaction with the oxygen neighbors and that by going from $YBa_2Cu_3O_7$ to $YBa_2Cu_4O_8$ one expects to see effects in the chain O position. In particular, the deviation of theoretical from experimental oxygen EFGs is about 12% for all sites.

For the apex O(1) and the planar O(2) and O(3) sites, the corresponding EFG tensors of $YBa_2Cu_3O_7$ and $YBa_2Cu_4O_8$ differ by only about 10%. However, a pronounced difference between both systems is observed for the chain O(4) site, whose asymmetry parameter changes from 0.48 in $YBa_2Cu_3O_7$ to 0.868 in $YBa_2Cu_4O_8$. Thus, the charge distribution in both structures is quite similar except for the chain oxygen site where the doubling of the chain affects the asymmetry of the charge distribution and causes a charge transfer from O(2) and O(3) to Cu(1) [20]. This exemplifies the interplay between the chains and the planes in these structures.

However, for the Cu(2) site in all these structures the theoretical values are too low, in case of $YBa_2Cu_4O_8$ by a factor of about 3. The origin of this discrepancy probably is that the local density approximation does not fully account for exchange-correlation effects between Cu-3*d* states [19]. In other words: the EFG calculations confirm the exceptional case played by the planar Cu and that the CuO_2 planes are mainly responsible for the superconductivity.

The dependence of the EFG at Cu sites on temperature and external pressure has been studied in several HTSC by observing the corresponding changes of the NQR frequencies, ν_Q. Since the ab initio calculations of the EFGs are not yet precise enough for a description of small changes of ν_Q one quite often resorts to the point charge model. The valence contribution, V_{zz}^{VAL}, originating from the holes in the open Cu 3*d* shell can be written in terms of the number of holes in the different *d* orbitals as follows:

$$V_{zz}^{VAL} = A\langle r^{-3}\rangle_{3d}[n_h(3d_{3z^2-r^2}) - n_h(3d_{x^2-y^2}) - n_h(3d_{xy}) + \tfrac{1}{2}n_h(3d_{xz}) - \tfrac{1}{2}n_h(3d_{yz})] \quad (1)$$

where A is a constant. This way one can estimate the hole concentration from the change of ν_Q.

A correlation between the EFG and the valence sum at the Cu(2) site in $Tl_2Ca_2Ba_2Cu_3O_y$, $Bi_2Ca_2Sr_2Cu_3O_y$, $Bi_2CaSrCuO_y$ and $(La, Sr)_2CuO_4$ has been found [25] which seems to emphasize the importance of the oxygen polarizability since the in-plane Cu-O bond lengths differ only slightly. The pressure dependence of the EFG allows one to draw conclusions about the change of the Cu valence state [26, 27] and about the charge transfer from chains to planes [28]. Some earlier results are discussed in [29]. Finally, EFG studies near T_c may answer the question whether the electronic transition is accompanied by some other type of transition. For $YBa_2Cu_3O_7$ and $YBa_2Cu_4O_8$, in our laboratory no evidence for a structural transition at T_c could be found [27] while Riesemeier et al. [30] reported for $YBa_2Cu_3O_7$ a discontinuous change of the NQR frequency in the vicinity of T_c which could be due to freezing out of a vibrational mode of the oxygen system.

3 Knight Shifts

Among the most important NMR parameters in HTSC studies which reflect *static* properties is the Knight shift, $K(T)$. It is defined as the relative displacement of the NMR frequency in a metal with respect to the frequency in a non-metallic reference compound. In the field of HTSC NMR it is customary to define $K(T)$ as that part of the total shift that arises from the interaction between the nuclear spin and the *spin* paramagnetism of the conduction electrons. The experiment yields a total relative resonance frequency shift given by

$$K_{tot}(T) = K_{chem} + K(T) + (1 - N)M(T)/B_0 \tag{2}$$

where K_{chem} denotes the chemical (or orbital) shift, M is the bulk magnetization, N the demagnetizing factor and B_0 is the external magnetic field. We have indicated that only the chemical shift, K_{chem}, is assumed to be temperature independent.

The importance of the Knight shift stems from the fact that it is proportional to the static electron spin susceptibility, χ_s, and the hyperfine interactions, A, of the nucleus. Thus, K may be expressed as

$$K(T) = \frac{1}{g\mu_B} \sum_i A_i \chi_s \tag{3}$$

where g is the Landé factor and μ_B the Bohr magneton. For a classical superconductor with s-wave spatial symmetry (as in the BCS theory), the Knight shift is constant in the normal state, declines in the superconducting state due to spin-singlet pairing and approaches zero with zero slope [31]. In several HTSC, however, K starts to decrease far above T_c and the shift is anisotropic.

3.1 Planar Cu and O and Y Shifts in Y-Ba-Cu-O

Before introducing the experimental facts, let us briefly discuss the different hyperfine interactions of nuclear spins in the CuO_2 plane of Y-Ba-Cu-O structures. Figure 2 shows the hybridization of the Cu-$3d_{x^2-y^2}$ and the O-$2p_x$ and $2p_y$ orbitals. To explain the Knight shift anisotropy and, as we will see later, the relaxation, Mila and Rice [32] proposed the following model of the hyperfine interactions. For the Cu(2) nucleus, there is a contribution A from the interaction with an electron of the on-site $3d_{x^2-y^2}$ orbital and a transferred interaction B with electrons of the $3d_{x^2-y^2}$ orbitals of the four neighboring copper ions. For the planar oxygen nuclei, there is a transferred hyperfine coupling C. Due to the weak hybridization of Cu-4s and O-2s states with Cu-d and O-p states, an isotropic hyperfine coupling between Cu and O may be anticipated. This has been confirmed by the large positive isotropic Knight shift at the planar Cu and O sites.

The Y ions are located between two CuO_2 planes and they can interact with eight Cu and O neighbors. The transferred hyperfine coupling D is expected to be mainly isotropic and to occur via the less distant O-2p orbitals rather than the Cu-3d orbitals.

The Knight shift must be extracted from the total shift according to Eq. (2); details of the evaluation are described for instance in [33, 34]. We denote by K_α and $K_\parallel$ ($K_\perp$) a Knight shift component measured with the external field, B_0, parallel to the crystallographic α axis ($\alpha = a, b, c$) or parallel (perpendicular) to the Cu-O bond, respectively. Upper indices are used to distinguish the isotopes. One then can determine the isotropic and axial (anisotropic) part of the Knight

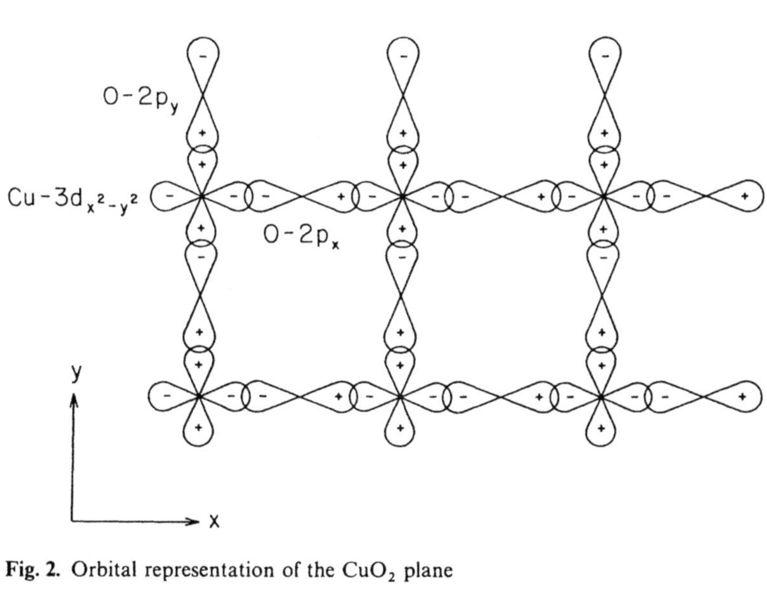

Fig. 2. Orbital representation of the CuO_2 plane

shift according to $^{63}K_{iso} = (^{63}K_a + {}^{63}K_b + {}^{63}K_c)/3$, $^{63}K_{ax} = (^{63}K_{ab} - {}^{63}K_c)/3$, $^{17}K_{iso} = (^{17}K_a + {}^{17}K_b + {}^{17}K_c)/3$ and $^{17}K_{ax} = (^{17}K_{\parallel} - {}^{17}K_{\perp})/3$. The anisotropic parts of the Cu and O shift probe the spin density on the Cu-*d* and O-*p* states, respectively, since the *s*-electrons contribute only to the isotropic Knight shift.

Figures 3 and 4 show the temperature dependence of various planar Cu and O Knight shift components in $YBa_2Cu_3O_{6.63}$ [35] and $YBa_2Cu_4O_8$ [36, 22, 37], respectively. K_c is zero due to a cancellation of on-site and transferred hyperfine interactions in both compounds. For each compound, all components exhibit the same temperature dependence with a decline which starts well above T_c; the behavior of both compounds is very similar. These facts support the "single spin fluid" model which states that the Cu-*d* holes and the doped holes (which mainly go into O-2*p* states) have one spin degree of freedom. Similar ^{17}O data for $YBa_2Cu_4O_8$ have been reported by Machi et al. [38] and similar Cu Knight shift data were obtained in $Bi_2Sr_2CaCu_2O_8$ [39].

Strong evidence for the single spin fluid model has been provided by ^{89}Y and ^{205}Tl Knight shift investigations. The group led by Alloul has pioneered the ^{89}Y NMR HTSC and has demonstrated that it allows one to probe the static (and dynamic) susceptibilities of the CuO_2 plane in $YBa_2Cu_3O_{6+x}$ [40, 41]

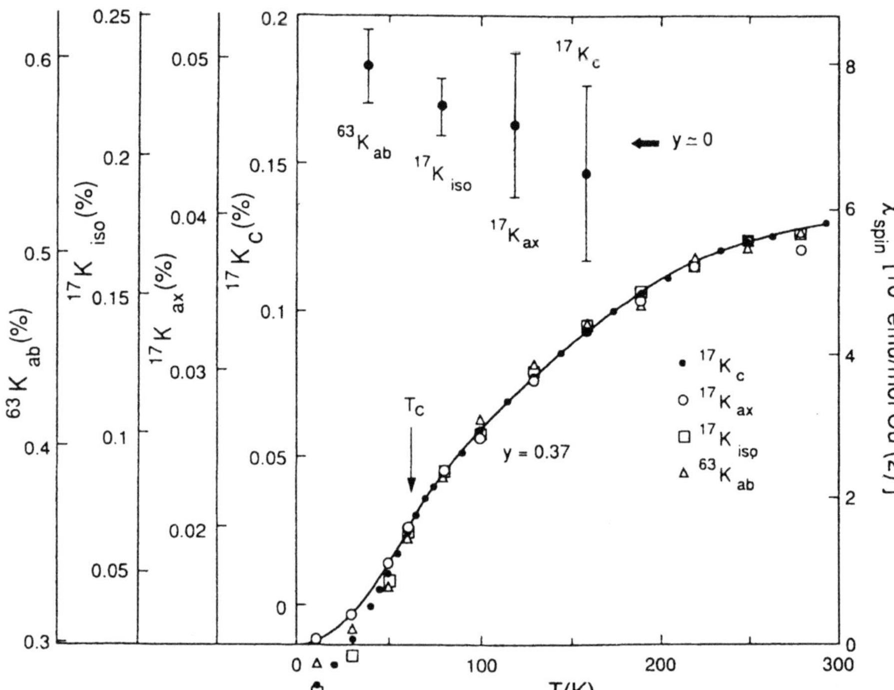

Fig. 3. Temperature dependence of various Knight shift components at Cu and O sites in $YBa_2Cu_3O_{6.63}$. From [35]

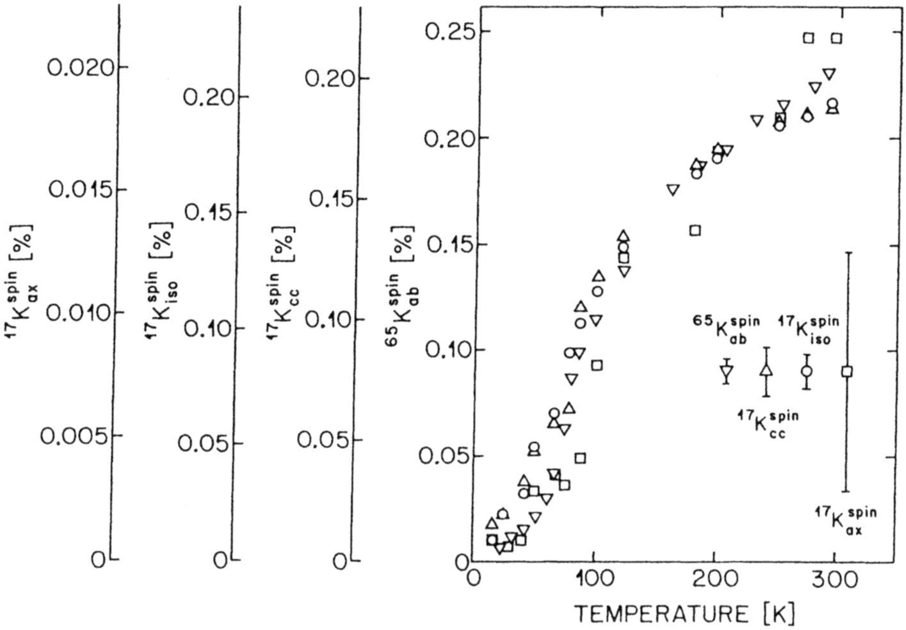

Fig. 4. Temperature dependence of various Knight shift components at Cu and O sites in $YBa_2Cu_4O_8$. From [37]

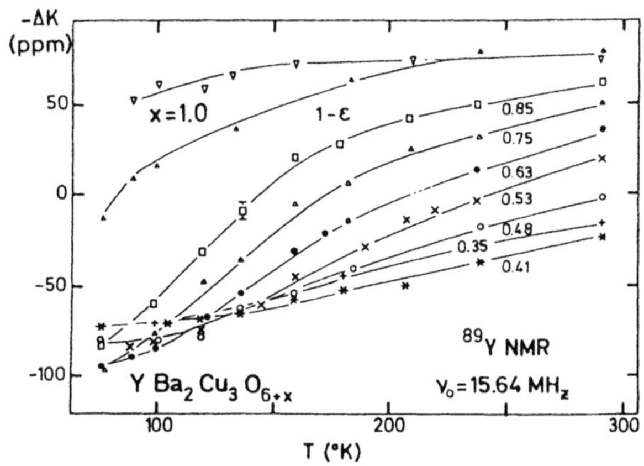

Fig. 5. Temperature dependence of Y Knight shift for a series of $YBa_2Cu_3O_{6+x}$ samples with YCl_3 used as reference. From [41]

via the Y Knight shift. Figure 5 shows the temperature dependence of the ^{89}Y total shift (including the chemical shift) for a series of $YBa_2Cu_3O_{6+x}$ samples ranging from the metallic to the semiconducting state. For $x = 1$ the shift is temperature independent as has already been shown for the Cu(2) shift in $YBa_2Cu_3O_7$ [9], but for smaller x values the shift has a large temperature dependence. For $x = 0.63$ the temperature dependence is found to scale with the planar ^{17}O Knight shift thus establishing the validity of the single spin fluid model.

We now turn to the very different temperature dependence of the Knight shift in over-doped samples where over-doped (under-doped) means doping that is over (under) the optimal doping for a maximum T_c. The C. Berthier group has performed detailed studies of the spin susceptibility as probed at Cu(2) and O(2, 3) sites in $YBa_2Cu_3O_{6+x}$ *single* crystals for the under-doped and over-doped regime (for a summary see [42]). As an example, Figs. 6 and 7 compare the principal components of the K_α tensor at the O(2, 3) sites in $YBa_{1.92}Sr_{0.08}Cu_3O_7$ and $YBa_2Cu_3O_{6.52}$ [42]. The slight increase of K_α in $YBa_{1.92}Sr_{0.08}Cu_3O_7$ with decreasing temperature is now recognized as the signature of slightly over-doped samples. In contrast, the K_α components in the under-doped $YBa_2Cu_3O_{6.52}$ decrease rapidly by lowering the temperature. This has been noticed also by other groups (for instance [35]) and also, as noted above, for the $YBa_2Cu_4O_8$ structure [36, 22, 37] which thus proves its low-doping character.

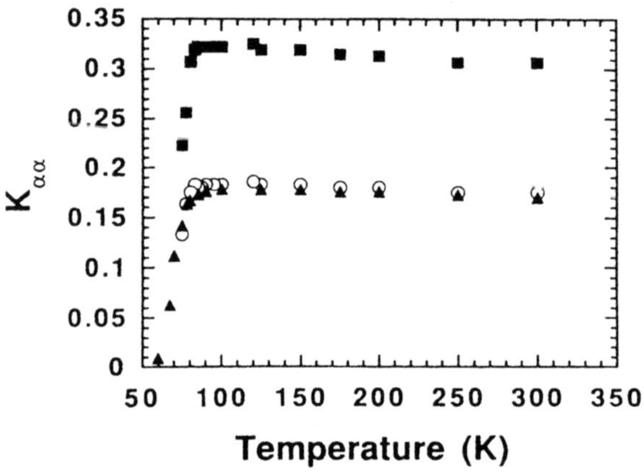

Fig. 6. Temperature dependence of the shift tensor at O(2, 3) sites in $YBa_{1.92}Sr_{0.08}Cu_3O_7$ with B_0 in the plane [$\parallel$ Cu-O-Cu bond (■), $\perp$ Cu-O-Cu bond (○)] and $B_0 \parallel c$ (▲). From [42]

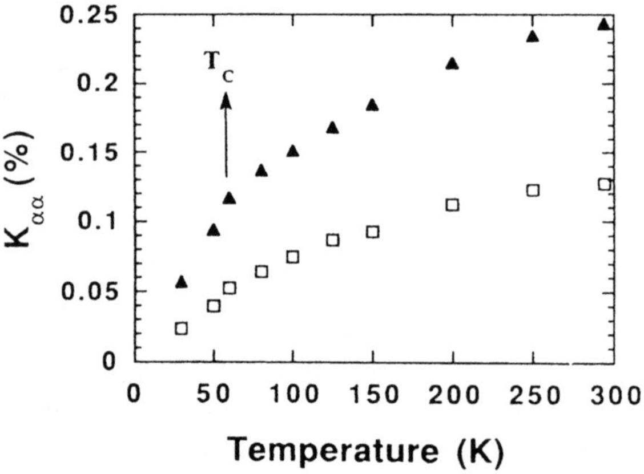

Fig. 7. Temperature dependence of the shift tensor at O(2, 3) sites in $YBa_2Cu_3O_{6.52}$ with B_0 in the plane [$\perp$ Cu-O-Cu bond (□), $\parallel$ Cu-O-Cu bond (▲)] and $B_0 \parallel c$. From [42]

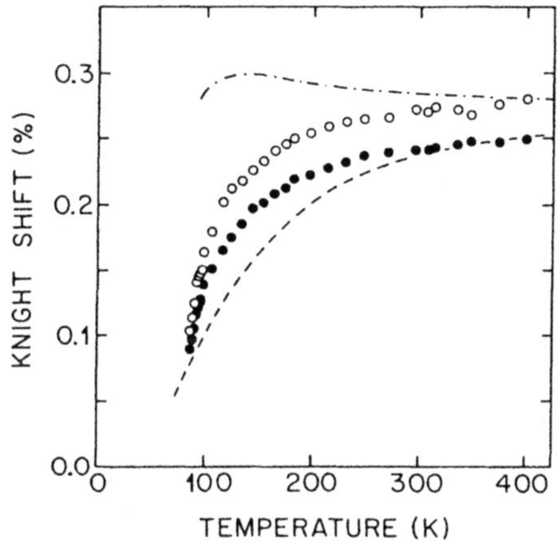

Fig. 8. Temperature dependence of the Knight shift of the two inequivalent planar Cu sites Cu(2) [$YBa_2Cu_3O_7$ block, *open circles*] and Cu(3) [$YBa_2Cu_4O_8$ block, *full circles*] in Y_2Ba_4·Cu_7O_{15}. The *dashed-dotted* and *dashed lines* represent the data measured in "pure" YBa_2Cu_3·O_7 and $YBa_2Cu_4O_8$, respectively. From [45]

A particular interesting Y-Ba-Cu-O structure is the "mixed-chain" compound $Y_2Ba_4Cu_7O_{15}$ which consists (along the *c* axis) of a sequence of alternating one-chain $YBa_2Cu_3O_7$ and double-chain $YBa_2Cu_4O_8$ blocks. The structure thus offers the possibility of studying the coupling between two adjacent *inequivalent* CuO_2 planes which have distinct carrier densities, a situation similar to the one encountered in (Bi, Pb)$_2Sr_2Ca_2Cu_3O_y$ [43, 44]. Figure 8 shows our results for the Knight shift at the two inequivalent Cu sites: Cu(2) in the $YBa_2Cu_3O_7$ block and Cu(3) in the $YBa_2Cu_4O_8$ block [45]. Apparently the coupling between the two inequivalent layers establishes a common (static) spin susceptibility. So to speak, the Cu(3)O_2 plane "imposes" its behavior with respect to the susceptibility in the Cu(2)O_2 plane.

3.2 Cu, O and Tl Shifts in La-Sr-Cu-O and Tl-Ba-Cu-O

It is well known that the T_c of HTSC is related to the hole concentration. Recent studies by the group of Asayama and Kitaoka at Osaka University considered the correlation between T_c and the AF spin fluctuations and therefore extended their investigations from the lightly-doped region near the magnetic insulator—superconducting transition in $La_{1.85}Sr_{0.15}CuO_4$ (La-Sr-Cu-O) to the heavily-doped region in $Tl_2Ba_2CuO_{6+y}$ (Tl-Ba-Cu-O) where T_c ranges from 85 to 0 K.

In La-Sr-Cu-O, both $^{17}K_c$ and $^{63}K_{ab}$ above T_c vary with temperature as the bulk susceptibility [46]. These behaviors are similar to those at the planar Cu and O sites and at Y in $YBa_2Cu_3O_{6.63}$. The hyperfine coupling constants $^{17}A_c$ and $^{63}A_{ab}$ estimated from the shift data, are almost the same as those in many Y-Ba-Cu-O compounds and hence there is no appreciable change in the static local character in the CuO_2 plane [47]. Since $^{63}A_{ab}$ is almost the same in La-Sr-Cu-O and $YBa_2Cu_3O_7$, it has been deduced that $\langle r^{-3}\rangle_{3d}$ is nearly the same in both. Thus, the local static structure of the $3d_{x^2-y^2}$ state in the plane is not significantly modified even though holes are doped into Y-Ba-Cu-O and La-Sr-Cu-O.

In Tl-Ba-Cu-O which has a single CuO_2 layer, the Cu Knight shift was found to be temperature independent in the normal state despite the inhomogeneity in the TlO_2 layers [48]. This reveals that the spin susceptibility in the CuO_2 layer is not significantly affected by such TlO_2 disorder. In Fig. 9, the Cu Knight shift components K_{ab} and K_c in Tl-Ba-Cu-O, La-Sr-Cu-O and $YBa_2Cu_3O_7$ are compared with respect to the hole content. Both components in the heavily-doped Tl-Ba-Cu-O increase with rising hole content. K_c in this compound is larger than those in lightly-doped structures. This is attributed not only to the larger spin susceptibility but also to an enhancement of the positive supertransferred hyperfine field B due to a slight increase of the weight factor of the 4s wave function in the 3d-2p-4s covalent bond [47]. This effect also plays an important role in relaxation.

Thus, the analysis of the Tl-Ba-Cu-O and La-Sr-Cu-O Knight shift data has shown that the mechanism of hyperfine coupling over the entire range of hole content in Y-Ba-Cu-O is well described by the Mila-Rice Hamiltonian involving almost localized spins.

Extensive studies by various groups have been performed on the Tl containing superconductors of the type $Tl_2Ba_2Ca_{n-1}Cu_nO_{2n+4}$ where the number of CuO_2 layers varies from $n = 1$ up to 4 associated with a variation of T_c. The maximum value of T_c, 125 K, is exhibited for $n = 3$. Many references are listed in [49].

In conclusion, the anisotropy of the Knight shift is satisfactorily explained by the Mila-Rice model. This model involves almost localized spins in the Cu(2)-$3d_{x^2-y^2}$ orbital. On doping, holes are introduced into the CuO_2 layer which are allowed to delocalize. A further support of this model is provided by the observation that the transferred hyperfine coupling constants do not change much over the range of oxygen content [35]. In other words: the Knight shift

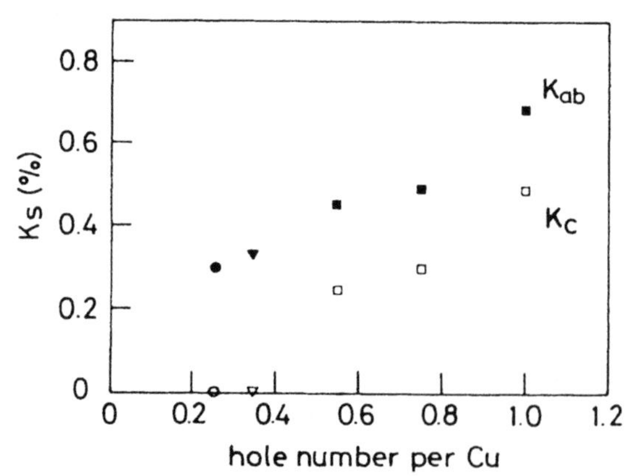

Fig. 9. Dependence of Cu Knight shift components on hole content in Tl-Ba-Cu-O [$B_0 \perp c$ (■), $B_0 \parallel c$ (□)], La-Sr-Cu-O [$B_0 \perp c$ (▼), $B_0 \parallel c$ (▽)] and Y-Ba-Cu-O [$B_0 \perp c$ (●), $B_0 \parallel c$ (○)] compounds. From [47]

results point to an unusual temperature dependence of the spin susceptibility. The important point is that the same temperature dependent spin susceptibility is observed in the spin–lattice relaxation rate to be discussed in the next Section.

4 Spin–Lattice and Spin–Spin Relaxation

The spin–lattice relaxation study of superconducting aluminum [1] yielded two important observations: the exponential dependence of the rate on inverse temperature and the occurrence of the so-called Hebel-Slichter coherence peak [8] just below T_c. These were the first direct proofs of the basic correctness of the BCS theory.

In HTSC, so far neither of these features has been observed. Although the first relaxation studies in $YBa_2Cu_3O_7$ [5, 6] produced energy gap values $2\Delta/k_b T_c$ which were larger than the classical BCS value of 3.5, they were not confirmed by subsequent more precise measurements. Instead, in most HTSC, the relaxation behavior below T_c follows a power law $1/T_1 \propto T^n$ with n between 3 and 4. If one plots for planar Cu sites in various HTSC, the normalized relaxation rates, i.e. the rate at temperature T, $1/T_1$, divided by the rate at T_c, $(1/T_1)_{T_c}$, as a function of T/T_c the data fall on a "universal" curve as shown in Fig. 10 [47]. This is a remarkable result in view of the differences in structure, composition, doping and T_c values. Actually, the power law reminds one of heavy fermion systems [8].

Rather than listing the many attempts to explain the relaxation features in HTSC below T_c we will start with a review of experimental relaxation studies in the normal conducting phase. Having done that we will deal with various current models to explain Knight shift and relaxation. After all, an understanding of NMR results of the normal conducting phase seems to be a prerequisite for any understanding of the superconducting phase.

It turned out that nearly all mechanisms which are responsible for spin-lattice relaxation in the HTSC are based on electron-nuclear interactions. The only exception seems to be the Ba relaxation in $YBa_2Cu_4O_8$ [23, 51] which is due to a two-phonon Raman process [23, 51]. Thus, for Cu, O and Y the relaxation is described by the Moriya formula [52, 53] which relates the relaxation rate to the hyperfine fields and the dynamical electron susceptibility. We write this formula in the following way [54]:

$$ {}^{i}\left(\frac{1}{T_1}\right)_{\alpha} = \frac{\gamma_i^2 k_B T}{2\mu_B^2} \sum_{\vec{q}, \alpha \neq \alpha'} |{}^{i}A_{\alpha'}(\vec{q})|^2 \frac{\chi''_{\alpha'\alpha'}(\vec{q}, \omega_i)}{\omega_i} \tag{4} $$

Here, i denotes the nuclear species and ω_i is its NMR frequency. The magnetic field B_0 is oriented along the α direction which coincides with one of the crystallographic axes (a, b, c). $\chi''_{\alpha'\alpha'}(\vec{q}, \omega_i)$ is the imaginary (dissipative) part of the complex electron susceptibility which depends on the excitation wave vector

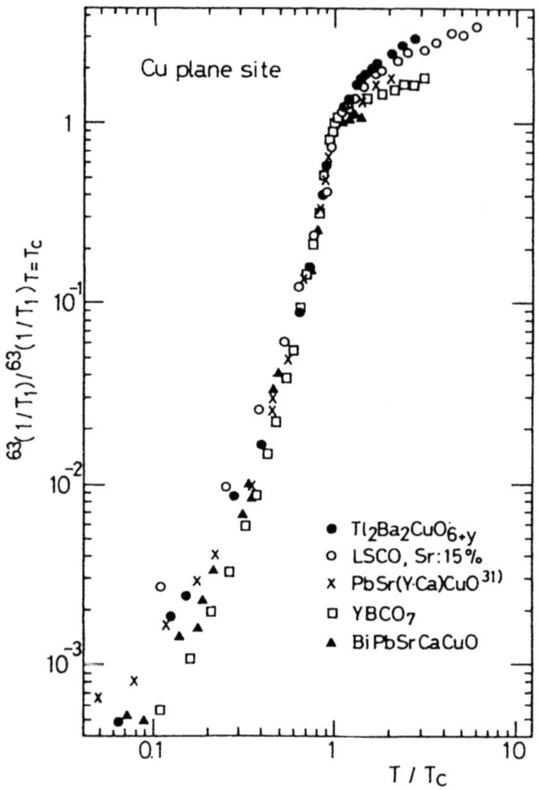

Fig. 10. Temperature dependence of planar Cu relaxation rates in various HTSC. From [47, 50]

$\vec{q}$ and the frequency ω. $A_{\alpha'}(\vec{q})$ are formfactors related to the hyperfine fields. Equation (4) demonstrates that the spin-lattice relaxation provides a tool to probe the average over the *q-dependent* electron susceptibility while the Knight shift yields information about the *static* susceptibility.

The success in applying Eq. (4) depends on the direct calculation of $\chi''(\vec{q}, \omega)$ which is very difficult. Models dealing with this problem will be discussed after we have presented a survey of the experimental facts about spin-lattice relaxation in some representative HTSC.

4.1 Temperature and Doping Dependence of T_1 in the Normal State

It is convenient to plot the quantity $(T_1 T)^{-1}$ as a function of temperature since this "relaxation rate per temperature unit" is a direct measure for the temperature dependence of the weighted susceptibilities and the anisotropy of the form factors. We start with a comparison of the $(T_1 T)^{-1}$ for planar Cu and O sites in several Y-Ba-Cu-O compounds which differ by their oxygen content.

4.1.1 Y-Ba-Cu-O

In Fig. 11, Cu data from single crystal studies of Berthier's group are compared [42]. For the over-doped compound $YBa_2Cu_3O_7$, $^{63}(T_1T)^{-1}$ increases when the temperature is lowered and then drops *at* T_c. For the two other samples with oxygen content $O_{6.52}$ and $O_{6.92}$ which are under-doped, $^{63}(T_1T)^{-1}$ reaches a maximum *above* T_c and then steadily decreases. While for the $O_{6.92}$ sample the decrease of $^{63}(T_1T)^{-1}$ becomes faster just below T_c, the $O_{6.52}$ sample exhibits a more "gentle" decrease.

Figures 12 and 13 make a similar comparison of the corresponding quantity $^{17}(T_1T)^{-1}$ for the planar oxygens in $YBa_{1.92}Sr_{0.08}Cu_3O_7$ and $YBa_2Cu_3O_{6.52}$ [42]. For the O_7 sample, there is again a continuous increase of $^{17}(T_1T)^{-1}$

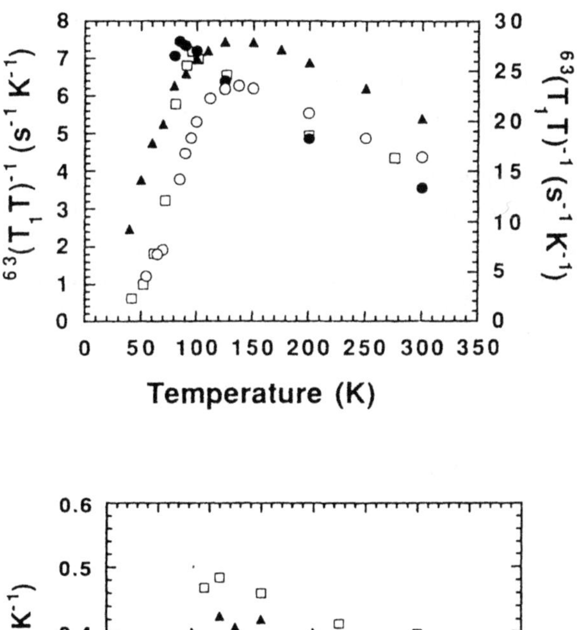

Fig. 11. Temperature dependence of $^{63}(T_1T)^{-1}$ for Cu(2) sites. Left-hand scale: □ $YBa_2Cu_3O_7$ (oriented powder); ○ $YBa_{1.93}Sr_{0.07}Cu_3 \cdot O_{6.92}$ (single crystal, $B_0 \parallel c$); ▲ $YBa_2Cu_3O_{6.52}$ (single crystal, $B_0 \parallel c$). Right-hand scale: ● $YBa_{1.92}Sr_{0.08}Cu_3O_7$ (single crystal, $B_0 \parallel a, b$ plane). From [42]

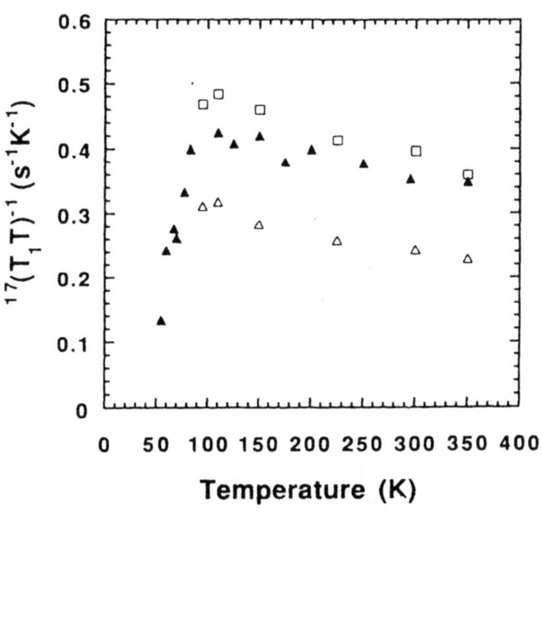

Fig. 12. Temperature dependence of $^{17}(T_1T)^{-1}$ for O(2,3) sites in $YBa_{1.92}Sr_{0.08}Cu_3O_7$ with $B_0 \parallel c$ (▲) and B_0 in the plane [$\perp$ Cu-O-Cu bond (□), $\parallel$ Cu-O-Cu bond (△)] From [42]

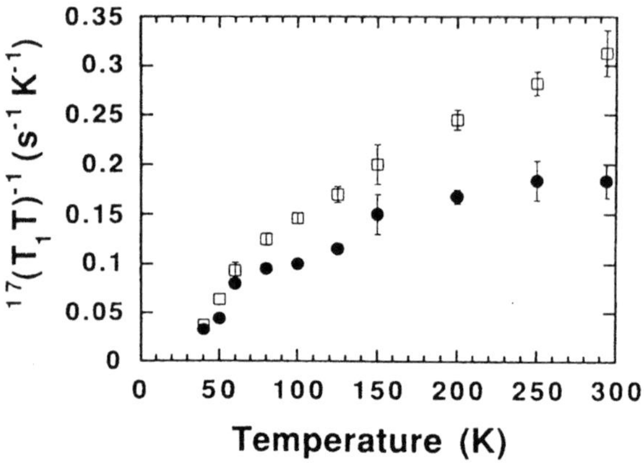

Fig. 13. Temperature dependence of $^{17}(T_1T)^{-1}$ for O(2, 3) sites in $YBa_2Cu_3O_{6.52}$ with $B_0 \parallel$ Cu-O-Cu bond (●) and $B_0 \perp$ Cu-O-Cu bond (□). From [42]

when the temperature is lowered, similar to the behavior of the Cu isotope although less pronounced. Previous measurements on powdered samples revealed only a temperature-independent behavior. Below T_c, the sharp decrease of $^{17}(T_1T)^{-1}$ is in agreement with other data from the literature [55, 56]. In particular, no Hebel-Slichter peak was found.

On the other hand the $O_{6.52}$ sample (Fig. 13) exhibits a continuous decrease of $^{17}(T_1T)^{-1}$ with decreasing temperature, similar to the behavior of the Knight shift. Let us denote by $^{17}(T_1T)_{\parallel}^{-1}$ and $^{17}(T_1T)_{\perp}^{-1}$ the quantities measured by aligning B_0 parallel and perpendicular, respectively, to the Cu-O-Cu bond (and in the CuO_2 plane). It then turns out that $^{17}(T_1T)_{\perp}^{-1}$ and $^{17}K_{\parallel}$ are proportional to each other. A similar behavior has been reported for $YBa_2Cu_3O_{6.63}$ [35] and for $YBa_2Cu_3O_{6.6}$ and $YBa_2Cu_3O_{6.8}$ [56]. However, $^{17}(T_1T)_{\parallel}^{-1}$ is not proportional to $^{17}K_{\parallel}$; in other words, the relaxation rate anisotropy is temperature dependent. Barriquand et al. were the first to observe such a temperature dependent anisotropy in $YBa_2Cu_3O_{6.5}$ [57].

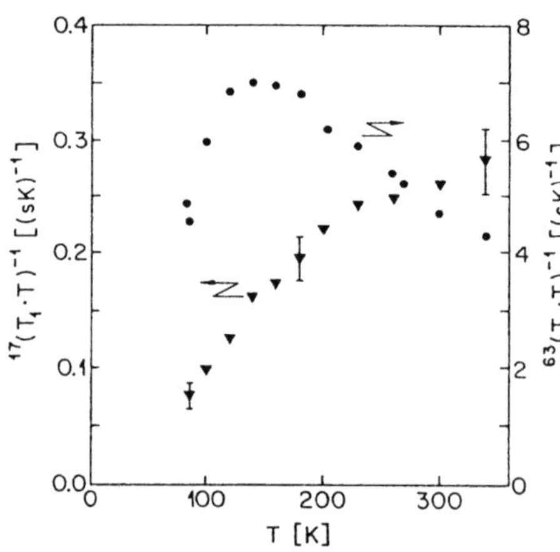

Fig. 14. Temperature dependence of the planar Cu and O $(T_1T)^{-1}$ in $YBa_2Cu_4O_8$. From [51]

So far we have discussed the one-chain Y-Ba-Cu-O compound $YBa_2Cu_3O_{6+x}$ with variable oxygen content x. The double-chain compound $YBa_2Cu_4O_8$ distinguished by its high thermal stability and precise stoichiometry, has intensively been investigated by our group and Yasuoka's group. In Fig. 14, we have plotted the temperature dependence of the quantity $(T_1T)^{-1}$ for both the planar Cu and O nuclei in the normal state of a $YBa_2Cu_4O_8$ sample with $T_c = 82\,K$ [51]. $^{63}(T_1T)^{-1}$ reaches a maximum around 130 K and rapidly decreases towards lower temperatures. On the other side, $^{17}(T_1T)^{-1}$ displays a monotonous decrease with lowering temperature. The behavior of both the oxygen and copper relaxation is similar to that of oxygen-deficient $YBa_2Cu_3O_{6+x}$ samples which again may be taken as a sign of under-doped material. It seems that for a fixed temperature, $^{17}(T_1T)^{-1}$ increases if one goes from $YBa_2Cu_3O_{6.6}$ to $YBa_2Cu_4O_8$ and to $YBa_2Cu_3O_{6.8}$. Zheng et al. [58] who measured $^{17}(T_1T)^{-1}$ in a $YBa_2Cu_4O_8$ sample with a *lower* T_c of 74 K, observed a less pronounced temperature dependence.

A valuable quantity for further discussions is the relaxation rate ratio $^{63/17}R = {}^{63}(T_1)^{-1}/{}^{17}(T_1)^{-1}$ which has been determined in several Y-Ba-Cu-O samples. In Fig. 15 we compare the $^{63/17}R$ values for $YBa_2Cu_4O_8$ [51] and $YBa_2Cu_3O_{6.52}$ [42]. Down to about 140 K, both ratios are the same and they increase with decreasing temperature; the $YBa_2Cu_4O_8$ material seems to diaplay the stronger temperature dependence. In $YBa_{1.92}Sr_{0.08}Cu_3O_7$, Berthier et al. [42] observed a ratio increasing with decreasing temperature although less pronounced as in $YBa_2Cu_4O_8$. Other groups have reported slightly different temperature behaviors for $YBa_{1.92}Sr_{0.08}Cu_3O_7$. The data of Hammel et al. [55] are shown in Fig. 16. The relaxation rates ^{63}W and ^{17}W at the planar Cu and O sites exhibit a different temperature dependence and magnitude in the normal state, while they show a similar temperature dependence below about 110 K (again without the Hebel-Slichter peak just below T_c). Thus, $^{63/17}R$ is constant below 110 K. Yoshinari et al. [56] have reported $^{63/17}R$ = constant only in the

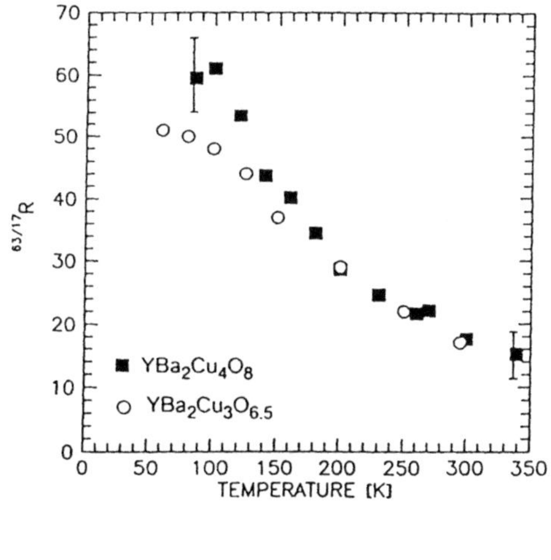

Fig. 15. Temperature dependence of the ratio $^{63/17}R$ for planar oxygen and copper sites in $YBa_2Cu_4O_8$ (□, $B_0 \parallel c$) [51] and in $YB_2Cu_3 \cdot O_{6.52}$ (○, $B_0 \parallel c$ for Cu(2) and $B_0 \perp$ Cu-O bond for O(2, 3)) [42]

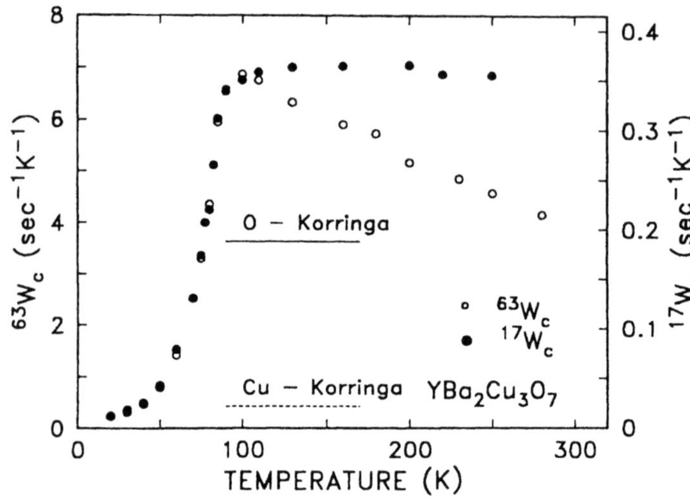

Fig. 16. Temperature dependence of the planar Cu and O relaxation rate in $YB_2Cu_3O_7$. From [59]

range 77 to 100 K. Berthier et al. [42] suspected that these discrepancies result from slightly sub-stoichiometric samples used by the other groups.

In Fig. 16, the so-called Korringa values of ^{63}W and ^{17}W are denoted by solid and dashed lines. The classical Korringa law states that for the free electron gas, the nuclear relaxation time and the isotropic part of the Knight shift are related in the following way:

$$T_1 T K_{iso}^2 \frac{4\pi k_B}{\hbar} \left(\frac{\gamma_n}{\gamma_e}\right)^2 = 1 \tag{5}$$

While ^{17}W is only moderately enhanced over the Korringa value, there is considerable enhancement for the ^{63}W value. We will return to the significance of this result.

We conclude the listing of Y-Ba-Cu-O studies by presenting the planar Cu relaxation data (Fig. 17) [51, 45] for the "mixed-chain" compound $Y_2Ba_4Cu_7O_{15}$ which consists of an alternating sequence of $YBa_2Cu_3O_7$ and $YBa_2Cu_4O_8$ blocks with adjacent CuO_2 planes of distinct carrier densities. The dotted and dashed lines represent the data measured in "pure" $YBa_2Cu_3O_7$ and $YBa_2Cu_4O_8$, respectively. For both planar Cu sites in the "mixed-chain" compound, the quantity $^{63}(T_1T)^{-1}$ reaches its maximum at the *same* temperature. As may be seen from the insert, the two relaxation rates are proportional to each other in the normal phase thus demonstrating that not only the static electron susceptibilities (as exemplified by the Knight shift) but also the *dynamic* susceptibilities are governed by the same temperature dependence. Obviously the two planes must be strongly coupled. At T_c, however, the proportionality between both rates ceases which may be due to the onset of a new relaxation mechanism and/or the change of the coupling.

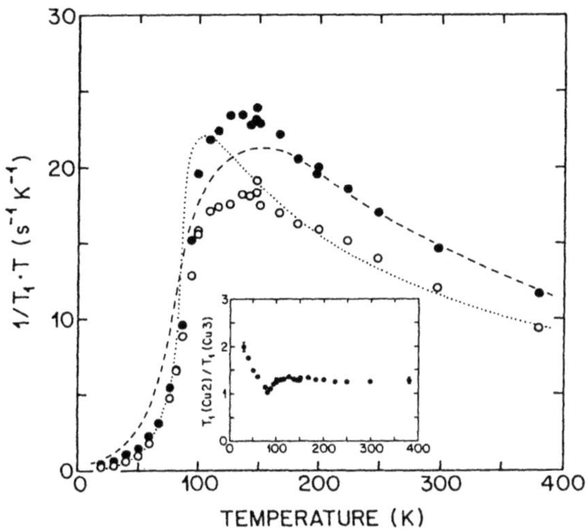

Fig. 17. Temperature dependence of $^{63}(T_1T)^{-1}$ for the two inequivalent planar Cu sites Cu(2) (○) and Cu(3) (●) in $Y_2B_4Cu_7O_{15}$. The insert shows the temperature dependence of the relaxation rates ratio. From [45]

4.1.2 La-Sr-Cu-O and Tl-Ba-Cu-O

Undoped La_2CuO_4 is an antiferromagnet (AF) with a Néel temperature, T_N, of about 300 K. Doping reduces T_N very rapidly and for $x \geqq 0.01$ a phase resembling a spin-glass phase occurs at low temperature. For $x \geqq 0.08$ all traces of AF ordering disappears and the system becomes metallic and superconducting. The interplay of superconductivity and AF has been of special interest in many NMR studies. References to earlier NMR/NQR investigations of La-Sr-Cu-O compounds and in particular to questions concerning the magnetism may be found for instance in [60].

In order to reveal the dependence of the Cu relaxation and hence the electron susceptibility on hole content, Kitaoka et al. [61] have measured the Cu relaxation in the normal and superconducting phases of $La_{2-x}Sr_xCuO_4$ for various doping levels, see Fig. 18. It is obvious that the relaxation rate in the normal phase is reduced by increasing the hole content, in other words: the enhancement of $^{63}(T_1T)^{-1}$ by the AF spin correlations is depressed by doping holes. Above T_c, the data can be described by a Curie-Weiss law, $(T_1T)^{-1} = AC/(T+\theta)$, where A is proportional to the square of the hyperfine field. One thus supposes that the staggered susceptibility at the zone boundary, $Q = (\pi/a, \pi/a)$, (where a is the lattice constant) is also described by a Curie-Weiss law: $\chi_Q(T) = c/(T+\theta)$. θ is zero for $x = 0.05$ and then increases linearly with x.

Starting from a similar point of view, the Borsa and Rigamonti group investigated $La_{2-x}Sr_xCuO_4$ [62]. One of the main question was concerned with the possible role of the dynamical microscopic magnetic properties in the mechanisms leading to the superconducting state. In particular, how does the Cu^{2+} spin dynamics evolve on going from the insulating paramagnetic state to the metallic and superconducting one as a function of doping? What is the

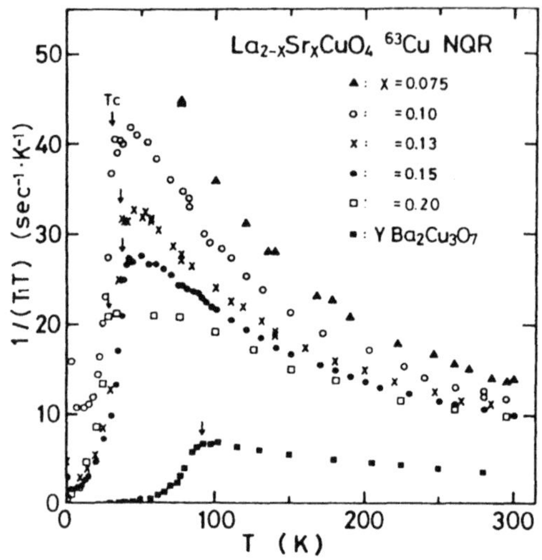

Fig. 18. Temperature dependence of $^{63}(T_1T)^{-1}$ in $La_{2-x}Sr_xCuO_4$ for various doping levels. From [63]

role of the holes in controlling the temperature and x dependence of the magnetic correlation length, $\xi(T)$? The group has studied the ^{139}La NQR relaxation in the spin-glass regime and the Cu NQR relaxation in the metallic and superconducting phases. NQR relaxation offers a unique method to probe the quasi-elastic fluctuations of the staggered magnetization in locally ordered regions off mesoscopic size.

As seen in Fig. 19, the ^{139}La relaxation rates in AF samples (with $x \leqq 0.05$) are strongly enhanced in the paramagnetic phase if the temperature approaches T_N (for the respective sample) from above [63, 64]. This indicates a collective critical behavior of the Cu^{2+} spin dynamics even in the x doping regime in which $\xi(T)$ is limited by the holes. In the parent compound La_2CuO_4 the $\xi(T)$ in the CuO_2 planes is very much larger than the distance, L, between charge defects. Thus, La NQR probes the $\omega = \omega_Q \approx 0$ quasi-elastic fluctuations of the staggered magnetization in the regions of size L with $a \ll L \ll \xi$. The interpretation of the data in terms of a local spin susceptibility will be discussed later. It is assumed that the only effect of doping is the local disruption of the AF correlation without affecting the intrinsic magnetic moment of Cu^{2+}.

The situation will change when for large doping ($x \geqq 0.08$) the extra holes affect the Cu spin and some sort of coupling of the carriers with the spin sets in. Borsa et al. [62] propose a simple extension of the model used for the explanation of the spin-glass data, to interpret the main physical aspects accompanying the superconducting transition.

The $Tl_2Ba_2CuO_{6+y}$ material offers the possibility to study the effect of *heavy* doping on the Cu and Tl relaxation. T_c can be varied between 0 and 85 K by reducing oxygen in the interstitial sites between the double TlO_2 layers. Kitaoka et al. [65, 47] measured both the ^{63}Cu (see Fig. 20) and the ^{205}Tl (see Fig. 21) relaxation for T_c values of 72, 40 and 0 K. Obviously the magnitude of $^{63}(T_1T)_{ab}^{-1}$

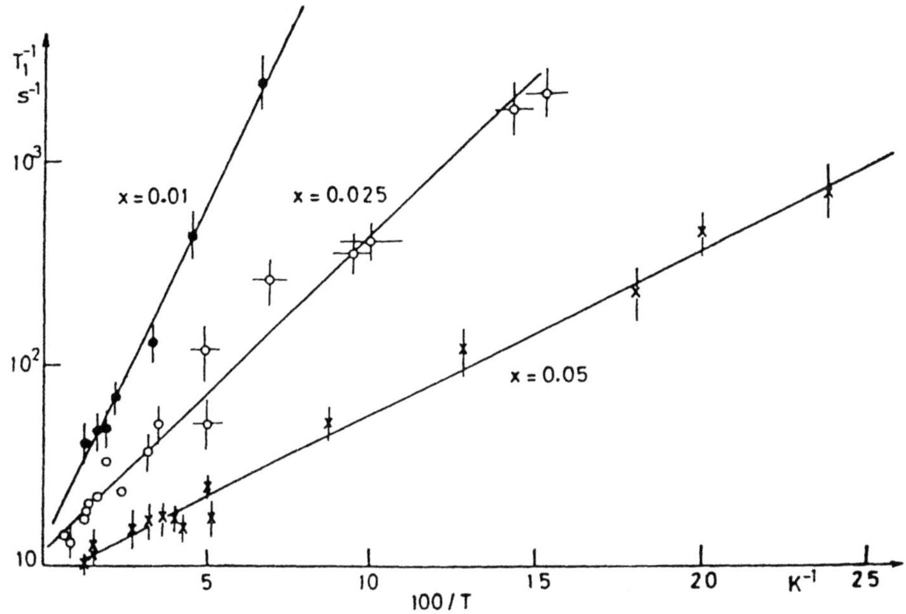

Fig. 19. Temperature dependence of ^{139}La relaxation rate in $La_{2-x}Sr_xCuO_4$ for various doping levels. From [64]

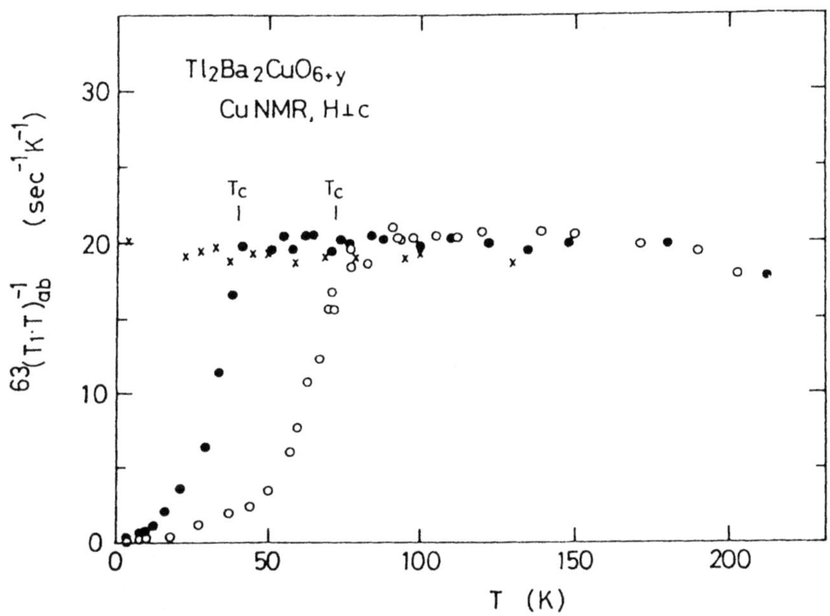

Fig. 20. Temperature dependence of $^{63}(T_1T)^{-1}$ in $Tl_2Ba_2CuO_{6+y}$ with $T_c = 72$ K (○), 40 K (●) and 0 K (×). From [47]

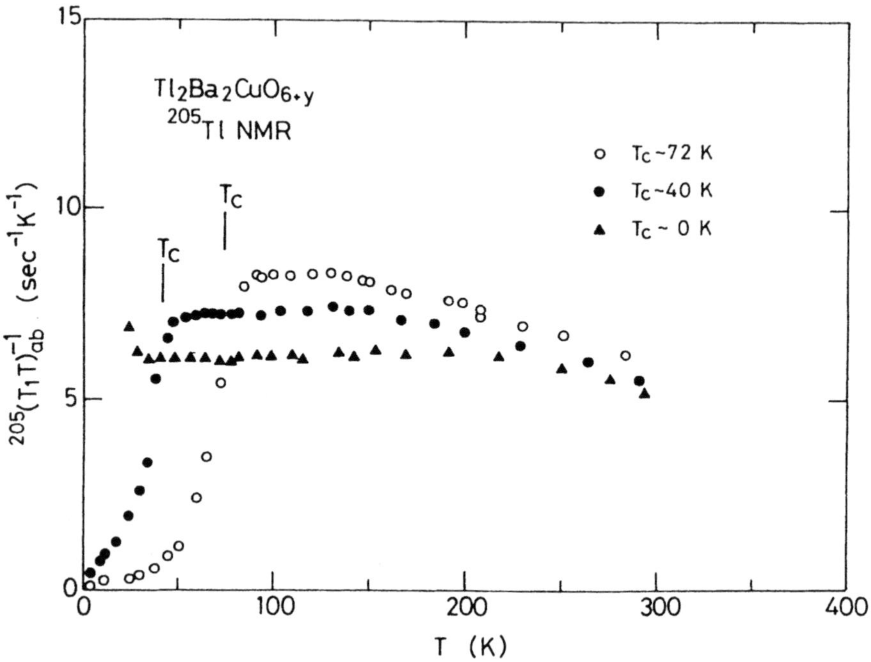

Fig. 21. Temperature dependence of $^{205}(T_1T)^{-1}$ in $Tl_2Ba_2CuO_{6+y}$ with $T_c = 72\,K$ (○), 40 K (●) and 0 K (▲). From [47]

at temperatures above T_c is not affected by doping. Furthermore, the temperature region where T_1T is constant depends on T_c. In contrast to the oxygen-deficient $YBa_2Cu_3O_{6+x}$ and to the low-doped $YBa_2Cu_4O_8$ the decrease of $^{63}(T_1T)^{-1}$ starts at temperatures only slightly above T_c.

However, in contrast to Kitaoka et al. [65], Vyaselev et al. [66] found that the temperature intervals where T_1T is constant, are in fact rather small. Actually, it seems that T_1T always follows the relation $T_1T = a + bT$ with a increasing and b diminishing as T_c is lowered.

4.1.3 Cation Doped Y-Ba-Cu-O

Among the many activities to "tailor" HTSC for a particular purpose is doping in order to increase T_c. For instance, Miyatake et al. [67] were able to enhance T_c of $YBa_2Cu_4O_8$ from 82 K to 90 K by doping with 10% Ca. To understand the enhancement it is crucial to know which crystallographic site the Ca^{+2} ion is occupying and whether the substitution changes the hole concentration.

Because of the controversy over this issue, we investigated this problem by measuring the Cu NQR spectrum at 100 K in $YBa_2Cu_4O_{8\pm y}$ doped with 10%

Ca ($T_c = 91$ K) [68]. The NQR lines of the plane and chain Cu sites appear at the same frequencies ν_Q as in the pure compound, however, both lines are broadend by static defects. Within the line widths no temperature dependence of ν_Q was observed and no indication for magnetic ordering at the Cu sites was found. At 20.600 MHz, near to the chain Cu(1) line, a resonance line has been detected which is absent in the pure $YBa_2Cu_4O_{8\pm y}$ spectrum. Because of its similarities with the Cu(1) spectrum we have identified the 20.600 MHz peak as a Cu(1) resonance line arising from those Cu(1) nuclei which are nearest neighbors of a lattice defect. A careful discussion of the signal intensities has revealed that the defect is a Ca ion occupying a Ba site. We thus concluded that at least a major fraction of Ca ions occupies Ba sites.

Does this type of substitution change the hole concentration? Since the spin-lattice relaxation time of both the Cu(1) and Cu(2) nuclei in $YBa_2Cu_4O_8$ are not changed by 10% Ca doping [69] we concluded, according to the results from Sect. 4.1.2, that the change in hole concentration is negligible. Furthermore, since ν_Q of Cu(2) in Ca doped $YBa_2Cu_4O_8$ remains unchanged with respect to pure $YBa_2Cu_4O_8$, the increase of the number of holes is estimated to be less than 0.013 per Cu. It therefore seems that the increase of T_c by Ca doping must be explained by another mechanism.

The opposite conclusion concerning this point has been reached by Machi et al. [70]. They claim that it is possible to separate in the relaxation data the contribution arising from the Brillouin zone corner which reflects the AF spin fluctuations. A reduction of this component due to Ca doping then suggests an increase of the hole concentration. The authors, however, do not comment on the additional Cu(1) line we have observed. Our NMR study of the ^{43}Ca nucleus itself in Ca doped $YBa_2Cu_4O_8$ is still in progress and hopefully it will decide between the two possibilities.

A somewhat opposite situation occurs by doping fully oxygenated $YBa_2Cu_3O_7$ with Zn or Al. Zn is thought to occupy primarily Cu(2) sites and produces a dramatic *decrease* in T_c [71] whereas Al ions go mainly into Cu(1) sites and have very little effect on T_c [72]. The effect of the dopants could be studied via ^{89}Y NMR. Balakrishnan et al. [73] noticed that Zn causes a temperature-dependent broadening of the line and an increase of T_1^{-1} while Al produces only a temperature independent broadening and has no appreciable effect on T_1. Alloul et al. [74] also observed the broadening effect of Zn which was taken as evidence for the generation of local magnetic moments on or near the Cu(2) sites [73, 74] since Zn unlike Al produces a large increase in Curie-like magnetism and therefore should affect the ^{89}Y NMR.

Warren Jr. et al. [75] and Walstedt and Warren Jr. [76] pursued the question whether NMR can tell the difference between $YBa_2Cu_3O_7$ samples whose T_c has been reduced either by Zn doping or by removing oxygen. Zn doping causes, among others, an increase of the Cu(2) line width which is believed to arise from a Ruderman-Kittel-Kasuya-Yosida (RKKY) effect, i.e. an indirect exchange interaction of local moments at Cu(2) sites via conduction electrons. Dipolar broadening by magnetic impurities cannot account for the magnitude of the effect. The interpretation is consistent with the previous analyses.

As to relaxation, Zn doping leads to an appreciable decrease of the ^{63}Cu spin-lattice relaxation rates and this for both sites without a change in the general character of the $1/T_1 - vs - T$ curve as has been observed in oxygen deficient Y-Ba-Cu-O compounds. The *increase* of the ^{89}Y relaxation rate [73] is attributed to a breaking of the magnetic symmetry around the Y site [76] which then allows (as we will see in Sect. 5) the AF fluctuations to enhance the Y relaxation rate. These results show clearly that relaxation is affected completely different in Y-Ba-Cu-O compounds where T_c is lowered by reduction in carrier density as compared to Y-Ba-Cu-O material where T_c is reduced by magnetic pair breaking effects alone. Similar studies of the planar copper relaxation and Knight shift in Zn doped Y-Ba-Cu-O material have been performed by Kitaoka et al. [47] who conclude that Zn impurities affect the local AF spin correlation therefore reducing T_c.

4.2 Anisotropy and Field Dependence of Spin–Lattice Relaxation

So far, we have summarized investigations of the spin-lattice relaxation which were concerned with its dependence on temperature and doping. Two other crucial parameters the rate may depend on, are the orientation in the external magnetic field, B_0, and the strength of this field. Because of the anisotropy of many physical properties of the HTSC one expected to find anisotropies in the relaxation which then would reflect e.g. the anisotropies in the AF spin fluctuations. A further question is how these fluctuations are affected by an external magnetic field. In Sect. 4.1.1, we came across conflicting results for the ratio $^{63/17}$R of the relaxation rates of planar copper and oxygen in Y-Ba-Cu-O. A ratio which is constant below 120 K in Y-Ba-Cu-O as measured by Hammel et al. [55], would suggest that the AF correlations do not change at T_c, whereas Yoshinari's results [56] point to the opposite. Since we know that $^{63}T_1$ depends strongly on B_0, one must wonder whether or not the oxygen relaxation time, ^{17}T$_1$, depends also on B_0. Until the B_0 variations of $^{17}T_1$ have been studied one must consider conclusions about $^{63/17}$R preliminary.

For the relaxation-rate anisotropy to be discussed now, it is convenient to introduce the quantity $r = W_{ab}/W_c$ where W_α is the spin-lattice relaxation rate at a planar site in a Y-Ba-Cu-O material with $\alpha = a, b, c$ specifying the orientation of B_0. The Urbana and Los Alamos NMR groups were the first to report systematic investigations of the quantity r.

Barrett et al. [77] noticed that r of Cu(2) is independent of temperature and equal to 3.74 in the normal phase of $YBa_2Cu_3O_7$ (see also Fig. 24) but undergoes a rapid change just below T_c. The group also studied ^{17}O(2, 3) and ^{89}Y. A straightforward application of the Millis-Monien-Pines (MPP) theory which we will discuss in Sect. 5, was not found to be consistent with the experimental results below T_c. Then, Martindale et al. [78] performed detailed studies of r

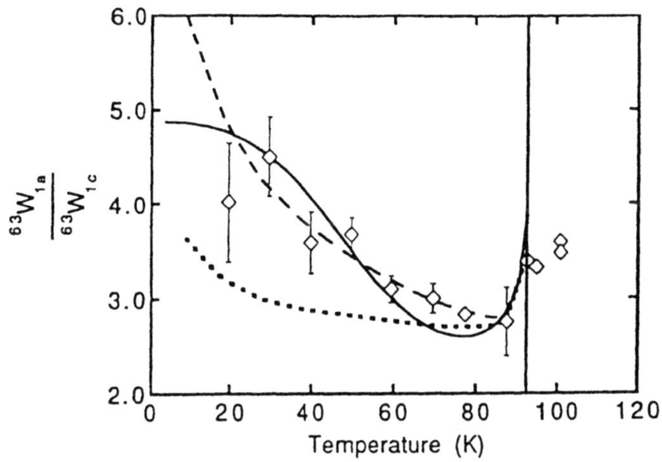

Fig. 22. Anisotropy of the Cu(2) relaxation rate in $YBa_2Cu_3O_7$. The various lines are discussed in Sect. 5. From [78]

and of the field dependence of W_α in the superconducting state. Since the change of r due to a magnetic field could be a secondary effect caused for example by fluxoid cores or by T_c suppression in a magnetic field, the high field experiments should be extended to low fields where the field induced anisotropy effects may be neglected. Figure 22 shows the temperature dependence of r in an oriented powder sample of $YBa_2Cu_3O_7$ [78]. The various lines are theoretical fits based on a BCS spin-singlet, orbital-d-wave pairing state to be discussed in Sect. 5. In addition, Martindale et al. found that in the superconducting state W_c, and to a lesser degree W_{ab}, become enhanced in a magnetic field, with the enhancement growing at lower temperature.

Takigawa et al. [79] independently measured the temperature-dependent Cu(2) anisotropy below T_c in $YBa_2Cu_3O_7$ using also low magnetic fields. Their results are given in Fig. 23 where we will compare them with our data for $YBa_2Cu_4O_8$. The $YBa_2Cu_3O_7$ results obtained by the Urbana and Los Alamos groups are similar.

Our group has reported similar investigations of the W_α anisotropy and its field dependence for Cu(2) in $YBa_2Cu_4O_8$ [80] which has a lower charge carrier concentration than $YBa_2Cu_3O_7$. In particular, the Cu(2) rate W_c, field independent in the normal state, shows a field-dependent enhancement in the superconducting state that begins already 13 K *above* T_c in a 5.17 T field. Consequently, the anisotropy r that is temperature- and field-independent above $T_c + 13$ K ($r = 3.3$), starts to diminish below this temperature as may be seen in Fig. 24.

Our recent low and high field data for the anisotropy are summarized in Fig. 25 [82]. Below T_c, both the high and low field r values decrease with the low field r passing through a pronounced minimum at 55 K. Within an

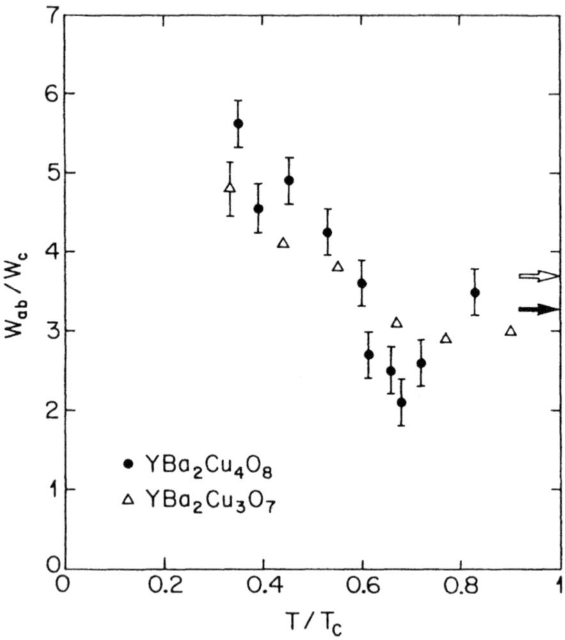

Fig. 23. Cu(2) relaxation rate anisotropy r vs the reduced temperature T/T_c in a weak magnetic field: ● $YBa_2Cu_4O_8$ in $B=0.58\,T$ (from [82]), △ $YBa_2Cu_3O_7$ in $B=0.44\,T$ (from [79])

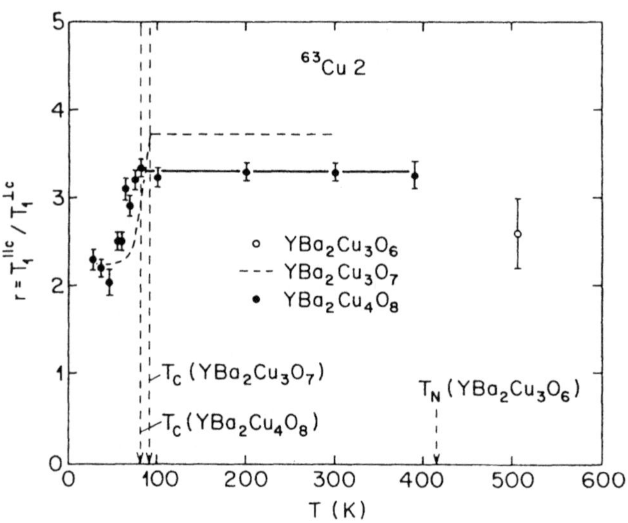

Fig. 24. Temperature dependence of the Cu(2) relaxation anisotropy r in $YBa_2Cu_4O_8$ (full circles, from [80]), $YBa_2Cu_3O_7$ (dashed line, from [77]) and in semi-conducting $YBa_2Cu_3O_6$ (open circle, from [81])

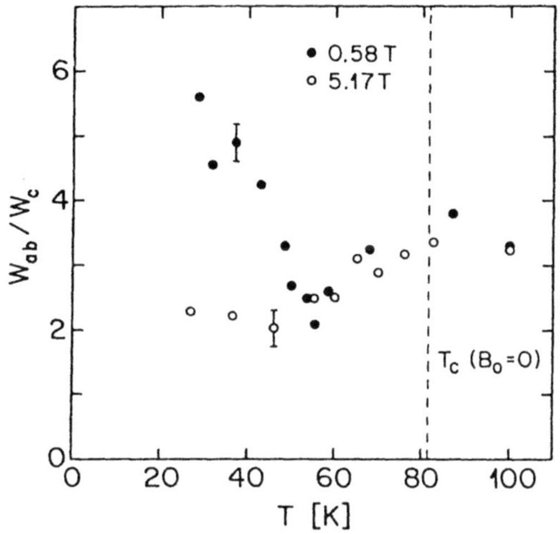

Fig. 25. Temperature dependence of Cu(2) relaxation anisotropy r for $YBa_2Cu_4O_8$ in two different magnetic fields. From [82]

experimental error of $\pm 10\%$, the weak field ratio $r = 3.2$ in the normal conducting phase agrees with our old high-field result $r = 3.3$ [80]. As Fig. 23 reveals, the low field r values for $YBa_2Cu_4O_8$ and $YBa_2Cu_3O_7$ of Takigawa et al. [79] are quite similar. However the minimum of r in $YBa_2Cu_4O_8$ seems to be deeper and is positioned about $0.1 T/T_c$ lower than in $YBa_2Cu_3O_7$. The upturn of r with decreasing temperature is much more pronounced as compared to $YBa_2Cu_3O_7$. Furthermore, just below T_c the $YBa_2Cu_4O_8$ data exhibit a much softer decrease of r; the derivative dr/dT is almost zero at T_c. It remains to be shown whether the different behaviour of the two structures can be reproduced by recent models based on d-wave pairing BCS theory to be discussed in the next Section.

The field dependence of the Cu(2) relaxation rate in $YBa_2Cu_4O_8$ is given in Fig. 26 for strong, weak and zero fields using an oriented powder sample. The same results normalized to the respective rate at T_c are re-plotted as a function of the reduced temperature $T/T_c(B)$ in Fig. 27. W_c depends stronger on field than W_{ab} does; the field dependence increases with decreasing temperature. On the other hand, down to about $T = 0.7 T_c$, W_{ab} is slightly enhanced by B_0 as observed in $YBa_2Cu_3O_7$ [78] and also recently reported by Borsa et al. [83]. However, below $0.7 T_c$, the enhancement gives way to a suppression that becomes more evident at lower temperatures; this contrasts with the behavior of W_{ab} in $YBa_2Cu_3O_7$.

The opposite response of W_{ab} and W_c to the application of a magnetic field seems to rule out the possibility hat fluxoid cores or thermally activated fluxoid motion cause the field dependence as it has been discussed for $YBa_2Cu_3O_7$ [78]. To account for the opposite response of W_{ab} and W_c an unexpected field related breaking of the spin-rotation invariance in the superconducting state has to be considered. Finally, we want to stress the fact that in $YBa_2Cu_4O_8$ the high field W_c is larger than the NQR rate from the lowest temperatures used in our

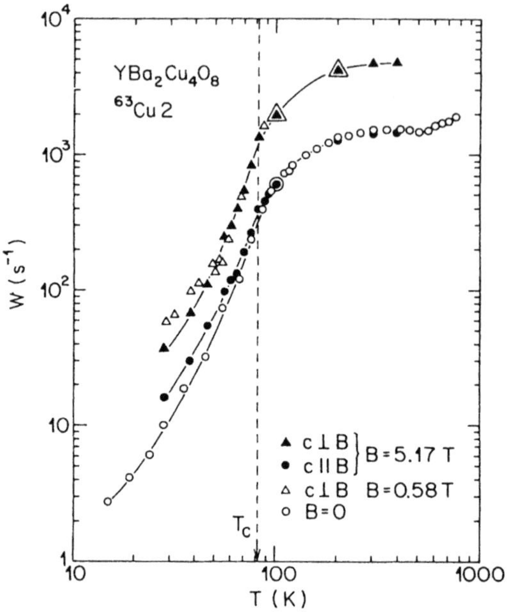

Fig. 26. Cu(2) spin–lattice relaxation rates vs temperature in $YBa_2Cu_4O_8$ for different magnetic fields and orientations. The *triangles* are for $B \perp c$ and the *circles* for $B \parallel c$. From [82]

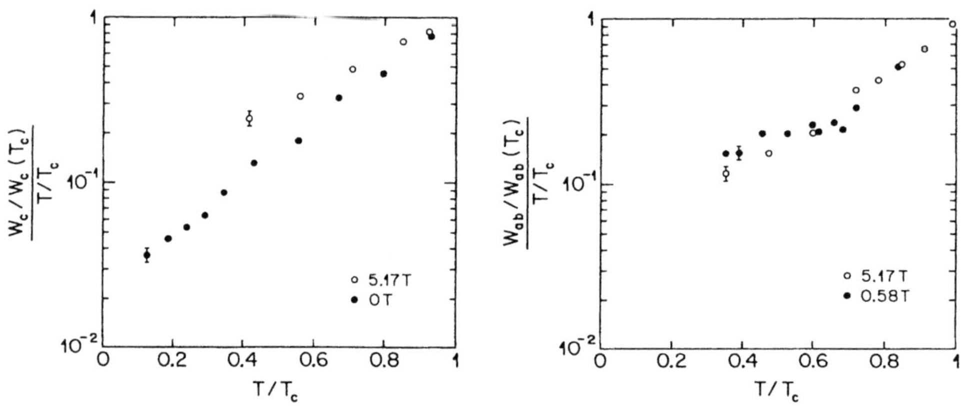

Fig. 27. Temperature dependence of the normalized Cu(2) spin–lattice relaxation rates W_c **(left)** and W_{ab} **(right)** divided by T/T_c in $YBa_2Cu_4O_8$. From [82]

experiment, up to $T_c + 13\,K$ (see Fig. 26). This is in contrast with a recent observation by Borsa et al. [83] who found in the temperature region just above T_c an opposite behaviour for $YBa_2Cu_3O_7$ and $La_{1.85}Sr_{0.15}CuO_4$.

4.3 Nuclear Spin–Spin Relaxation

Another very strong experimental constraint on the static spin susceptibility, $\chi'(\vec{q})$, at non-zero wave vector $\vec{q}$ can be deduced from the transverse relaxation obtained by the NMR/NQR spin echo method [84]. The transverse relaxation

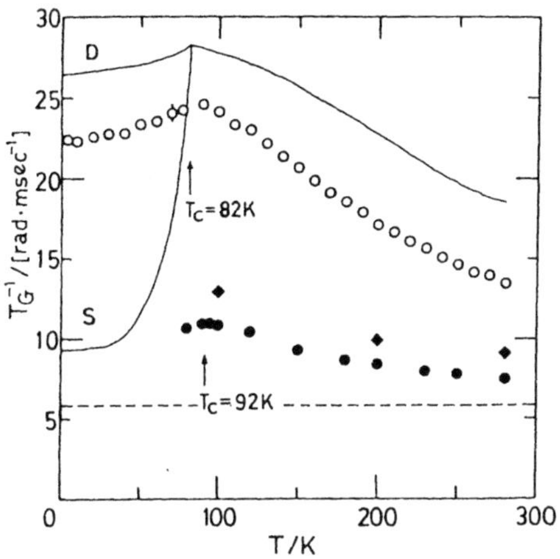

Fig. 28. Temperature dependence of the Gaussian decay rate, T_{2G}^{-1}, of planar Cu in $YBa_2Cu_3O_{6.98}$ (*full circles*), $YBa_2^{63}Cu_3O_{7-\delta}$ (*rhombus*), $YBa_2Cu_4O_8$ (*open circles*). The *solid curve* is a calculation of Bulut *et al.* [89] within a 2D Hubbard model. The symbol S (D) refers to a s-(d-) wave energy gap. The *dashed line* indicates the estimated T_{2G}^{-1} based on the direct nuclear dipole–dipole interaction. From [86]

ordinarily originates from spin–lattice relaxation and/or direct as well indirect nuclear spin–spin coupling. The first mechanism leads to an exponential decay of the transverse magnetization, whereas the second one, under certain circumstances, produces a Gaussian decay [85]. It is the measurement of this Gaussian component, $^{63}T_{2G}^{-1}$, at the planar Cu site that provides quantitative information concerning the static spin susceptibility, $\chi'(\vec{q})$, for non-zero $\vec{q}$ [85–88].

The absolute value and the temperature dependence of $^{63}T_{2G}^{-1}$ could be understood as the combined contribution from (1) temperature-independent direct nuclear dipole–dipole interaction, and (2) temperature-dependent indirect nuclear spin–spin coupling mediated by AF-correlated Cu(2) electron spins. From the temperature dependence of $^{63}T_{2G}^{-1}$ for $YBa_2Cu_4O_8$ and $YBa_2Cu_3O_7$ [86, 87] Yasuoka's and Slichter's group show that, above T_c, $\chi'(\vec{q})$ satisfies a Curie–Weiss law around $Q = (\pi/a, \pi/a)$. Below T_c in $YBa_2Cu_4O_8$, $\chi'(\vec{q})$ remains enhanced (see Fig. 28) exhibiting the same temperature dependence as calculated by Bulut et al. [89] within a 2D Hubbard model in random-phase approximation (RPA) for a d-wave superconductivity.

The combined analysis of $^{63}(T_1T)^{-1}$ and $^{63}T_{2G}^{-1}$ in $YBa_2Cu_4O_8$ [86] undoubtedly reveals the formation of a pseudo gap in the AF fluctuation spectrum at temperature T^* which is appreciably higher than T_c. There is however, little evidence for pseudo-gap behavior in $YBa_2Cu_3O_7$ [87].

5 Interpretation of Knight Shift and Spin–Lattice Relaxation

The ultimate goal of any theory of HTSC must be to explain the behavior of typical NMR parameters like Knight shift and spin–lattice relaxation in both the normal and the superconducting state. To our knowledge such a comprehensive theory is not yet available. However, many models have been proposed

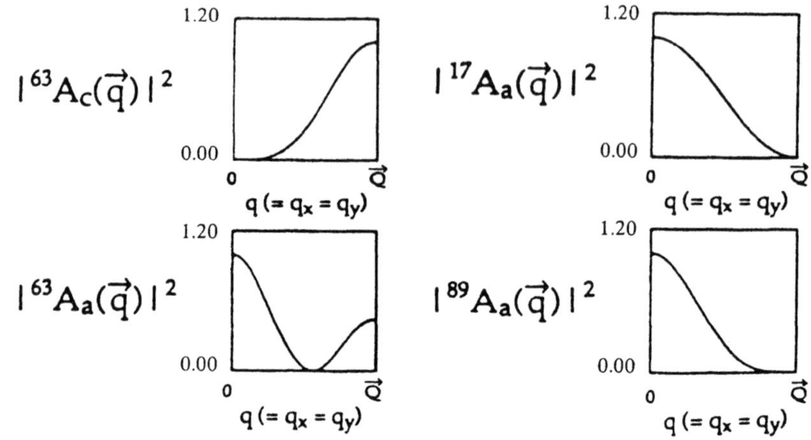

Fig. 29. Dependence of form factors on wave vector. From [54]

to explain at least part of the properties of the HTSC. In this Section we will present some of these new approaches.

As mentioned several times, it is now widely believed that magnetic spin–lattice relaxation in several HTSC imply the existence of AF fluctuations and correlations. Inelastic neutron scattering experiments in $YBa_2Cu_3O_{6+x}$ (see, e.g., [90–93]) have shown how these dynamic spin correlations survive once doping has destroyed the long-range magnetic ordering. The fluctuations peak near $\vec{Q} = (\pi/a, \pi/a)$ which is the AF ordering point in $\vec{q}$-space.

Given these AF fluctuations, the relaxation rates of Cu, O and Y nuclei will follow the Moriya equation, Eq. (4), which requires knowledge of the formfactors $A_{\alpha'}(\vec{q})$ and the imaginary (dissipative) part of the complex electron susceptibility $\chi''_{\alpha'\alpha'}(\vec{q}, \omega_i)$. The dependence of the formfactors on $\vec{q}$ is shown in Fig. 29. Note that at $\vec{Q}$ the oxygen formfactors vanishes while the copper formfactor has a maximum. Remember hat the single spin fluid model is well established experimentally. Hence, in order to explain the different relaxation behavior of Cu and O in, e.g., $YBa_2Cu_3O_7$ (see Fig. 16) one has to postulate a strong $\vec{q}$ dependence of $\chi''(\vec{q}, \omega)$ and temperature-dependent AF spin correlations. The problem is to calculate $\chi''(\vec{q}, \omega)$.

5.1 *Millis–Monien–Pines Model*

Millis, Monien and Pines have put forward a phenomenological model (in short MMP) [94] which together with its variants [95, 96], has quantitatively described the normal state relaxation of Cu, O and Y in $YBa_2Cu_3O_7$, $YBa_2Cu_3O_{6.63}$ and $La_{1.85}Sr_{0.15}CuO_4$. Most recent NMR data have raised questions with respect to the applicability of the MMP model. However, before discussing these questions we will sketch out the MMP model following closely Ref. [94] and [95].

The model is based on six major assumptions:

(1) There is only one electron spin $\vec{S}$ per unit cell composed of the planar Cu and O spins. $\vec{S}$ has direct and transferred hyperfine couplings denoted by A, B, C and D, with the nucleus at the same site and the neighboring nuclei, respectively, as described in Sect. 3.1.

(2) The electron spins $\vec{S}$ interact anti-ferromagnetically with a finite temperature-dependent correlation length $\xi(T)$ which increases with decreasing temperature.

(3) The electron susceptibility consists of two parts:

$$\chi(\vec{q},\omega) = \chi'(\vec{q},\omega) + i\chi''(\vec{q},\omega) = \chi_{QP}(\vec{q},\omega) + \chi_{AF}(\vec{q},\omega) \tag{6}$$

where the first part describes the quasi-particle (normal Fermi-liquid-like) contribution and the second the AF correlations.

(4) Using the abbreviation $\chi_0 = \chi'_{QP}(0,0)$, one sets for small values of ω:

$$\mathrm{Im}(\chi_{QP}) \approx \frac{\pi\omega}{\Gamma}\chi_0 \tag{7}$$

which is assumed to be independent of $\vec{q}$: Γ is the characteristic spin fluctuation energy for χ_{QP}.

(5) The AF part of $\chi(\vec{q},\omega)$ is written as

$$\chi_{AF}(\vec{q},\omega) = \chi_0 \frac{(\xi/\xi_0)^2}{1 + \xi^2(\vec{Q} - \vec{q})^2 - i(\omega/\omega_{SF})} \tag{8}$$

$1/\xi_0$ is the wave vector at which χ_{AF} starts to dominate χ_{QP}. $\hbar\omega_{SF}$ is a typical energy scale for the AF paramagnon that describe the AF spin dynamics.

(6) $\xi(T)$ has a Curie-Weiss like temperature dependence:

$$\left(\frac{\xi(T)}{a}\right)^2 = \left(\frac{\xi}{a}\right)^2_{T=0} \frac{|T_x|}{T_x + T} \tag{9}$$

where T_x denotes a temperature scale for the variatons of $\xi(T)$.

The MMP model gives a quantitative account of relaxation data at Cu, O and Y sites in $YBa_2Cu_3O_7$ [94]. Also the Cu(2) relaxation data in $YBa_2Cu_4O_8$ could be fitted with a similar ξ [26]. At O and Y sites, χ_{QP} makes the dominant contribution to the relaxation rate since the O and Y formfactor vanishes at $q = Q$ and thus no temperature dependence of the rate is expected. For the Cu sites, on the other hand, χ_{AF} is dominating and because of the temperature-dependent ξ, this leads to non-Korringa behavior.

There is, however, a serious shortcoming of this attractive model: the assumption of a temperature dependent correlation length has not been confirmed by neutron-scattering investigations [91]. These experiments measure a scattering cross-section which is proportional to the dynamical structure factor

$S(\vec{q},\omega)$ which in turn is related to $\chi''(\vec{q},\omega)$ via the equation:

$$S(\vec{q},\omega)=\frac{1}{\pi}\frac{1}{1-\exp(-\hbar\omega/kT)}\chi''(\vec{q},\omega) \tag{10}$$

Millis and Monien [97] re-examined the MMP model and extended it to temperatures below T_c. Instead of Eq. (8) they considered a Gaussian χ'' which falls off more rapidly with q. By re-examining the fits done in [94,95) for the normal state data they concluded that—quoting from [97]—"the NMR data require a spin susceptibility strongly peaked at or near the zone corner, and that although the uncertainties in the value and temperature dependence of the AF correlation length implied by the data are large, the available NMR data are more consistent with a temperature-dependent correlation length than a temperature-independent one." They emphasized that the crucial experiment to test the existence of a temperature-dependent ξ is the ratio of the oxygen and yttrium relaxation rates, $^{17/89}R$. Thus, more precise experimental values for this ratio are needed. The authors believe that "the presently available neutron data are not consistent with any interpretation of the NMR data based on the Mila-Rice Hamiltonian."

5.2 *Spin–Gap Model*

Millis and Monien [97] also discussed $\chi''(\omega,\vec{q})$ for finite frequencies as measured in neutron scattering and they derived relations which may be compared with neutron scattering data. The basic assumption was that there is only one important frequency scale at each q, called Γ_q. It could be given for instance by

$$\Gamma_q=\frac{\Gamma}{\beta^{1/2}}\frac{1+q^2\xi^2}{\xi^2} \tag{11}$$

where Γ has been defined above and β is a parameter measuring the relative strength of the AF fluctuations to the zone-center fluctuations. Millis and Monien showed that the assumption of *one* frequency scale cannot be correct in under-doped compounds like $YBa_2Cu_3O_{6.63}$ and $Bi_2Sr_2CaCuO_8$. In fact, they found that the anomalous temperature dependence of the relaxation rate in these materials must be traced back to the temperature dependence of Γ_q with the result that the spectral weight in the spin fluctuations decreases as the temperature is lowered. This missing spectral weight must reappear at a higher energy in the form of transitions over a "spin pseudo-gap".

The spin-gap effect manifests itself in the relaxation rates in the following way. The effect competes with the AF fluctuations which increase with descending temperature. Below a temperature T^* where the maximum of $(T_1T)^{-1}$ occurs, the spin-gap effect predominates and thus $(T_1T)^{-1}$ is decreasing. In the under-doped structures, T^* lies well above T_c while in over-doped compounds T^* nearly coincides with T_c.

It was Rice [98] who noted from a theoretical point of view the widespread nature of the spin-gap effect and Rossat-Mignod et al. [90] were the first to discover the effect in $YBa_2Cu_3O_{6+x}$ by neutron scattering. The effect is attracting considerable theoretical attention; however, the microscopic origin of the effect seems to be still unknown. First reports about the spin-gap effect in NMR appeared in Refs. [99–102] although with no quantitative description of the data. More recent discussions may be found in [76], [42] and [62].

What is the analytical expression to describe the NMR relaxation rate and Knight shift data, at least in the normal conducting state? In the absence of a detailed theory one may proceed as follows. Tranquada et al. [93] found that their neutron-scattering data in the presence of a spin-gap could be described by a phenomenological expression for the susceptibility of the form

$$\chi''(\omega) = A\left[\tanh\left(\frac{\hbar(\omega-\omega_g)}{2k_BT}\right) + \tanh\left(\frac{\hbar(\omega+\omega_g)}{2k_BT}\right)\right] \tag{12}$$

where A is a constant and ω_g is the experimental gap frequency. If one assumes that Eq. (12) is applicable to NMR data where $\hbar\omega \ll k_BT$, one obtains

$$\chi''(\omega \approx 0) = BT^{-1}\left[1 - \tanh^2\left(\frac{\Delta_\pi}{2T}\right)\right] \tag{13}$$

where we have introduced $\Delta_\pi = \hbar\omega_g/k_B$ with the index π denoting the zone-corner. A greater fit flexibility is obtained by introducing an additional temerature dependence via a factor $T^{-\alpha}$. Then the "relaxation rate per temperature unit" may be written as

$$\frac{1}{T_1T} = CT^{-\alpha}\left[1 - \tanh^2\left(\frac{\Delta_\pi}{2T}\right)\right] \tag{14}$$

where C is a constant.

Some applications of Eq. (14) to Knight shift and relaxation rates are discussed by Mehring [103]. For instance, the Cu relaxation rates in the normal state of $YBa_2Cu_4O_8$ [21, 36] are well described by Eq. (14) with parameters $\Delta_\pi = 280\,K$ and $\alpha = 1.5$. For fitting the Cu(2) Knight shift data in the same compound by the formula

$$\chi(0) = \chi_0\left[1 - \tanh^2\left(\frac{\Delta_0}{2T}\right)\right] \tag{15}$$

Mehring obtained a value of $\Delta_0 = 233\,K$. If the temperature variation of the Knight shift is really due to the opening of a spin gap and the relations used are appropriate for analyzing the data, the results imply that a spin gap, Δ_0, opens also at $q = (0, 0)$ and that this gap is smaller than the gap Δ_π at $Q = (\pi/a, \pi/a)$.

Figure 30 shows the fit of Eq. (14) to the Cu(2) and Cu(3) relaxation rates of the "mixed-layer" compound $Y_2Ba_4Cu_7O_{15}$ [45]. The parameters for both fits are $\Delta_\pi = 240 \pm 20\,K$ and $\alpha = 1.25$ and thus agree within the error limits with $\Delta_\pi = 260 \pm 10\,K$ and $\alpha = 1.25$ we obtained for our Cu(2) data in $YBa_2Cu_4O_8$.

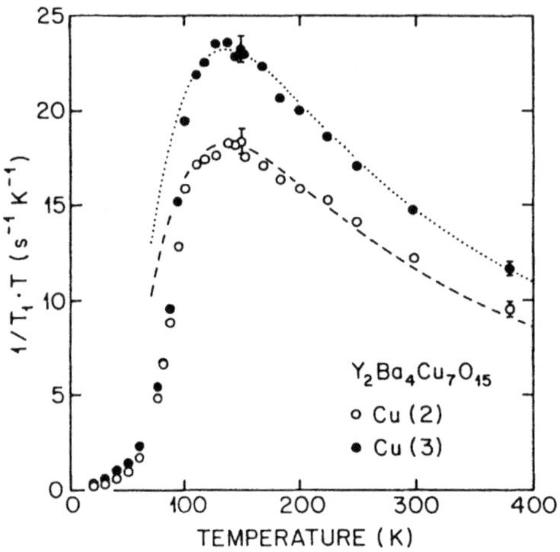

Fig. 30. Fit of Eq. (14) to the Cu(2) and Cu(3) relaxation rates in $Y_2Ba_2Cu_7O_{15}$. The parameters of both fits are $\Delta_\pi = 240\,K$ and $\alpha = 1.25$. From [45]

This result again demonstrates the dominating role played by the double-chain block in $Y_2Ba_4Cu_7O_{15}$. On the other side, an interpretation of the Knight shift by the spin-gap effect yields $\Delta_0 = 150 \pm 20\,K$ for both blocks while we obtain $\Delta_0 = 190 \pm 20\,K$ from new and more precise data of "pure" $YBa_2Cu_4O_8$. Thus, the differences between Δ_π and Δ_0 in $YBa_2Cu_4O_8$ and $Y_2B_4Cu_7O_{15}$ are about the same.

Similar differences between Δ_π and Δ_0 were reported by Winzek and Mehring [49] for the spin-gap effect of the Tl relaxation and Knight shift in $Tl_2Ba_2CaCu_2O_{8-\delta}$ studied for various values of δ. While Δ_0 did not depend on δ, Δ_π seems to decrease with increasing T_c which itself increases with increasing hole concentration.

While there is no doubt that NMR is ideally suitable for probing the static and dynamical correlations of the Cu^{2+} spins, one of the most debated issues is whether these correlations are directly related to the mechanism of the superconducting transition. There seems to be a lack of a direct relationship between the transition temperature and the spin dynamics. This is for example shown in a study by Borsa et al. [62] who have compared ^{89}Y NMR shift and ^{63}Cu relaxation measurements in $YBa_2Cu_4O_8$ obtained by various groups. T_c of these samples has been changed either by varying the oxygen stoichiometry or by doping with Ca. The Knight shifts have about the same values and the same temperature dependence and the relaxation rates exhibit a remarkably small scatter of data. It is, however, feasible that at higher energies not accessible to NMR/NQR techniques, magnetic excitations play a role in the pairing mechanism of the quasi-particles.

We conclude this paragraph by emphasizing the fact that is not sufficient to fit data by a formula. Eremin et al. [104] have proposed an approach to explain the temperature dependence of the Cu(2) spin–lattice relaxation and Knight shift in $YBa_2Cu_4O_8$ by employing charge excitation in a two-component

Hubbard description. For the relaxation rate, an expression is derived which is mathematically identical to Eq. (14) (with $\alpha = 1$) with the exception that Δ is replaced by the chemical potential.

5.3 *Test of Pairing State Symmetry*

In this Section we are dealing with NMR/NQR studies which tried to answer the question about the spatial symmetry of the pairing state of the quasi-particles. We will be concerned with Knight shift and spin–lattice anisotropy investigations.

As discussed by Leggett [105], the spin susceptibility in the superconducting state, χ_s, is related to the normal state susceptibility, χ_n, via

$$\chi_s = Y_l(T)\chi_n \tag{16}$$

where

$$Y_l(T) = \int_{-\infty}^{\infty} N(E)\left(-\frac{\partial f}{\partial E}\right)\mathrm{d}E \tag{17}$$

is a function which depends on the angular momentum, l, involved in the pairing. $N(E)$ is the superconducting density of states and f is the Fermi distribution function. The temperature dependence of $Y_l(T)$ determines the variation of the Knight shift with temperature, see Eq. (3). For $l = 0$, $Y_0(T)$ is called the Yosida function [31].

Let us start with Cu Knight shift data in $YBa_2Cu_3O_7$. Figure 31 shows the (isotropic) shift for the chain Cu(1) nuclei in the superconducting state with $B_0 \parallel c$ [54]. The dashed line is the Yosida function as calculated for the weak-coupling, spin-singlet, $l = 0$ pairing case of the BCS theory. Obviously

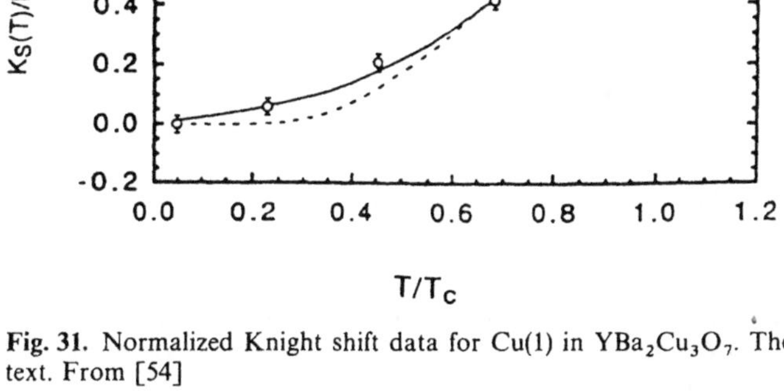

Fig. 31. Normalized Knight shift data for Cu(1) in $YBa_2Cu_3O_7$. The *curves* are explained in the text. From [54]

there is disagreement at lower temperatures. In contrast, the solid line is based on a d-wave model ($l = 2$) calculated by J. P. Lu [106] under the assumption that the superconducting properties of the chains are derived by a proximity effect with the planes. For the wave vector, $\vec{k}$, dependent energy gap function, Lu uses the tight-binding expressions

$$\Delta(\vec{k}, T) = \Delta(T)[\cos k_x - \cos k_y] \tag{18}$$

where

$$\Delta(T) = \Delta(T = 0) \tanh \alpha[(T_c/T) - 1]^{1/2} \tag{19}$$

The best fits correspond to $\Delta(T = 0)/k_B T_c = 2.3$ and $\alpha = 1.74$ and there seems to be a preference for the d-wave model.

The situation is different for the planar Cu(2) sites in $YBa_2Cu_3O_7$. In Figs. 32 and 33 the experimental data (for $B_0 \perp c$) are fitted by a strong-coupling Yosida function and spin-singlet pairing but with either s-wave (Fig. 32) or d-wave (Fig. 33) pairing [107, 54]. The analysis was based on the assumption that the ^{89}Y Knight shift (used as a reference) obeyed a BCS weak-coupling Yosida function. The fits yield $\Delta(T = 0)/k_B T_c$ parameters of 1.90 (s-wave, Fig. 32) and 3.13 (d-wave, Fig. 33). However, the experimental points change only slightly if the one-component model is adopted, i.e. a proportionality between Cu and Y shifts [54]. Thus, the experimental precision does not allow us to distinguish between s-wave and d-wave pairing.

A second possibility to check the spatial symmetry of the pairing is the temperature dependence of the spin–lattice relaxation rate anisotropy. As mentioned above, the various curves in Fig. 22 (taken from [54]) are fits based on BCS spin-singlet orbital d-wave pairing states calculated by Bulut and Scalapino [108, 109], Lu [110], and Lu and Pines [111]. Bulut and Scalapino explained the data within a BCS framework which included AFM correlations by a mean-field type approach. A complex spin susceptibility is introduced:

$$\chi(\vec{q}, \omega) = \frac{\chi_0(\vec{q}, \omega)}{1 - \bar{U}\chi_0(\vec{q}, \omega)}$$

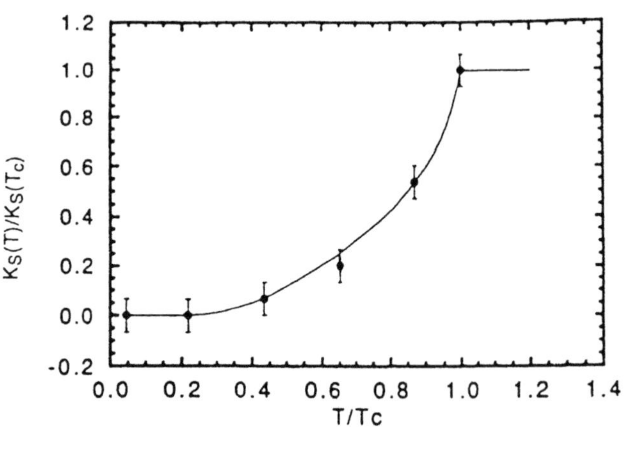

Fig. 32. Normalized Cu(2) Knight shift in $YBa_2Cu_3O_7$. Fit with s-wave pairing model; see text

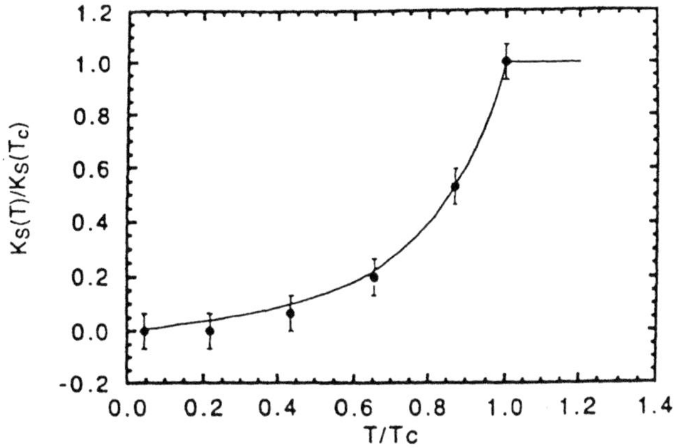

Fig. 33. Normalized Cu(2) Knight shift in $YBa_2Cu_3O_7$. Fit with *d*-wave pairing model; see text

where $\chi_0(\vec{q}, \omega)$ is given by the BCS expression and $\Delta(\vec{k}, T)$ and $\Delta(T)$ are given by Eqs. (18) and (19), respectively. In order to fit $W_\alpha(T)$ below T_c, an appropriate lifetime broadening is introduced into $\chi_0(\vec{q}, \omega)$ by means of a complex ω.

The essential results for the relaxation rate anisotropy are the following [54]. The Hebel-Slichter coherence peak which is not observed experimentally, can only be made to disappear by using *d*-wave pairing. The anisotropy ratio, *r*, *drops* just below T_c for both *s*-wave and *d*-wave pairing. However, the increase of *r* at lower temperatures can only be accomplished by using *d*-wave pairing. The various curves in Fig. 22 differ with respect to the degree of "AF enhancement" used in the calculation. Both, the solid curve (by Bulut and Scalapino, [108]) and the dashed curve (by Lu and Pines, [111]) follow from an inclusion of such an enhancement (which is "strong" for the dashed curve) while the dotted curve (by Lu, [110]) is calculated for "no enhancement". A further test of these theories will be the application to the corresponding data for the $YBa_2Cu_4O_8$ compound [82].

6 Conclusions

We have discussed several illustrative experimental studies in some representative members of high-temperature superconductors, mainly from the Y-Ba-Cu-O family but supplemented by examples from La-Sr-Cu-O and Tl-Ba-Cu-O compounds. Lack of space permitted to present mainly electric field gradient, Knight shift and relaxation studies. We summarize some of the major results.

The ab initio computation of electric field gradients at different atomic sites in Y-Ba-Cu-O compounds have turned, except for the planar Cu site, quite successful, reaching an agreement between theoretical and experimental results within about 10%.

The Knight shift data besides proving the spin-singlet superconducting pairing, represents a firm support for the "single spin-fluid" model in which Cu-*d* holes and the doped holes (predominately dwelling in the O-2*p* states) possess only one spin degree of freedom.

The NMR/NQR experiments deliver a clear evidence that anti-ferromagnetic spin correlations and fluctuations strongly influence the plane spin dynamics, however, their influence abates with rising doping levels.

Furthermore, the spin-lattice and spin–spin relaxation as well the Knight shift in the under-doped material reveal the formation of a pseudo gap in the anti-ferromagnetic fluctuation spectrum far above T_c, a phenomena that still awaits an adequate non-phenomenological theoretical treatment.

It appears that the anisotropy of planar Cu relaxation with its unusual re-enhancement at very low temperatures, is a sensitive indicator of the orbital pairing state of the superconducting quasi-particles. The experiments show that there is no coherence peak in the relaxation just below T_c nor the relaxation weakens exponentially with decreasing temperature in the superconducting state as observed in the conventional *s*-wave superconductors. Instead, the temperature dependence of the relaxation is rather power-law like, i.e., $T_1^{-1} \propto T^n$ with n between 3 and 4. In order to explain all these unconvention NMR/NQR features of the superconducting state, orbital *d*-wave pairing had to be introduced.

Indeed, the NMR/NQR experiments as well the interaction between experiment and theory have contributed to a better understanding of the electronic properties of high-temperature superconductors. We believe that in the future NMR/NQR will continue to demonstrate its power in this field.

Acknowledgements. We are grateful to various colleagues for permission to reproduce figures from their publications. We also wish to thank the Swiss National Science Foundation for continued and generous support of our work.

7 References

1. Hebel LC, Slichter CP (1957) Phys. Rev. 107: 901
2. Bednorz JG, Müller KA (1986) Z. f. Physik B64: 189
3. Lütgemeier H, Pieper MW (1987) Solid State Commun. 64: 267
4. Riesemeier H, Grabow C, Scheidt EW, Müller V, Lüders K, Riegel D (1987) Solid State Commun. 64: 309
5. Mali M, Brinkmann D, Pauli L, Roos J, Zimmermann H, Hulliger J (1987) Physics Lett. A 124: 112
6. Walstedt RE, Warren Jr, WW, Bell RF, Brennert GF, Espinosa GP, Remeika JP, Cava RJ, Rietman EA (1987) Phys. Rev. B36: 5727
7. Tachiki M, Muto Y, Syono Y (eds) (1991) Proc. of Third Intern. Conf. on Materials and Mechanisms of Superconductivity, High-Temperature Superconductors, Kanazawa, Japan, Physica C, p 185
8. MacLaughlin DE (1976) In: Solid state physics, vol. 31, Academic New York, p. 1

9. Pennigton CH, Slichter CP (1990) In: Ginsberg DM (ed) Physical properties of high-temperature superconductors, vol. II. World Scientific, New Jersey
10. Walstedt RE, Warren Jr WW (1990) Science 248: 1082
11. Kuzmany H, Mehring M, Fink J (eds) (1990) Electronic properties of high-T_c superconductors and related compounds. Springer Series in Solid-State Science 99. Springer, Berlin Heidelberg New York
12. Mehring M (guest ed) (1992) Appl. Magn. Reson. 3
13. Hewat AW, Capponi JJ, Chaillout C, Marezio M, Hewat EA (1987) Solid State Commun. 64: 301
14. Beno MA, Soderholm L, Capone DW, Hinks DG, Jorgensen JD, Schuller IK, Segre CU, Zhang K, Grace JD (1987) Appl. Phys. Lett. 51: 57
15. Fischer P, Karpinski J, Kaldis E, Jilek E, Rusiecki S (1989) Solid State Commun. 69: 531
16. Sahoo N, Markert S, Das TP, Nagamine K (1990) Phys. Rev. B 41: 220
17. Sulaiman SB, Sahoo N, Das TP, Donizelli O, Torikai E, Nagamine K (1991) Phys. Rev. B44: 7028
18. Blaha P, Schwarz K, Herzig P (1985) Phys. Rev. Lett. 54: 1192
19. Schwarz K, Ambrosch-Draxl C, Blaha P (1990) Phys. Rev. B 42: 2051
20. Ambrosch-Draxl C, Blaha P, Schwarz K (1991) Phys. Rev. B 44: 5141
21. Zimmermann H, Mali M, Brinkmann D, Karpinski J, Kaldis E, Rusiecki S (1989) Physica C 159: 681
22. Mangelschots I, Mali M, Roos J, Brinkmann D, Rusiecki S, Karpinski J, Kaldis E (1992) Physica C 194: 277
23. Lombardi A, Mali M, Roos J, Brinkmann D, Yakubowskii A (to be published)
24. Yakubowskii A, Egorov A, Lütgemeier H (1992) Appl. Magn. Reson. 3: 665
25. Riesemeier H, Gärtner S, Müller V, Lüders K (1992) Appl. Magn. Reson 3: 641
26. Zimmermann H (1991) Ph.D. Thesis, University of Zürich
27. Brinkmann D (1992) Appl. Magn. Reson 3: 483
28. Zimmermann H, Mali M, Mangelschots I, Roos J, Brinkmann D, Karpinski J, Rusiecki S, Kaldis E (1990) J. Less-Common Metals 164–165: 132
29. Brinkmann D (1991) In: Jonas J (guest ed) High pressure NMR, NMR basic principles and progress 24, Springer, Berlin Heidelberg New York, p 1
30. Riesemeier H, Kamphausen H, Scheidt EW, Stadermann G, Lüders K, Müller V (1990) In: Kuzmany H, Mehring M, Fink J (eds) Electronic properties of high-T_c superconductors and related compounds: Springer Series in Solid-State Sciences 99. Springer, Berlin Heidelberg New York, p 225
31. Yosida K (1958) Phys. Rev. 110: 769
32. Mila F, Rice TM (1989) Physica C157: 561
33. Barrett SE, Durand DJ, Pennington CH, Slichter CP, Friedman TA, Rice JP, Ginsbeg DM (1990) Phys. Rev. B 41: 6283
34. Takigawa M, Hammel PC, Heffner RH, Fisk Z (1989) Phys. Rev. B 39: 7371
35. Takigawa M, Reyes AP, Hammel PC, Thompson JD, Heffner RH, Fisk Z, Ott KC (1991) Phys. Rev. B 43: 247
36. Zimmermann H, Mali M, Mangelschots I, Roos J, Pauli L, Brinkmann D, Karpinski J, Rusiecki S, Kaldis E (1990) J. Less-Common Metals 164 & 165: 138
37. Bankay M, Mali M, Roos J, Brinkmann D (1993) (to be published)
38. Machi T, Tomeno I, Miyatake T, Koshizuka N, Tanaka S, Imai T, Yasuoka H (1990) Physica C 173: 32
39. Walstedt RE, Bell RF, Mitzi DB (1991) Phys. Rev. B 44: 7760
40. Alloul H, Mendels P, Collin G, Monod P (1988) Phys. Rev. Lett. 61: 746
41. Alloul H, Ohno T, Mendels P (1989) Phys. Rev. Lett. 63: 1700
42. Berthier C, Berthier Y, Butaud P, Clark WG, Gillet JA, Horvatič M, Ségransan P, Henry JY (1992) Appl. Magn. Reson. 3: 449
43. Trokiner A, Le Noc L, Schneck J, Pouguet AM, Mellet R, Primot J, Savary H, Gao YM, Aubry S (1991) Phys. Rev. B 44: 2426
44. Howes AP, Dupree R, Paul DMcK, Male S (1992) Physica C 193: 189
45. Stern R, Mali M, Mangelschots I, Roos J, Brinkmann D, Heinmaa I, Genoud JY, Graf T, Muller J (1993) (to be published)
46. Ishida K, Kitaoka Y, Zheng Gq, Asayama K (1991) Phys. Soc. Jpn 60: 3526
47. Kitaoka Y, Ishida K, Ohsugi S, Fujiwara K, Zheng Gq. Asayama K (1992) Appl. Magn. Reson. 3: 549

48. Kitaoka Y, Ohsugi S, Ishida K, Asayama K (1991) Phys. Soc. Jpn 60: 2351
49. Winzek N, Mehring M (1992) Appl. Magn. Reson. 3: 535
50. Kohara T, Ueda K, Takenaka T, Kohori Y, Oda Y (1990) Physica B 165 & 166: 1307
51. Mangelschots I, Mali M, Roos J, Stern R, Bankay M, Lombardi A, Brinkmann D (1993) In: Müller KA (ed) Phase separation in cuprate superconductors. World Scientific, Singapore, p 262
52. Moriya T (1963) J. Phys. Soc. Jpn 18: 516
53. Moriya T, Takahashi Y, Ueda K (1990) J. Phys. Soc. Jpn 59: 2905
54. Slichter CP, Barrett SE, Martindale JA, Durand DJ, Pennington CH, Klug CA, O'Hara KE, DeSoto SM, Imai T, Rice JP, Friedmann TA, Ginsberg DM (1992) Appl. Magn. Reson. 3: 423
55. Hammel PC, Takigawa M, Heffner RH, Fisk Z, Ott KC (1989) Phys. Rev. Lett. 63: 1992
56. Yoshinari, Y, Yasuoka H, Ueda Y, Koga K, Kosuge K (1990) J. Phys. Soc. Jpn 59: 3698
57. Barriquand F, Odier P, Jerome D (1991) Physica C 177: 230
58. Zheng Gq, Kitaoka Y, Asayama K, Kodama Y, Yamada Y (1992) Physica C 193: 154
59. Takigawa M (1992) Appl. Magn. Reson. 3: 495
60. Imai T, Yoshimura K, Uemura T, Yasuoka H, Kosuge K (1990) J. Phys. Soc. Jpn 59: 3846
61. Kitaoka Y, Ishida K, Ohsugi S, Fujiwara K, Asayama K (1991) Physica C 185–189: 98
62. Borsa F, Caretta P, Corti M, Rigamonti A (1992) Appl. Magn. Reson. 3: 509
63. Borsa F, Corti M, Rega T, Rigamonti A (1989) Il Nuovo Cimento 11D: 1785
64. Rigamonti A, Borsa F, Corti M, Rega T, Ziolo J, Waldner F (1990) In: Bednorz JG, Müller KA (eds) Earlier and recent aspects of superconductivity. Springer, Berlin Heidelberg New York, p 441
65. Kitaoka Y, Fujiwara K, Ishida K, Asayama K, Shimakawa Y, Manako T, Kubo Y (1991) Physica C 179: 107
66. Vyaselev OM, Kolesnikov NN, Kulakov MP, Schegolev IF (1992) Physica C 199: 50
67. Miyatake T, Gotoh S, Koshizuka N, Tanaka S (1989) Nature 341: 41
68. Mangelschots I, Mali M, Roos J, Zimmermann H, Brinkmann D, Rusiecki S, Karpinski J, Kaldis E, Jilek E (1990) Physica C 172: 57
69. Mangelschots I, Mali M, Roos J, Zimmermann H, Brinkmann D, Karpinski J, Kaldis E, Rusiecki S (1990) J. Less-Common Metals 164 & 165: 78
70. Machi T, Tomeno I, Miyatake T, Tai K, Koshizuka N, Tanaka S, Yasuoka H (1991) Physica C 185–189: 1147
71. Borges HA, Wells GL, Cheong SW, Kwok RS, Thompson JD, Fisk Z, Smith JL (1987) Physica B 148: 411
72. Siegrist T, Sunshine S, Murphy D, Cava RJ, Zahurak SM (1987) Phys. Rev. B 35: 7137
73. Balakrishnan G, Caves LJW, Dupree R, McPaul D, Smith ME (1989) Physica C 161: 9
74. Alloul H, Mendels P, Casalta H, Marucco JF, Arabski J (1991) Phys. Rev. Lett. 67: 3140
75. Warren Jr WW, Walstedt RE, Bell RF, Schneemeyer LF, Waszczak J, Dupree R: preprint
76. Walstedt RE, Warren Jr WW (1992) Appl. Magn. Reson 3: 469
77. Barrett SE, Martindale JA, Durand DJ, Pennington CH, Slichter CP, Friedmann TA, Rice JP, Ginsberg DM (1991) Phys. Rev. Lett 66: 108
78. Martindale JA, Barrett SE, Klug CA, O'Hara KE, DeSoto SM, Slichter CP, Friedmann TA, Ginsberg DM (1992) Phys. Rev. Lett 68: 702
79. Takigawa M, Smith JL, Hults WL (1991) Phys. Rev. B 44: 7764
80. Zimmermann H, Mali M, Bankay M, Brinkmann D (1991) Physica C 185–189: 1145
81. Mali M, Mangelschots I, Zimmermann H, Brinkmann D (1991) Physica C 175: 581
82. Bankay M, Mali M, Roos J, Mangelschots I, Brinkmann D (1992) Phys. Rev. B 46: 11228
83. Borsa F, Rigamonti A, Corti M, Ziolo J, Hyun O, Torgeson DR (1992) Phys. Rev. Lett. 68: 698
84. Pennington CH, Slichter CP (1991) Phys. Rev. Lett. 66: 381
85. Pennington CH, Durand DJ, Slichter CP, Rice JP, Bukowski ED, Ginsberg DM (1989) Phys. Rev. B39: 274
86. Itoh Y, Yasuoka H, Fujiwara Y, Ueda Y, Machi T, Tomeno I, Tai K, Koshizuka N, Tanaka S (1992) J. Phys. Soc. Jpn 61: 1287
87. Imai T, Slichter CP, Pailikas AP, Veal B (1992) Appl. Magn. Reson. 3: 729
88. Song Y-Q, Halperin WP (1992) Physica C 191: 131
89. Bulut N, Scalapino DJ (1991) Phys. Rev. Lett. 67: 2898
90. Rossat-Mignod J, Regnault LP, Vettier C, Burlet P, Henry JY, Lapertod G (1991) Physica B 169: 58
91. Rossat-Mignod J, Regnault LP, Vettier C, Bourges P, Burlet P, Bossy J, Henry JY, Lapertod G (1991) Physica B 185–189: 86

92. Bourges P, Gehring PM, Hennion B, Moudden AH, Tanquada JM, Shirane G, Shamoto S, Sato M (1991) Phys. Rev. B 43: 8690
93. Tranquada JM, Gehring PM, Shirane G, Shamoto S, Sato M (1992) Phys. Rev. B 46: 5561
94. Millis AJ, Monien M, Pines D (1990) Phys. Rev. B 42: 167
95. Monien M, Pines D, Takigawa M (1991) Phys. Rev. B 43: 258
96. Monien M, Monthoux P, Pines D (1991) Phys. Rev. B 43: 275
97. Millis AJ, Monien H (1992) Phys; Rev. B 45: 3059
98. Rice TM (1992) In: Iye Y (ed) The physics and chemistry of oxide superconductors, Springer, Berlin Heidelberg New York
99. Warren Jr WW, Walstedt RE, Brennert GF, Cava RJ, Tycko R, Bell RF, Dabbagh G (1989) Phys. Rev. Lett. 62: 1193
100. Horvatić M, Ségransan P, Berthier C, Berthier Y, Butaud P, Henry JY, Couach M, Chaminade JP (1989) Phys. Rev. B 39: 7332
101. Horvatić M, Butaud P, Ségransan P, Berthier Y, Berthier Y, Butaud C, Henry JY, Couach M (1990) Physica C 166: 151
102. Kitaoka Y, Berthier Y, Butaud P, Horvatić M, Ségransan P, Berthier C, Katayama-Yoshida H, Okabe Y, Takahashi T (1990) Physica C 162–164: 265
103. Mehring M (1992) Appl. Magn. Reson. 3: 383
104. Eremin MV, Markendorf R, Zavidonov AY, Brinkmann D, Bankay M, Mali M, Mangelschots I, Roos J, Stern R (1993) (submitted to Solid State Commun)
105. Leggett AJ (1975) Rev. Mod. Phys. 47: 331
106. Lu JP (private communication to C. P. Slichter)
107. Barrett SE (1991) Ph.D. Thesis, University of Illinois at Urbana-Champaign
108. Bulut N, Scalapino DJ (1992) Phys. Rev. Lett. 68: 706
109. Bulut N, Scalapino DJ (1992) Phys. Rev. B45: 2371
110. Lu JP (1992) Modern Physics Letters B 6: 547
111. Lu JP, Pines D (1992) (private communication to C. P. Slichter)

Author Index Volumes 21–31

MIX
Papier aus verantwortungsvollen Quellen
Paper from responsible sources
FSC® C105338

If you have any concerns about our products,
you can contact us on
ProductSafety@springernature.com

In case Publisher is established outside the EU,
the EU authorized representative is:
Springer Nature Customer Service Center GmbH
Europaplatz 3, 69115 Heidelberg, Germany

Printed by Libri Plureos GmbH
in Hamburg, Germany